META-ANALYSIS IN
MEDICINE AND
HEALTH POLICY

Biostatistics: A Series of References and Textbooks

Series Editor
Shein-Chung Chow

President, U.S. Operations
StatPlus, Inc.
Yardley, Pennsylvania

Adjunct Professor
Temple University
Philadelphia, Pennsylvania

1. *Design and Analysis of Animal Studies in Pharmaceutical Development*, edited by Shein-Chung Chow and Jen-pei Liu
2. *Basic Statistics and Pharmaceutical Statistical Applications*, James E. De Muth
3. *Design and Analysis of Bioavailability and Bioequivalence Studies, Second Edition, Revised and Expanded*, Shein-Chung Chow and Jen-pei Liu
4. *Meta-Analysis in Medicine and Health Policy*, edited by Dalene K. Stangl and Donald A. Berry
5. *Generalized Linear Models: A Bayesian Perspective*, edited by Dipak K. Dey, Sujit K. Ghosh, and Bani K. Mallick

ADDITIONAL VOLUMES IN PREPARATION

Medical Biostatistics, Abhaya Indrayan and S.B. Sarmukaddam

META-ANALYSIS IN MEDICINE AND HEALTH POLICY

edited by

Dalene K. Stangl
Duke University
Durham, North Carolina

Donald A. Berry
University of Texas M. D. Anderson Cancer Center
Houston, Texas

CRC Press
Taylor & Francis Group
Boca Raton London New York

CRC Press is an imprint of the
Taylor & Francis Group, an **informa** business

CRC Press
Taylor & Francis Group
6000 Broken Sound Parkway NW, Suite 300
Boca Raton, FL 33487-2742

First issued in paperback 2019

© 2000 by Taylor & Francis Group, LLC
CRC Press is an imprint of Taylor & Francis Group, an Informa business

ISBN-13: 978-0-8247-9030-1 (hbk)
ISBN-13: 978-0-367-39873-6 (pbk)

Visit the Taylor & Francis Web site at
http://www.taylorandfrancis.com

and the CRC Press Web site at
http://www.crcpress.com

Series Introduction

The primary objectives of the Biostatistics series are to provide useful reference books for researchers and scientists in academia, industry, and government, and to offer textbooks for undergraduate and/or graduate courses in the area of biostatistics. This book series will provide comprehensive and unified presentations of statistical designs and analyses of important applications in biostatistics, such as those in biopharmaceuticals. A well-balanced summary will be given of current and recently developed statistical methods and interpretations for both statisticians and researchers/scientists with minimal statistical knowledge who are engaged in applied biostatistics. The series is committed to providing easy-to-understand state-of-the-art references and textbooks. In each volume, statistical concepts and methodologies will be illustrated through real examples.

Meta-analysis is a commonly employed systematic reviewing strategy for addressing research or scientific questions in health-related research. It is especially useful when results from several studies disagree with regard to direction of effect, or sample sizes are individually too small to detect an effect, or a large trial is too costly and time-consuming to perform. It has been a concern whether the results from a meta-analysis covering a number of studies, which may or may not be conducted under the same study protocol, are statistically valid. The validity of a meta-analysis depends on the selection of studies and the heterogeneity among studies. As a result, the United States Food and Drug Administration (FDA) suggests that the issues of publication bias (or selection bias) and study-by-treatment interaction should be carefully evaluated before a meta-analysis is conducted for evaluation of safety and efficacy of a pharmaceutical compound in clinical research and development. Meta-analyses for uncombinable studies should be avoided for good clinical practice. This volume not only introduces important statistical concepts, designs, and methodologies of meta-analysis but also provides applications in clinical research through practical examples. The book will serve as a bridge among biostatisticians, health-related researchers/scientists, and regulatory agents, by providing a good understanding of key statistical concepts regarding design, analysis, and interpretation in health-related research, especially in medicine and health policy.

Shein-Chung Chow

Preface

Enhancements in research methodology and statistical computing, along with the demand for more accountable decision-making, have made the collection and analysis of data an integral component of every aspect of health (i.e., prevention, diagnosis, treatment, and policy). In the United States alone, annual spending on quantitative health-related research is in the billions of dollars. Replication of studies is often mandated and necessary to ensure validity and generality of results. However, there is also a great need for improved methods of meta-analysis to integrate research findings. This book reviews current methods of meta-analysis and introduces cutting-edge methods that improve quantitative meta-analysis and enable better decision-making.

This book is written for applied statisticians, students of statistics and biostatistics, and others who use statistical methods in their professional life. Our objective is to teach, so we rely heavily on examples. The authors present a common problem, develop a methodology to address the problem, and follow up with one or more examples. The level of chapters ranges from elementary to advanced; however, each chapter starts from first principles and proceeds to state-of-the-art techniques about which there are many open research questions suitable for graduate projects and dissertations.

Several chapters address controversies and make appropriate and practical recommendations. These recommendations are derived with the notion that statisticians must be able to persuade nonstatisticians as to the appropriateness of their conclusions. To this end, we accent pictorial presentations that are backed up by mathematical analyses.

This book is primarily a reference. However, it is ideal as a supplemental text for master's- and Ph.D.-level statistics and biostatistics courses. An undergraduate course in statistical theory and methods will provide the necessary background for most of the chapters. Each chapter could serve as a basis for a student project in which the student can present the analysis, think through the pros and cons of the methods, investigate other evidence that bears on the medical or policy question, and suggest improvements or further steps that could be carried out.

Dalene K. Stangl
Donald A. Berry

Contents

Contributors

Keith R. Abrams, Ph.D. Department of Epidemiology and Public Health, University of Leicester, Leicester, England

Donald A. Berry, Ph.D. Department of Biostatistics, University of Texas M.D. Anderson Cancer Center, Houston, Texas

Scott M. Berry, Ph.D. Department of Statistics, Texas A & M University, College Station, Texas

James Brophy, M.D., F.R.C.P.(C), Ph.D. Department of Medicine, University of Montreal, Montreal, Quebec, Canada

Francesca Dominici, Ph.D. Department of Biostatistics, Johns Hopkins University, Baltimore, Maryland

William DuMouchel, Ph.D. Statistics Research, AT&T Shannon Labs, Florham Park, New Jersey

Guido Francini, M.D. Division of Medical Oncology, Siena University, Siena, Italy

Geof H. Givens, Ph.D. Department of Statistics, Colorado State University, Fort Collins, Colorado

Leon J. Gleser, Ph.D. Department of Statistics, University of Pittsburgh, Pittsburgh, Pennsylvania

Martin Hellmich, Dr.rer.medic.   Department of Medical Statistics, Informatics, and Epidemiology, University of Cologne, Cologne, Germany

David R. Jones, Ph.D.   Department of Epidemiology and Public Health, University of Leicester, Leicester, England

Lawrence Joseph, Ph.D.   Department of Epidemiology and Biostatistics, McGill University, Montreal, Quebec, Canada

Paul C. Lambert, M.Sc.   Department of Epidemiology and Public Health, University of Leicester, Leicester, England

Daniel T. Larose, Ph.D.   Department of Mathematical Sciences, Central Connecticut State University, New Britain, Connecticut

Chantal Milan, Ph.D.   Fondation Française de Cancérologie Digestive (F.F.C.D.), Centre d'Epidémiologie et de Population de Bourgogne, Dijon, France

Sharon-Lise Normand, Ph.D.   Department of Health Care Policy, Harvard Medical School, Boston, Massachusetts

Ingram Olkin, Ph.D.   Department of Statistics, Stanford University, Stanford, California

Giovanni Parmigiani, Ph.D.   Institute of Statistics and Decision Sciences, Duke University, Durham, North Carolina

Donna K. Pauler, Ph.D.   Biostatistics Center, Massachusetts General Hospital, Boston, Massachusetts

Nargis Rahman, Ph.D.   Department of Mathematics, Imperial College of Science, Technology and Medicine, London, England

Bruno Sansó, Ph.D.   Centro de Estadísticas y Software Matemático, Universidad Símon Bolívar, Caracas, Venezuela

Daniel J. Sargent, Ph.D.   Department of Health Sciences Research, Mayo Clinic, Rochester, Minnesota

Chris Shaw, Ph.D. Department of Epidemiology and Public Health, University of Leicester, Leicester, England

Richard Simon, D.Sc. Biometric Research Branch, National Cancer Institute, Bethesda, Maryland

David D. Smith, Ph.D. Center for Drug Evaluation and Research, Office of Biostatistics, U.S. Food and Drug Administration, Rockville, Maryland

Dalene K. Stangl, Ph.D. Institute of Statistics and Decision Sciences, Duke University, Durham, North Carolina

Michael A. Stoto, Ph.D. Department of Epidemiology and Biostatistics, George Washington University School of Public Health and Health Services, Washington, D.C.

Alexander J. Sutton, M.Sc. Department of Epidemiology and Public Health, University of Leicester, Leicester, England

Valter Torri, M.D. Department of Oncology, Istituto di Ricerche Farmacologiche Mario Negri, Milano, Italy

Richard L. Tweedie, Ph.D., D.Sc. Division of Biostatistics, School of Public Health, University of Minnesota, Minneapolis, Minnesota

Jon Wakefield, Ph.D. Department of Statistics and Biostatistics, University of Washington, Seattle, Washington

Benny C. Zee, Ph.D. Department of Community Health and Epidemiology, Queen's University, Kingston, Ontario, Canada

1

Meta-analysis: Past and Present Challenges*

Dalene K. Stangl
Duke University, Durham, North Carolina

Donald A. Berry
University of Texas M.D. Anderson Cancer Center, Houston, Texas

I. META-ANALYSIS: A PARADIGM SHIFT

Most broadly defined, meta-analysis includes any methodology for combining information across sources. Of most interest to statisticians are quantitative approaches to summarize all relevant information pertaining to a research question. Introductory texts on the subject include Refs 1–8. Each of these books focuses on methods for deriving a common estimate of effect. In contrast, Ref. 9 emphasizes the need to move this focus to one of quantifying and reporting the heterogeneity between studies. This message reflects the methodological development seen in the statistics field during the last decade. Statistical methodology for meta-analysis has been moving away from approaches focused only on fitting a common estimate toward approaches that include estimating the extent and sources of heterogeneity among studies. This chapter will review this progression, highlight complexities encountered, summarize how the chapters in the present volume address heterogeneity between studies, and suggest future directions for methodological development.

*This work was partially supported by grant SBR-9809267 from the National Science Foundation.

1

A. Progression from Estimation of Effect to Estimation of Heterogeneity

The earliest standardized methodologies for meta-analysis include combining p-values and combining effect-size estimates. Examples of the former methodology are summarized in Refs 4 and 10. They include Tippett's minimum-p method (11) and Fisher's product of p-values (12). Tippett's minimum-p test rejects the null hypothesis, that each of the k individual-study effects is equal to zero, if any of the p-values is less than α, where $\alpha = 1 - (1 - \alpha^*)^{1/k}$, and α^* is the apriori determined significance level for the combined significance test. Fisher's product of p-values compares

$$-2 \sum_{i=1}^{k} \log(p_i)$$

with the $100(1 - \alpha^*)\%$ critical value of the chi-squared distribution with $2k$ degrees of freedom. These methods require little input, but cannot give estimates of effect sizes or correct the publication biases toward statistically significant effects, and suffer the problems inherent in multiple testing.

While requiring more thorough journal reporting, combining estimates of effect size across studies overcomes some of the problems encountered in combining p-values. Examples of combined effect size (13) include combined standardized mean differences, correlations, differences between proportions, and odds ratios. These procedures are restricted in that they are based on large-sample theory, and while standard errors and tests of homogeneity are possible, these tests have low power. Hence, in practice, the focus remains on the pooled estimates of effect rather than giving due emphasis to the degree of heterogeneity between studies. Reference 14 suggests that: ". . . preparing and presenting a single estimate as the distillation of all that is known has drawn the most criticism." This distillation is responsible for much of the misunderstanding and controversy in meta-analysis today. Some of these misunderstandings are summarized in this volume by Simon (Chap. 12), and one particular controversy is addressed by Berry (Chap. 3). Berry describes how the single-estimate distillation has inappropriately resulted in discrediting of meta-analysis because of perceived disagreements between meta-analyses and the results of large clinical trials. He shows how these perceived disagreements are errors in reasoning that are caused by failure to explicitly account for study heterogeneity.

In both the frequentist and Bayesian paradigms, the recent shift from presenting single-estimate summaries parallels the transition in popularity from fixed-effects models to random-effects models. While a brief review and comparison of these models will be presented here, nearly every chapter in this volume recognizes and promotes this shift. Further explanation and applications are given by Abrams et al. (Chap. 2), Berry (Chap. 3), Brophy and Joseph (Chap. 4), Dominici and Parmigiani (Chap. 5), DuMouchel and Normand (Chap. 6), Larose (Chap. 8), Pauler and Wakefield (Chap. 9), Rahman and Wakefield (Chap. 10), Sargent et al. (Chap. 11), and Smith et al. (Chap. 13).

1. Fixed-effects Models

Fixed-effects models for meta-analysis assume that the studies being modeled are homogeneous. There are no differences in underlying study populations, no differences in patient-selection criteria that might affect response to therapy, and the therapies are applied in the same way. Technically, patients who are assigned the same treatment but in different studies are taken to be exchangeable (or partially exchangeable in the case of covariates).

Let Y_i be a sufficient statistic for the effect of interest. For large individual-study sample sizes, the response, the individual-study effect, is approximately normal:

$$Y_i \sim \mathrm{N}(\mu, \sigma_i^2),$$

where σ_i is the standard deviation of the response. If we assume that the σ_i^2 are known, the minimum-square-error estimate of all linear estimators for μ is

$$\hat{\mu} = \frac{\sum_{i=1}^{I} \frac{1}{\sigma_i^2} Y_i}{\sum_{i=1}^{I} \frac{1}{\sigma_i^2}},$$

which has distribution

$$\hat{\mu} \sim N\left(\mu, \left(\sum_{i=1}^{I} \frac{1}{\sigma_i^2}\right)^{-1}\right).$$

The sole source of variability is assumed to be the within-study variation. A test for homogeneity of study effects compares

$$Q = \sum_{i=1}^{I} \frac{1}{\sigma_i^2}(Y_i - \hat{\mu})^2$$

to the chi-squared distribution with $I - 1$ degrees of freedom.

Reference 4 presents a thorough overview of fixed-effects models. References 9 and 15 provide shorter, but more recent overviews. There are many articles in the literature that compare and contrast the main methods for fixed-effects models. Reference 16 summarizes and compares the main methods for binary-response fixed-effects models, including those of Mantel and Haenszel (17), Woolf (18), Mantel, Haenszel, and Peto (19), and logistic regression. The Mantel–Haenszel and Woolf methods use weighted averages of the maximum-likelihood estimates of the log-odds ratios and odds ratios in each study, respectively. The Mantel–Haenszel–Peto method uses a score and Fisher information statistics from the conditional likelihoods for study-specific effects to estimate pooled effects, and logistic regression derives maximum-likelihood estimates from a full binomial likelihood. Analogous methods are available for continuous outcomes.

Fixed-effects models continue to be the most common method of meta-analysis. However, the assumption of homogeneity is usually unrealistic, given variability among studies and/or research and evaluation protocols. Indeed it is this assumption that underlies the controversy between the relative value of large clinical trials versus meta-analysis, which is addressed by Berry (Chap. 3). Equally important, the fixed-effects model underestimates variability and hence may lead to erroneous conclusions of statistical significance.

2. Random-effects Models

The random-effects formulation avoids the homogeneity assumption by modeling a random effect, θ_i for study i. Each θ_i is assumed to be selected from a distribution of study effects. The response at study i is

$$Y_i \sim N(\theta_i, \sigma_i^2)$$

and the individual-study effects are exchangeable with a normal distribution:

$$\theta_i \sim N(\mu_\theta, \tau^2).$$

Here μ_θ represents the mean of the study effects, and τ^2 represents the between-study variability. If τ^2 is known, then μ_θ is estimated by

$$\hat{\mu}_\theta = \frac{\sum_{i=1}^{I} (\sigma_i^2 + \tau^2)^{-1} y_i}{\sum_{i=1}^{I} (\sigma_i^2 + \tau^2)^{-1}}.$$

Reference 20 provides methods of estimation if τ^2 is unknown. If there are study covariates, then one can model the study-specific effects as

$$\theta_i \sim N(x_i' \beta, \tau^2),$$

where study-specific effects are assumed to be exchangeable for the partition defined by x_i.

In the random-effects model, the study effects represent samples from a population. These models "borrow strength" across studies in making estimates of both study-specific effects, θ_i, as well as an estimate of the population effect, μ. The estimate for θ_i is a weighted average of the study-specific effect estimate and the estimate of the population average. DuMouchel and Normand (Chap. 6) provide formulas for these estimates in their appendix. References 16 and 21–24 compare fixed-effects, random-effects, empirical-Bayes, and fully Bayesian models.

The primary difference in how we will define empirical-Bayes models and fully Bayesian models lies in estimating parameters of the distribution of study effects. In an empirical-Bayes model, these parameters are estimated from the "current" data, while in a fully Bayesian model another level is added, and the parameters of the distribution of study effects are given a prior distribution. In the empirical-Bayes model, the only prior information incorporated is choice of a parametric family for the distribution of study effects. The relevant experimental unit is the study; therefore, when the number of studies is small, this estimation is imprecise. Also, the uncertainty embedded in estimation of these parameters is not included in the uncertainty estimates for study effects (25).

In the fully Bayesian model, prior distributions are placed on the parameters of the distribution of the study effects, rather than estimating these parameters from the "current" data. While experts and previous studies may provide information about these hyper-parameters, typically as we move upward in the hierarchy, less and less information is known about parameters, so priors become more diffuse. In fully Bayesian models, this uncertainty is incorporated into estimates of study effects.

Authors tend to vary in their use of the label "random-effects model." For example, DuMouchel and Normand (Chap. 6) prefer to reserve the label for empirical-Bayes models while other authors use the label to refer to both empirical-Bayes and fully Bayesian models.

a. A Frequentist Perspective

Interpreting a random-effects model from a frequentist perspective is problematic unless there are many studies. First, relying on maximum-likelihood analysis and asymptotic theory requires large samples. Second, the classical test for homogeneity between studies has low statistical power (26). Third, negative estimates of variability are possible; see, e.g., Ref. 20. And fourth, a single estimate for the variability between studies may be unsatisfactory.

Some researchers object to random-effects models for philosophical reasons. For example, in a comprehensive study of the treatment of early breast cancer (27), random-effects models were not used, because:

> The statistical assumptions needed for such statistical methods to be of direct medical evidence are unlikely to be met. In particular, the different trial designs that were adopted would have to have been randomly selected from some underlying set of possibilities that includes the populations about which predictions are to be made. This is unlikely to be the case, since trial designs are adopted for a variety of reasons, many of which depend in a complex way on the apparent results of earlier trials. Moreover, selective factors that are difficult to define may affect the types of patients in trials, and therapeutic factors that are also difficult to define may differ between trials, or between past trials and future medical practice.

These authors failed to understand that using a fixed-effects model makes a more rigid key assumption that is even more unlikely to be met. Viewing the random-effects formulation from a Bayesian perspective avoids many of these problems.

b. A Bayesian Perspective

From a Bayesian perspective, all parameters are random in that they have probability distributions. These distributions depend on all available information and not just on the data at hand. The Bayesian paradigm is synonymous with meta-analysis. For both, the goal is to incorporate all information to predict with as much accuracy as possible some future event, and the uncertainty associated with it, and to present this prediction in a manner that leads to coherent decisions.

According to DV Lindley, "Meta-analysis is a natural for the Bayesian . . . " (28). Suppose that the meta-analyst is interested in some effect θ, common to all studies. After updating with observations from the first study, x_1, the posterior distribution is

$$f(\theta|x_1) \propto f(x_1|\theta)f(\theta),$$

where $f(\theta)$ is the prior distribution of the effect θ, and $f(x_1|\theta)$ is the likelihood of the data given the effect. Now the meta-analyst can use $f(\theta|x_1)$ as a prior for the next analysis, producing

$$f(\theta|x_1, x_2) \propto f(x_2|\theta)f(\theta|x_1)$$

and so on. This updating scheme is common in all applications of Bayesian methods and demonstrates how all Bayesian analysis can be seen as meta-analysis. A particularly clear example of such updating is presented by Brophy and Joseph (Chap. 4).

Testing for heterogeneity of effects between studies requires estimates of between-study variability. So, the greater the number of studies included in the meta-analysis the better, as long as the quality of included studies is high. Many meta-analyses include less than a dozen studies, which means that the estimates are not very precise. In fact, many authors in this volume are concerned about precisely this issue. While some Bayesians advocate the use of Bayes factors to test for heterogeneity, if the number of studies is small, some of the same problems arise as with frequentist tests of significance. Reference 29 uses Bayes factors to test for two types of heterogeneity, additive and interactive, in meta-analysis of 2×2 contingency tables. Chapters by Pauler and Wakefield (Chap. 9) and Abrams et al. (Chap. 2) both demonstrate the use of Bayes factors to test for heterogeneity. Some limitations of Bayes factors are discussed in Refs 30–32.

Using a fully Bayesian perspective for meta-analysis has several advantages. First, the Bayesian paradigm provides a method for synthesizing all available information in a formal, consistent, and coherent manner. Second, it explicitly incorporates model and parameter uncertainty. A third and very important aspect of the Bayesian approach is that one can average over the current distribution of unknown parameters to find a predictive distribution for future observations. For a future study outcome, x^*, the predictive distribution is

$$f(x^*|x_1, x_2) = \int f(x^*|\theta) f(\theta|x_1, x_2) d\theta.$$

This predictive distribution is a quantity of central interest to the decision maker.

References that cover the topic of decision making from an applied perspective include Refs 33–36, while Refs 37 and 38 take more theoretical perspectives. Applications using a decision-theoretic perspective which could be easily extended to the meta-analysis context may be found in Refs 39–44. These applications cover the development of clinical recommendations, determining the effectiveness of vaccines, analyzing multi-center clinical trial data, and determining who should remediate contaminated geographic areas. Rahman & Wakefield (Chap. 10) demonstrate the value of predictive distributions for decision making in a meta-analysis of pharmacokinetic data. Data from four phase I studies are combined via a hierarchical model that is nonlinear at the first stage. The authors derive overall and study-specific predictive distributions for drug clearance, half life, and volume parameters as well as for drug concentration as a function of time. These distributions are useful for decisions such as determining optimal dosages and designing future studies. Simon (Chap. 12) and Stoto (Chap. 14) review the role of meta-analysis in medical and health-policy decision making respectively, and both authors discuss controversies that ensue.

Bayesian methods offer flexible modeling schemes. One incorporates information contained in the data at hand with available prior information. The posterior distribution of the variance components as well as posterior and predictive distributions for study-specific effects are used to investigate heterogeneity in effects across studies. A typical Bayesian hierarchical model for meta-analysis is (45, 21, 16):

Level I: $Y_i \sim N(\theta_i, \sigma_i^2)$

Level II: $\theta_i \sim N(\mu, \tau^2)$

Level III: $\mu \sim \pi(\mu), \tau^2 \sim \pi(\tau^2)$.

Here Y_i is the effect from the ith study, and σ_i represents the standard deviation for the effect. Although σ_i is often assumed known, a prior for the σ_i may be incorporated as well. As with the empirical-Bayes models these models "borrow strength" across studies in estimating both study-specific effects, θ_i, and the population effect, μ. The posterior mean for θ_i is a weighted average of the study-specific effect estimate and the posterior mean of the population effect. Study-specific effects are shrunk toward the overall population mean, μ, while more accurate estimates of uncertainty are derived for the study-specific effects and the population effect. Formulas for posterior means and variances are provided in the appendix of DuMouchel and Normand (Chap. 6).

c. A Brief Literature Review of Empirical-Bayesian and Fully Bayesian Approaches

Early development and application of random-effect (empirical-Bayes and fully Bayesian) meta-analyses are included in the following publications. Bayesian hierarchical models for meta-analysis were introduced in Ref. 46. Empirical-Bayesian approaches are used in Ref. 47 (and elsewhere). Reference 49 considers general parametric approaches for meta-analysis of clinical trials. Sampling-based methods to hierarchical-Bayesian models for normally distributed data are applied in Ref 21. Reference 49 compares Bayesian and empirical-Bayes methods for 2×2 tables and demonstrates that empirical-Bayes methods underestimate the variance of the pooled estimate. References 3 and 50–52 introduced the Confidence Profile Method, a software package for carrying out Bayesian meta-analyses evaluating health-care interventions. Reference 53 developed a random-effects dose–response meta-analysis model incorporating correlation between observations within studies and including study-level covariates. Results from controlled and uncontrolled studies using random-effects models for meta-analysis are combined in Ref. 54. Studies involving variation in response classification are examined in Ref. 55. The use of meta-analysis in pharmaceutical studies is investigated in Ref. 56; more specifically Ref. 57 investigates meta-analysis for dose–response models. References 22 and 24 apply Bayesian meta-analytic models to lung-cancer studies.

Random-effects, asymptotic-Bayesian and exact-Bayesian methods are compared in Ref. 58. A meta-analysis for 2×2 tables for vaccine efficacy using empirical-Bayes methods is presented in Ref. 59. Reference 60 estimates and adjusts for selection bias in Bayesian meta-analysis. Bayesian approaches to model discrimination in meta-analysis are investigated in Ref. 61. Variability in the underlying population risk across studies in meta-analysis of clinical trials is also investigated.

Empirical-Bayesian and fully-Bayesian approaches applicable to meta-analysis of time-to-event data are presented in Refs 43 and 62–69.

In this volume, Sargent et al. (Chap. 11) present a meta-analysis of randomized trails of chemotherapy for colon cancer. Using individual patient survival data in a random-effects proportional-hazards model, Sargent et al. demonstrate the importance of carefully choosing a parametrization. They show how subtle model changes can change results. An example is whether the model is parametrized so that shrinkage occurs within treatment groups, or in the relative treatment effect, or in both.

II. COMPLEXITIES ADDRESSED IN THIS VOLUME

Difficulties in implementing either the Bayesian or classical paradigm include defining a method for choosing studies, incorporating inconsistency between study designs and outcomes, assessing and including measures of study quality, matching outcome scales of measurement, adjusting for publication bias, accommodating missing data, checking for model fit, incorporating study-level covariates, and finding or developing appropriate software. A thorough review of recent research on these complexities is available in Ref. 70. Solutions or partial solutions to many of these complexities are presented in this volume.

A. Inconsistent Outcomes and Result Reporting

Outcomes are rarely identical across studies. While endpoints may be named the same, different measures, different implementations, and different populations may prevent direct comparison and pooling of data. Even when measures, implementation, and populations are very close, researchers may have chosen to present results in different ways. To compare two groups on a binary outcome, the researcher may choose measures such as differences in proportions, relative risks, odds ratios, risk differences, or the number needed to treat. References 71–72 provide detailed comparisons and discussion of the advantages and disadvantages of these measures, and DuMouchel and Normand (Chap. 6) briefly do so. Reference 70 provides a concise summary and discusses transforming between measures when results are inconsistently reported. If enough information is given in individual studies, transforming between any of these measures is possible. For example, if the odds ratio and the incidence of the event in the control group are reported, then the odds ratio can be transformed to the relative risk. The amount of additional information reported will dictate the possible transformations. The problem of combining across different measures is not restricted to binary data. Solutions for analogous problems with continuous data are presented by Abrams et al. (Chap. 2) and with mixed, binary, and continuous data, by Dominici and Parmigiani (Chap. 5).

Abrams et al. (Chap. 2) vary, between their studies, both the outcome measures used and the way results are reported. The six studies included in the meta-analysis use four different instruments to measure anxiety level in two groups of patients. For each group, four of the six studies reported anxiety levels at baseline and follow-up, while the other two studies reported only the mean change from baseline. The authors present both frequentist and Bayesian analyses that take these differences into account. In the frequentist analysis, results from each study are transformed to standardized group differences by assuming that the within-subject correlation is zero for studies that do not report the correlation. Both fixed- and random-effects models are then fitted to these group differences. In the Bayesian analysis, a three-level hierarchy is introduced which places a distribution not only on the standardized group differences and the means of these differences, but also on the variances of the standardized group differences, the overall pooled effect, and the variance in results across studies. This Bayesian model is fitted using a range of values for the within-subject correlation. The model is then extended so that the within-subject correlation is incorporated as a random parameter. Bayes factors are used to examine competing models and to average across plausible models.

A second chapter that addresses variability in reporting of results is that of Dominici and Parmigiani (Chap. 5). Their focus is conducting meta-analyses when some outcomes are continuous and others binary. Using assumptions similar to those in Ref. 73, they develop a hierarchical-Bayesian latent-variables approach. Rather than dichotomizing continuous responses, they assume an underlying latent continuous variable for the discrete outcomes. They apply their approach to a meta-analysis of efficacy of calcium-blockers for preventing migraine headaches.

B. Model Uncertainty

Like any sophisticated statistical analysis, meta-analysis is tricky. Getting the model correct or at least "good enough" requires careful thought and a delicate separation of the wheat from the chaff. In the hierarchical models promoted in much of this volume, the problem is multiplied because there are several levels of the model: one for within-study variability, one for between-study variability, and often one for the priors on parameters of the between-study variability. Practical graphical techniques for model selection are described by DuMouchel and Normand (Chap. 6). For binary data, they describe plots to assess heterogeneity in risk differences, risk ratios, and odds ratios. This determination is important in choosing which parametrization to analyze. They explain how various plots can be used to help detect publication bias and the presence of study-level covariates that may be related to effect size. Extending Ref. 56, they discuss the use of cross-validated residuals to check for model fit and the presence of outliers. They suggest sensitivity analysis to the prior for the between-study variance component.

Many chapters in this volume address choosing between fixed- and random-effects models: Abrams et al. (Chap. 2), Berry (Chap. 3), Brophy and Joseph (Chap. 4), DuMouchel and Normand (Chap. 6), Larose (Chap. 8), Pauler and Wakefield (Chap. 9), Rahman and Wakefield (Chap. 10), Sargent et al. (Chap. 11), and Smith et al. (Chap. 13). Most of these chapters use residual plots and/or Bayes factors to examine the heterogeneity between studies and to compare the fit of the fixed- and random-effects models. Several chapters demonstrate the difficulty of this task in view of the typically small number of studies included in the meta-analysis. Cautionary warnings on the use of Bayes factors are presented in subsection I.A.2. Sargent et al. (Chap. 11) argue that the decision to fit a fixed-effects versus a random-effects model should be made prior to the analysis.

C. Assessment and Prior Specifications

For most problems, investigators will not have the luxury of postponing decisions until enough data are available to make the decision insensitive to all possible prior distributions. Decisions must be made under uncertainty, and effort needs to go into modeling prior information as well as assessing the impact of possible priors post data analysis. Eliciting a prior distribution is challenging in most Bayesian analyses, and meta-analysis is no exception. In meta-analysis, elicitation is not just about assessing distributions of model parameters. Choosing which studies to include, whether to weight studies depending upon quality of design and implementation, and related decisions are part of the elicitation process. These topics are covered in Ref. 2, and a brief review of elicitation of prior distributions will be provided here.

Most applications using Bayesian models rely on prior distributions that are either estimated from data or are reference priors (eg., Jeffreys', constant, or unit-information). In many problems, resorting to reference priors is essentially the same as doing a frequentist analysis, although the interpretations differ. Reference 32 presents a review of these priors and discusses their limitations. Graphical methods as described in Refs 74–75, quantile prediction of outcomes as described in Refs 76–80, and providing a range of priors representing beliefs from "skeptical" to "sold" (81–82) are likely to become more popular as software is disseminated and made more user-friendly. Alternate methods are examining classes of priors (83–84) and partitioning priors into subspaces which either support or do not support particular decisions (85–87). Common practice mandates a sensitivity analysis that checks the robustness of results to the prior specification.

In meta-analysis, one of the most important elicitations is the prior distribution of the between-study variability. This distribution controls how much shrinkage occurs across studies, and will have an impact on the variance of the posterior distribution of the overall population effect. At times, there may be little expert opinion or previous empirical evidence to guide the choice of this prior distribution. When the number of studies included in a meta-analysis is small, as is typical, this prior distribution can be very influential on the conclusions. DuMouchel and Normand (Chap. 6) explain the impact of this prior on shrinkage, derive shrinkage factors, and make suggestions for this choice of prior. Pauler and Wakefield (Chap. 9) also address this issue and make suggestions.

D. Missing Data

Many types of missing data can occur in meta-analyses. One type is when entire studies are excluded, because they were not known to the meta-analysts. Publication bias refers to the possibility that the results of these studies may differ from those of published studies. Because publication bias is one of the most widely researched and most important topics of meta-analysis, it merits separate discussion and will be considered in a later section in this introductory chapter. At the study level, missing data may mean that the relevant summary statistics, such as effect sizes and standard errors, are not available for all studies being considered, that the same covariate information is not available for all studies, or that different subsets of effects are available in each study. At the individual-subject level, all data are usually missing in meta-analysis. While rapidly improving data storage and transfer may change this in the future, few meta-analyses use the individual-subject data.

Missing data are the norm in meta-analysis. Several chapters in this volume address missing data of one type or another, although this may not be the primary focus. Standard approaches for dealing with missing data include complete-case analysis, use of missing-value indicator variables, single-value imputation (either an unconditional mean, conditional mean based on other covariates, or conditional means based on other covariates and the outcome variable), maximum likelihood, and multiple imputation. Bayesian methods for missing covariates include data augmentation and the use of hierarchical submodels. These methods assume the missing data to be a random variable, with an underlying stochastic mechanism. When using data augmentation, one imputes missing data from the variable's predictive distribution. When using hierarchical submodels, missing data are considered to be model parameters.

Gleser and Olkin (Chap. 7) address missing-data problems that arise in meta-analyses when multiple interventions are being compared and different subsets of interventions are included in each study. In the context of a meta-analysis, addressing the effectiveness of three antihypertension therapies for preventing heart disease, they develop and apply an approximation method to estimate effectiveness when each study examines only one or two of three possible therapies. All studies include a control group that receives none of the three therapies. While it is extremely helpful that each study includes this control group, the analysis must adjust for the fact that the comparisons of the therapies with the control are positively correlated within studies. Their approach approximates these correlations and uses weighted least squares to estimate effect sizes and simultaneous confidence intervals. They demonstrate their method both when effects are estimated as increments in proportions and as log-odds ratios. The former may use an arcsine variance-stabilizing transformation that eliminates the need for sample variance estimates.

E. Covariates

In addition to combining results across studies, meta-analyses allow for examining the impact of covariates in a way that individual studies cannot. Differences in the dosage of intervention administered, and differences in the severity of subjects' impairment are but two examples of covariates that may vary more across studies than within studies. In a carefully conducted meta-analysis, researchers can examine the impact of these covariates, and the resultant knowledge is greater than the sum of the parts. Chapters by Brophy and Joseph (Chap. 4), Larose (Chap. 8), DuMouchel and Normand (Chap. 6), Pauler and Wakefield (Chap. 9), and Sargent et al. (Chap. 11) present models for incorporating study-level covariates and provide examples. Larose examines the impact of the study-level covariate—duration of estrogen exposure—on development of endometrial cancer. Duration of estrogen exposure is incorporated through a Bayesian random-effects model. The outcome variable is the logarithm of the relative risk of contracting cancer for estrogen users relative to those who have never used estrogen. Similarly, DuMouchel and Normand (Chap. 6) use a Bayesian hierarchical model to examine the impact of two study-level covariates: route of nicotine-replacement therapy (gum versus patch) and the intensity of support (high versus low) on rates of smoking cessation. They also demonstrate how standard errors of individual-study effects can be included as a covariate to determine whether study size is correlated with outcome. This is also demonstrated by Pauler and Wakefield (Chap. 9).

In a similar vein, Brophy and Joseph (Chap. 4) adjust for two study-level covariates that threaten to bias the results of their meta-analysis. These authors look at clinical trials that compare the ability of two thrombolytic agents to reduce mortality following acute myocardial infarction. The covariates were rates of revascularization (angioplasty and/or by-pass surgery) and method of drug administration. Because this meta-analysis combines only three studies, incorporating these covariates in the regression manner of Larose (Chap. 8) or DuMouchel and Normand (Chap. 6) was not possible. Instead, the authors create prior distributions for the increased mortality rates due to decreased revascularization and less-effective administration protocols used in some studies, and then they recalculate posterior distributions using these priors.

In yet another example, Sargent et al. (Chap. 11) incorporate two study-level covariates, treatment duration and drug dosage, within a random-effects Cox proportional-hazards regression model. They use stratified Cox models within a hierarchical random-effects structure, grouping treatment effects by level of covariate—that is, studies with covariates in common were given a single prior. Results showed that dosage, but not duration, had an impact on parameter estimates.

F. Publication Bias

A serious threat to the validity of any meta-analysis is publication bias. Publication bias can occur at pre-publication (called the "file-drawer problem"), journal review, and/or post-publication (88). This bias arises when submission, review, or publication of meta-analyses is restricted to studies having a particular type of result. Investigators, reviewers, and editors often base decisions regarding submission or acceptance of manuscripts for publication on whether the study shows a "statistically significant" effect. Published studies are the most accessible. Careless meta-analysis compiles a set of biased studies, overestimates effects, and underestimates standard errors. Unwary users of such analyses are presented with highly convincing arguments that can lead to inappropriate decision making. For example, Ref. 89 examined a particular treatment for ovarian cancer. This demonstrated that a meta-analysis of only the published trials led to a conclusion of significant improvement, while inclusion of all studies registered with the International Cancer Research Data Bank did not. The scientific literature in most fields includes documentation that publication bias exists. Discussions and literature reviews on publication bias can be found in Refs 90–91. Both cite studies in the natural sciences, social sciences, and medicine demonstrating the pervasiveness of this problem. Other references include Refs 92–96.

Reference 97 reviews methods for identifying publication bias. It explains why sample sizes provide an important clue for detecting publication bias, describes the "funnel graph" (5), which plots sample size against effect size across studies to determine the potential of publication bias, and presents statistical tests that help formally assess the presence of publication bias. Reference 98 proposes methods to quantify the funnel graph, structuring the association between bias and sample size through a model which assumes no correlation between the effects and sample size. Graphical methods for detecting publication bias are also reviewed by DuMouchel and Normand (Chap. 6).

The impact of publication bias can be mediated in several ways. Meta-analyses can be restricted to sampling studies contained in certain pre-defined sampling frames, such as "complete" registers. However, this method is restrictive, and more appropriate methods have been developed. In Ref. 99, a pooled z-score from published studies is calculated. Then the number of zero-effect studies that would be required to deem this pooled z-score no longer significant is compared with an estimate of the number of studies that have gone unpublished. This "file-drawer" method is easy to implement, but missing results are centered at the null hypothesis of no effect, it does not adjust for degree of publication bias, and does not provide a corrected estimate. DuMouchel and Normand (Chap. 6) adjust for publication bias by incorporating standard errors of study-specific effects as a covariate. Their example on assessing the impact of nicotine-replacement therapy on smoking cessation shows large differences in effect estimates after adjusting for possible publication bias. Pauler and Wakefield (Chap. 9) also use this approach in the context of a meta-analysis of 13 randomized trials comparing drugs to reduce hypertension.

Early efforts to model publication bias include Refs. 100–102. References 100–101 examine publication bias where a study is published only if it yields significant results. Their methods use weight functions that assume observations in certain parts of a distribution are more likely to be observed. Reference 102 uses an approach to include specific fixed monotonic families of weight functions, and Ref. 103 further extends publication-bias models to account for heterogeneity as well as bias. More sophisticated work with weight functions appear in Refs 104–105. Both consider publication-bias models in which the weight function is estimated by maximum likelihood. Using Bayesian selection models, Refs 106–107 model the selection mechanisms of published results. Data augmentation is used to estimate and adjust for publication bias in Ref. 60. Synthesizing and building on these works, Refs 108–109 show that statistical models describing publication bias can be constructed quite naturally using weighted distributions. This method models bias by adjusting the probabilities of actual event occurrence to arrive at the probabilities that events are observed and recorded. Reference 110 uses hierarchical selection models, with parametric and step weight functions, to address selection bias along with heterogeneity between study effects and sensitivity of results to any unobserved study effects. Most of these authors adjust for publication bias based on a single factor, such as significance level. Smith et al. (Chap. 13) extend the data-augmentation methods of Ref. 60 by developing a strategy that adjusts for significance level and also for the number of and outcomes of studies which may be missing, within strata defined by research quality. They apply their method to a meta-analysis of studies of cervical cancer rates associated with use of oral contraceptives.

Another type of publication bias, called first-report bias, is discussed by Simon (Chap. 12). Within the context of clinical trials, he argues that the first study should often be interpreted differently than confirmatory studies. He presents a rationale based on calculating the posterior odds of an effective treatment given the design (detectable treatment effect, power, and desired significance) and the prior probability that the treatment represents a medically important improvement.

G. Software

Several computer-software packages and collections of program macros are available for doing meta-analysis. A comparative review of three packages—DSTAT, TRUE EPISTAT, and FAST*PRO—is available in Ref. 111, where a review covers data requirements, model assumptions, and data input/output. As noted in the article, two of the three companies had released new versions which addressed all criticisms made in the article by the time the article was published. This, along with the review of Sutton et al. (Chap. 15), demonstrates the speed with which meta-analysis software is becoming available and adapting to user needs. Sutton et al. briefly highlight others' reviews of DSTAT, TRUE EPISTAT, and FAST*PRO and present a thorough review of seven newer packages: four stand-alone packages (Review Manager, EasyMA, MetaGraphs, and Descartes), two collections of macros that run on SAS and Stata, and one general statistical package with meta-analysis options (Arcus). The authors point out the improvements and versatility of the available software, but also note that these resources are still deficient in providing model checking, incorporating publication bias, and performing sensitivity analysis.

III. OTHER CHALLENGES ADDRESSED IN THIS VOLUME

Most chapters in this volume propose improving meta-analysis methods by increasing the level of quantitative sophistication. A chapter of a different sort is Stoto (Chap. 14). Based upon first-hand experience with two Institute of Medicine projects, he addresses the role of quantification versus professional-group judgement in meta-analysis, and proposes pragmatic guidelines. While not dismissing the need for improved methods, his development contrasts with the increasingly technical advancements proposed in most of the other chapters. Stoto is in accord with most chapters on the importance of the review of heterogeneity and systematic variation between studies. He also argues that (a) group judgements require both quantitative and qualitative aspects, but reporting of results within a few standard qualitative categories (e.g., no association, inadequate/insufficient, limited/suggestive, and sufficient) was most useful, most reliable, and does not make finer distinctions about the quality of evidence than the data support; (b) professional groups should be comprised of "non-biased experts," naïve on the topic in question, that adopt a neutral starting point in evaluating evidence; and (c) formal quantitative meta-analysis approaches are often impractical because of the heterogeneity in studies, the difficulty in extracting data from published literature, and the focus on causation rather than statistical association. These conclusions challenge the increasingly sophisticated methods proposed in most of the volume concerning the role of the statistician in conducting meta-analyses.

IV. THE FUTURE OF META-ANALYSIS

So what lies ahead? What will be the fate of quantitative methods for meta-analysis? The key may lie in our ability to develop and use increasingly sophisticated statistical models to produce more reasoned conclusions.

Improved technologies will allow increasing sophistication. We are likely to see much more emphasis on analyses that make use of the original raw data rather than the few summary statistics that fit on journal pages. Computing and communication advances will make data collection more complete, data transfer much simpler, and data analysis across studies much more sophisticated. Research will continue to develop methods for combining studies with different types of outcome and different sampling and data collection designs. We are also likely to see increasing use of observational data, available via international registries, to supplement clinical trials.

The models presented in this volume will be able to capitalize on these improved technologies. However, hurdles remain. Three such hurdles are small samples, unclear definition of the unit of analysis, and choice of parametrization. To derive good estimates of between-study heterogeneity and to be able to check model fit, a sufficient number of studies must be included. If the number of studies is small—say less than 10—but the studies themselves are large, our estimate of variability may be satisfactory. But assessing model fit will still be problematic for a small number of large studies. We need to continue searching for adequate ways of checking that a sample of studies represents the population of studies, or for ways to adjust if it does not.

Related to the problem of estimating variability is the unit-of-analysis question: what constitutes a trial or study in a meta-analysis? The meta-analysis by Brophy and Joseph (Chap. 4) included three clinical trials each sampling more than 20,000 patients with several outcome events each. Within each of these trials, patients were recruited from many treatment centers. The authors explored the heterogeneity across the three clinical trials, but the heterogeneity between treatment centers within trials, and between clinicians within treatment centers, is not explored. The question of concern is: at what level of disaggregation can we stop worrying about heterogeneity?

Several chapters demonstrate that the choice of parametrization makes a difference. This has implications for decision making. For example, in the Cox proportional-hazards models of Sargent et al. (Chap. 11), parametrization determined whether shrinkage occurred within each treatment group or in the relative treatment effect. Shrinkage of survival proportions within each treatment group will not give the same answer as shrinking the difference between these two proportions. Most authors chose decision models analogous to hypothesis testing, and used standard significance criteria. Given that the choice of parametrization can affect whether confidence and posterior intervals cover the region of the null hypothesis, it is clear that we must do a better job of educating consumers and decision makers about this problem and its implications. This brings us to the biggest challenge for the new millennium.

Will improved but more sophisticated methods have any more impact than previous methods? Stoto argues that simple is better. How do we employ increasingly sophisticated statistical models to produce increasingly relevant output? We need to focus on decision making. For example, consider the choice between two treatments for cancer. Which is more useful: a predictive survival curve for each treatment or a log hazards ratio? If we want to test a point null hypothesis, then the latter is more useful. If we want to make a decision about the relative utility of the two treatments, the former is more useful. The former is in units which the decision maker can directly and easily consider, while the latter is not. Using a predictive distribution, the decision maker can assess the various decisions under a variety of utility functions, while the units of a log hazards ratio may be irrelevant for making decisions. The shift from fixed- to random-effects models has been a desirable one. But, do we need to take it a step further by providing predictive distributions for the outcomes of interest?

References

1. TD Cook, H Cooper, DS Cordray, H Hartman, LV Hedges, RJ Light, TA Louis, F Mosteller. Meta-Analysis for Explanation. New York: Russell Sage Foundation, 1992.
2. H Cooper, LV Hedges. The Handbook of Research Synthesis. New York: Russell Sage Foundation, 1994.
3. DM Eddy. The confidence profile method: a Bayesian method for assessing health technologies. Operations Research 37:210–228, 1989.
4. LV Hedges, I Olkin. Statistical Methods for Metaanalysis. Orlando, FL: Academic Press, 1985.
5. RJ Light, DB Pillemer. Summing Up: The Science of Reviewing Research. Cambridge, MA: Harvard University Press, 1984.
6. DB Petitti. Meta-Analysis, Decision Analysis, and Cost-Effectiveness Analysis: Methods for Quantitative Synthesis in Medicine. Oxford University Press, 1994.
7. R Rosenthal. Meta-Analytic Procedures for Social Research. Beverly Hills: Sage Publications, 1984.
8. KW Wachter, ML Straf. The Future of Meta-Analysis. New York: Russell Sage Foundation, 1990.
9. D Draper, D Gaver, P Goel, J Greenhouse, L Hedges, C Morris, C Waternaux. On combining information: statistical issues and opportunities for research. A report of a panel convened by the National Academy of Sciences, 1992.
10. BJ Becker. Combining significance levels. In: H Cooper and LV Hedges, eds. The Handbook of Research Synthesis. New York: Russell Sage Foundation, 1994, pp 215–230.
11. LHC Tippett. The Method of Statistics. London: Williams and Norgate, 1931.

12. RA Fisher. Statistical Methods for Research Workers. London: Oliver and Boyd, 1932.

13. WR Shadish, CK Haddock. Combining estimates of effect size. In: H Cooper and LV Hedges, eds. The Handbook of Research Synthesis. New York: Russell Sage Foundation, 1994, pp 261–282.

14. JC Bailar III. The promise and problems of meta-analysis. New England Journal of Medicine 337:559–561, 1997.

15. LV Hedges. Fixed effects models. In: H Cooper and LV Hedges, eds. The Handbook of Research Synthesis. New York: Russell Sage Foundation, 1994, pp 285–300.

16. TC Smith, DJ Spiegelhalter, A Thomas. Bayesian approaches to random-effects meta-analysis: A comparative study. Statistics in Medicine 14:2685–99, 1995.

17. N Mantel, W Haenszel. Statistical aspects of the analysis of data from retrospective studies of disease. Journal of the National Cancer Institute 22:719–748, 1959.

18. B Wolfe. On estimating the relation between blood group and disease. Annals of Human Genetics 19:251–253, 1955.

19. S Yusef, R Peto, J Lewis, R Collins, P Sleight. Beta blockade during and after myocardial infarction: An overview of the randomized trials. Progress in Cardiovascular Diseases 27:335–371, 1985.

20. R DerSimonian, N Laird. Meta-analysis in clinical trials. Controlled Clinical Trials 7:177–188, 1986.

21. Morris, C., Normand, S.L. Hierarchical models for combining information and for meta-analyses with discussion. In: J. Bernardo, J. Berger, A. Dawid, and A. Smith, eds. Bayesian Statistics 4. London: Oxford University Press, 1992, pp 321–344.

22. BJ Biggerstaff, RL Tweedie, KL Mengersen. Passive smoking in the workplace: classical and Bayesian meta-analyses. Int. Arch Occup. Environ. Health 66:260-77, 1994.

23. XY Su, ALW Po. Combining event rates from clinical trials: Comparison of Bayesian and classical methods. Annals of Pharmacotherapy 30:460–465, 1996.

24. RL Tweedie, DJ Scott, BJ Biggerstaff, KL Mengersen. Bayesian meta-analysis, with application to studies of ETS and lung cancer. Lung Cancer 14:171–94, 1996.

25. R Kass, D Steffey. Approximate Bayesian inference in conditionally independent hierarchical models. Journal of the American Statistical Association, 84(407):717–726, 1989.

26. JL Fleiss. Measures of effect size for categorical data. In: H Cooper and LV Hedges, eds. The Handbook of Research Synthesis. New York: Russell Sage Foundation, 1994, pp 245–260.

27. Early Breast Cancer Trialists' Collaborative Group. Treatment of Early Breast Cancer. Volume 1: Worldwide Evidence, 1985–1990. Oxford: Oxford University Press, 1990.

28. DV Lindley. A discussion of: Hierarchical models for combining information by C.N. Morris and S.L. Normand. In: JM Bernardo, JO Berger, AP Dawid, AFM Smith, eds. Bayesian Statistics IV. London: Oxford Science Publications, 1992, pp 341–342.

29. S Berry. Understanding and testing across 2×2 tables: Application to meta-analysis. Statistics in Medicine 17:2353–2369, 1998.

30. M Lavine, M Schervish. Bayes factors: What they are and what they are not. American Statistician 53(2):119–122, 1999.

31. JB Kadane, JM Dickey. Bayesian decision theory and the simplification of models. In: J Kmenta, J. Ramsey, eds. Evaluation of Econometric Models, Academic Press, 1980, pp 245–268.

32. D Stangl. The use of reference priors and Bayes factors in the analysis of clinical trials. *Proceedings of the IMA Conference on Statistical Methods in the Health Sciences*, Minneapolis, 1997, pp 237–250.

33. HC Sox, MA Blatt, MS Higgins, KI Marton. Medical Decision Making. Boston: Butterworth and Heinemann, 1988.

34. DA Berry. Decision analysis and Bayesian methods. In: PF Thall, ed. Recent Advances in Clinical Trial Design and Analysis. Boston: Kluwer Academic Publishers, 1995, pp 125–154.

35. V Hasselblad, D McCrory. Meta-analysis Tools for Medical Decision Making: A Practical Guide. Medical Decision Making 15:81–96, 1995.

36. R Lewis, DA Berry. Decision theory. In: P Armitage and T Colton, eds. Encyclopedia of Biostatistics. New York: John Wiley & Sons, 1998.

37. MH Degroot. Optimal Statistical Decisions. McGraw Hill Book Company, 1970.

38. JO Berger. Statistical Decision Theory and Bayesian Analysis. 2nd ed. New York: Springer-Verlag, 1985.

39. G Parmigiani, MS Kamlet. A cost–utility analysis of alternative strategies in screening for breast cancer. In: C Gatsonis, J Hodges, R Kass, N Singpurwalla, eds. Case Studies in Bayesian Statistics, New York: Springer-Verlag, 1993, pp 390–402.

40. G Parmigiani, M Ancukiewicz, D Matchar. Decision models in clinical recommendations development: The stroke prevention policy model. In: DA Berry, DK Stangl, eds. Bayesian Biostatistics. New York: Marcel Dekker, 1996, pp 207–233.

41. DA Berry, MC Wolff, D Sack. Public health decision making: A sequential vaccine trial with discussion. In: JM Bernardo, JO Berger, AP Dawid, AFM Smith, eds. Bayesian Statistics. Oxford, England: Oxford University Press, 1992, pp 79–96.

42. DA Berry, MC Wolff, D Sack. Decision making during a phase III randomized controlled trial. Controlled Clinical Trials 15:360–379, 1994.
43. D Stangl. Prediction and decision making using Bayesian hierarchical models. Statistics in Medicine 14:2173–2190, 1995.
44. L Wolfson, J Kadane, M Small. Expected utility as a policy-making tool: An environmental health example. In: D Berry and D Stangl, eds. Bayesian Biostatistics. New York: Marcel Dekker, 1996.
45. W DuMouchel. Bayesian meta-analysis, In DA Berry, ed. Statistical Methodology in the Pharmaceutical Sciences. New York: Marcel Dekker, 1990, pp 509–529.
46. WH DuMouchel, JE Harris. Bayes methods for combining the results of cancer studies in humans and other species with discussion. Journal of the American Statistical Association 78:293–315, 1983.
47. SW Raudenbush, AS Bryk. Empirical Bayes metaanalysis. Journal of Educational Statistics 10:75–98, 1985.
48. JB Carlin. Meta-analysis for 2×2 tables: A Bayesian approach. Statistics in Medicine 11:141–158, 1992.
49. A Whitehead, J Whitehead. A general parametric approach to the meta-analysis of randomized clinical trials. Statistics in Medicine 10:1665–1677, 1991.
50. DM Eddy, V Hasselblad, R Schachter. A Bayesian method for synthesizing evidence. International Journal of Technical Assistance in Health Care 6:31–55, 1990.
51. DM Eddy, V Hasselblad, R Shachter. An introduction to a Bayesian method for meta-analysis: the confidence profile method. Medical Decision Making 10:15–23, 1990.
52. DM Eddy, V Hasselblad, R Shachter. Meta-analysis by the Confidence Profile Method. The Statistical Synthesis of evidence. Academic Press Inc., 1992.
53. SL Normand. Random effects methods for dose-response meta-analyses. Technical Series in Statistics, Report 1–1994, Dept. of Health Care Policy, Harvard Medical School, 1994.
54. Z Li, CB Begg. Random effects models for combining results from controlled and uncontrolled studies in a meta-analysis. Journal of the American Statistical Association 89:1523–1527, 1994.
55. A Whitehead, NMB Jones. A meta-analysis of clinical trials involving different classifications of response into ordered categories. Statistics in Medicine 13:2503–2515, 1994.
56. WH DuMouchel. Predictive cross-validation in Bayesian meta-analysis. In: JM Bernardo, JO Berger, AP Dawid, AFM Smith, eds. Bayesian Statistics 5. Oxford: Clarendon Press, 1996, pp 107–124.
57. W DuMouchel. Meta-analysis for dose-response models. Statistics in Medicine 14:679–685, 1995.

58. A Rogatko. Bayesian Approach for meta-analysis of controlled clinical trials. Communications in Statistics, Part A (Theory and Methods) 21:1441–1462, 1992.
59. CS Berkey, DC Hoaglin, F Mosteller, DA Colditz. A random-effects model for meta-analysis. Statistics in Medicine 14:395–411, 1995.
60. GH Givens, DD Smith, RL Tweedie. Bayesian data-augmented meta-analysis that account for publication bias issues exemplified in the passive smoking debate, Statistical Science 12:221–250, 1997.
61. K Abrams, B Sanso. Discrimination in meta-analysis—a Bayesian perspective. Technical Report 95-03. Department of Epidemiology and Public Health, University of Leicester, U.K., 1995.
62. DG Clayton. A Monte Carlo method for Bayesian Inference in frailty models. Biometrics 47:467–485, 1991.
63. CB Begg, L Pilote. A model for incorporating historical controls into a meta-analysis. Biometrics 47:899–906, 1991.
64. RT Clemen. Making Hard Decisions. Boston: PWS-Kent Publishing Company, 1991.
65. R Gray. A Bayesian analysis of institutional effects in a multicenter cancer clinical trial. Biometrics 50:244–253, 1994.
66. C Morris, C Christiansen. Fitting Weibull duration models with random effects. Lifetime Data Analysis 1:347–359, 1995.
67. SK Sahu, DK Dey, H Aslanidou, D Sinha. A Weibull regression model with gamma frailties for multivariate survival data. Lifetime Data Analysis 3:123–137, 1997.
68. D Stangl. Modeling heterogeneity in multicenter clinical trials using Bayesian hierarchical models. Ph.D. Dissertation, Carnegie Mellon University, 1991.
69. D Stangl. Hierarchical analysis of continuous-time survival models. In: D Berry, D Stangl, eds. Bayesian Biostatistics. New York: Marcel Dekker, 1996, pp 429–450.
70. AJ Sutton, KR Abrams, DR Jones, TA Sheldon, F Song. Systematic reviews of trials and other studies. Health Technology Assessment 2(19), 1999.
71. JC Sinclair, MB Bracken. Clinically useful measures of effect in binary analyses of randomized trials. Journal of Clinical Epidemiology 47:881–889, 1994.
72. JL Fleiss. The statistical basis of meta-analysis. Statistical Methods in Medical Research 2:147–160, 1993.
73. A Whitehead, AJ Bailey, D Elbourne. Combining summaries of binary outcomes with those of continuous outcomes in a meta-analysis. Journal of Biopharmaceutical Statistics 9:1–16, 1999.
74. K Chaloner. Elicitation of Prior Distributions. In: D Berry, D Stangl, eds. Bayesian Biostatistics. New York: Marcel Dekker, 1996, pp 141–156.

75. K Chaloner, T Church, JP Matts, TA Louis. Graphical elicitation of a
     prior distribution for an AIDS clinical trial. The Statistician 42:341–353,
     1993.
76. JB Kadane. Predictive and structural methods for eliciting prior distribu-
     tions. In: A. Zellner, ed. Studies in Bayesian Analysis in Econometrics and
     Statistics In Honor of Harold Jeffreys. Amsterdam:North-Holland, 1980,
     pp 89–93.
77. JB Kadane, JM Dickey, R Winkler. Interactive elicitation of opinion for a
     normal linear model. Journal of the American Statistical Association
     75:845–854, 1980.
78. JB Kadane, RL Winkler. DeFinetti's methods of elicitation. In: R. Viertl,
     ed. Probability and Bayesian statistics. New York: Plenum: 97–110, 1987.
79. JB Kadane, LJ Wolfson. Priors for the design and analysis of clinical trials.
     In: D Berry, D Stangl, eds. Bayesian Biostatistics. New York: Marcel
     Dekker, 1996, pp 157–186.
80. JB Kadane, LJ Wolfson. Experiences in elicitation. The Statistician
     47(4):3–19, 1997.
81. DJ Spiegelhalter, LS Freedman, MK Parmar. Bayesian approaches to
     randomized trials. J. Royal Statistical Soc. Ser. A 157:357–416, 1994.
82. DJ Spiegelhalter, LS Freedman, MK Parmar. Bayesian approaches to
     randomized trials. In: D Berry, D Stangl, eds. Bayesian Biostatistics.
     New York: Marcel Dekker, 1996, pp 67–108.
83. L Wasserman. Recent methodological advances in robust Bayesian infer-
     ence. In: JM Bernardo, MH DeGroot, DV Lindley, AFM Smith, eds.
     Bayesian Statistics IV. London: Oxford University Press, 1992, pp 483–
     502.
84. JB Greenhouse, L Wasserman. Robust Bayesian methods for monitoring
     clinical trials. Statistics in Medicine 14:1379–1391, 1995.
85. BP Carlin, KM Chaloner, TA Louis, FS Rhame. Elicitation, monitoring,
     and analysis for an AIDS clinical trial. In: C Gatsonis, J Hodges, R Kass,
     N Singpurwalla, eds. Case Studies in Bayesian Statistics, vol. 2. New York:
     Springer, 1995, pp 48–84.
86. BP Carlin, TA Louis. Bayes and Empirical Bayes Methods for Data
     Analysis. London: Chapman and Hall, 1996.
87. D Sargent, BP Carlin. Robust Bayesian design and analysis of clinical trials
     via prior partitioning with discussion. IMS Lecture Note Series: Second
     International Workshop on Bayesian Robustness, 1996.
88. TC Chalmers, CS Frank, D Reitman. Minimizing the three stages of pub-
     lication bias. Journal of the American Medical Association 263: 1392–
     1395, 1990.
89. RJ Simes. Publication bias: the case for an international registry of clinical
     trials. Journal of Clinical Oncology 4:1529–1541, 1986.

90. CB Begg, JA Berlin. Publication bias: a problem in interpreting medical data with discussion. Journal of the Royal Statistical Society, Ser. A 151:419–463, 1988.

91. D Larose. Bayesian approaches to meta-analysis. Ph.D. dissertation, University of Connecticut, 1996.

92. K Dickersin. The existence of publication bias and risk factors for its occurrence. Journal of the American Medical Association 263:1385–1389, 1990.

93. K Dickersin, YI Min, CL Meinert. Factors influencing publication of research results, Journal of the American Medical Association 267:374–378, 1992.

94. PJ Easterbrook, JA Berlin, R Gopalan, DR Matthews. Publication bias in clinical research. Lancet 337:867–872, 1991.

95. MJ Mahoney. Publication prejudices: an experimental study of confirmatory bias in the peer review system. Cognitive Therapy Research 1:161–175, 1977.

96. TD Sterling, WL Rosenbaum, JJ Weinkam. Publication decisions revisited: the effect of the outcome of statistical tests on the decision to publish and vice versa. American Statistician 49:108–112, 1995.

97. CB Begg. Publication Bias. In: H Cooper and LV Hedges, eds. The Handbook of Research Synthesis. New York: Russell Sage Foundation, 1994, pp 399–410.

98. JA Berlin, CB Begg, TA Louis. An assessment of publication bias using a sample of published clinical trials. Journal of the American Statistical Association 84:381–392, 1989.

99. R Rosenthal. The file drawer problem and tolerance for null results. Psychological Bulletin 86:638–641, 1979.

100. DM Lane, WP Dunlap. Estimating effect size: Bias resulting from the significance criterion in editorial decisions. British Journal of Mathematical and Statistical Psychology 31:107–112, 1978.

101. LV Hedges. Estimation of effect size under nonrandom sampling: the effects of censoring studies yielding statistically significant mean differences. Journal of Educational Statistics 9:61–85, 1984.

102. S Iyengar, JB Greenhouse. Selection models and the file drawer problem. Statistical Science 3:109–135, 1988.

103. GP Patil, C Taillie. Probing encountered data, meta-analysis and weighted distribution methods. In: Y Dodge ed. Statistical Data Analysis and Inference. Elsevier Science Publishers B.V., North-Holland, 1989.

104. LV Hedges. Modeling publication selection effects in meta-analysis. Statistical Science 7:246–255, 1992.

105. KBG Dear, CB Begg. An approach for assessing publication bias prior to performing a meta-analysis. Statistical Science 7:237–245, 1992.

106. MJ Bayarri, MH DeGroot. The analysis of published significant results. In: W Racugno, ed. Rassegna di Metodi Statistici ed Applicazioni. Bologna: Pitagora, 1993, pp 19–41.

107. MJ Bayarri, MH DeGroot. A "BAD" view of weighted distributions and selection models. In: J Bernardo, J Berger, A Dawid, and A Smith, eds. Bayesian Statistics 4. London: Oxford University Press, 1992, pp 17–33.

108. DT Larose, DK Dey. Publication bias using weighted distributions in a Bayesian setting. ASA Proceedings of the Section on Bayesian Statistical Science, 1995, pp 121–126.

109. DT Larose, DK Dey. Weighted distributions viewed in the context of model selection: a Bayesian perspective. Test 5:227–246, 1996.

110. NP Silliman. Hierarchical selection models with applications in meta-analysis. Journal of the American Statistical Association 92(439):926–936, 1997.

111. SL Normand. Meta-analysis software: A comparative review. The American Statistician 49(3):298–309, 1995.

2

Meta-analysis of Heterogeneously Reported Study Results: A Bayesian Approach

Keith R. Abrams, Paul C. Lambert, and Chris Shaw
University of Leicester, Leicester, England

Bruno Sansó
Centro de Estadísticas y Software Matemático, Universidad Símon Bolívar, Caracas, Venzuela

Abstract

This chapter considers the quantitative synthesis of published comparative-study results when the outcome measures used in the individual studies *and* the way in which they are reported varies between studies. While the former difficulty may be overcome, at least to a limited extent, by the use of standardized effect sizes, the latter is often more problematic. Two potential solutions to this problem are: sensitivity analyses and a fully Bayesian approach, in which pertinent background information is included. Both approaches are illustrated using the results of a systematic review exploring the impact of screening for disease on levels of psychological morbidity.

I. INTRODUCTION

Often in epidemiological and health-service research settings, a variety of different outcome measures and instruments are used. An example is assessing quality of life. Heterogeneity of research instruments and measures poses problems when the quantitative synthesis of a number of individual study results is required (Greenland, 1987; Jones, 1992). The use of standardized effect sizes has been advocated as a means of overcoming this diversity of instruments, especially in a comparative setting. A further complication is the fact that individual studies may report their results in a manner which makes calculation of a standardized effect size problematic, or even impossible. Ideally, individual-patient data could be obtained, so that re-analysis of the studies could be undertaken to calculate the standardized effects in the individual studies (Stewart and Parmar, 1993). However, the process of obtaining individual-patient data is laborious, and frequently infeasible or impractical. Indeed, it may even yield a further complication, in that only individual-patient data for a proportion of the studies in the meta-analysis may be obtained (Sutton et al., 1998; Sutton et al., 1999). We therefore consider a variety of methods for the synthesis of heterogeneously reported studies, using a variety of outcome measures. In a similar manner, Dominici and Parmigiani (Chap. 3) discuss the synthesis of studies in which the same outcome measure has *either* been reported on a continuous *or* a dichotomous scale.

A number of studies have considered the effect of diagnostic screening, for a variety of chronic diseases, on the level of psychological morbidity experienced by individuals post screening (Shaw et al., 1999). A number of authors have reported changes in levels of anxiety and depression in individuals who have either been screened to target primary prevention or to implement therapy (Haynes et al., 1978; Marteau, 1993). However, such costs to the individual are frequently ignored in health-policy decision making, and this therefore was the rationale for undertaking a systematic review of the literature, i.e., to assess the costs, in terms of psychological morbidity, attributable to the process of screening and the subsequent results.

Here we concentrate on a subset of six studies in the systematic review which investigated the effect of screening, in terms of testing either positive or negative, on levels of long-term anxiety, i.e., at least 6 months after notification of the test result. The six studies identified used four different instruments, in the form of self-completed questionnaires, to measure anxiety: the hospital anxiety and depression scale (HAD) (Zigmond and Snaith, 1983), the state-trait anxiety inventory scale (STAI) (Spielberger et al., 1970), Kellner (Kellner and Sheffield, 1973), and Hopkins symptom checklist (HSCL) (Derogatis et al., 1974). To account for the four different instruments used, it was decided to calculate the standardized difference for the comparison of long-term anxiety levels of individuals who tested negative with those who tested positive, adjusting for initial, i.e., pre-screening, baseline anxiety levels. However, while two of the studies reported mean change from baseline for the two groups, the remaining four studies reported anxiety levels separately at baseline and follow-up only. Therefore, for these latter studies, it was impossible, based on the reported statistics, to fully account for within-subject correlation. Without performing a sensitivity analysis, it is difficult to assess the extent to which ignoring such correlation may bias the overall result of the meta-analysis.

In the original reporting of the meta-analysis (Shaw et al., 1999) a sensitivity analysis was performed in which a range of possible values for the within-subject correlations was assumed for the four studies, in which mean change was not reported. Unfortunately, the two studies that did adjust for baseline levels, had explicitly taken the correlation into account and reported the results in such a way that made it impossible to calculate the correlation. While a sensitivity analysis such as this assesses the impact of ignoring within-subject correlation, it does not give explicitly an overall pooled estimate of effect size that takes into account the uncertainty inherent in such meta-analysis problems.

A possible solution to this problem is to utilize a Bayesian approach. Bayesian methods for meta-analysis generally have been advocated by a number of authors (Louis and Zelterman, 1994; Abrams and Sansó, 1998; Sutton et al., 1999) and are reviewed by Stangl and Berry (Chap. 1). They have a number of potential advantages. Firstly, they enable an analysis to take account of all parameter uncertainty, a feature that classical methods have up until recently lacked (Hardy and Thompson, 1996; Biggerstaff and Tweedie, 1997). A second advantage occurs when there are a number of potentially competing models. Adopting a classical approach often entails choosing between such models, e.g., fixed- and random-effect models, which when they are finely balanced may not be considered appropriate, but certainly fails to account for the inherent model uncertainty (Abrams and Sansó, 1995). Bayesian approaches based on the posterior probability of the various models enable an overall pooled estimate to be obtained averaged over all models under consideration, and which takes into account the model uncertainty (Kass and Raftery, 1995; Raftery and Richardson, 1994). Finally, adoption of a Bayesian approach enables potentially pertinent background information to be included in a meta-analysis, which might otherwise be ignored (Abrams and Jones, 1995; Higgins and Whitehead,

1996). In the case of the meta-analysis considered here, other study results that did consider the within-subject correlation could be used to derive a prior distribution for the correlation parameter in a meta-analysis model.

The remainder of this chapter is organized with Sect. 2 discussing the illustrative example in further detail. Sect. 3 presents initial classical analyses, including a sensitivity analysis, while Sect. 4 considers an initial Bayesian approach to the problem, including whether fixed- or random-effect models should be adopted. Sect. 5 presents a Bayesian analysis, which takes into account both the uncertainty associated with the within-patient correlation and relevant background information. Finally, Sect. 6 discusses some of the limitations and possible extensions to the analyses presented.

II. PSYCHOLOGICAL MORBIDITY ASSOCIATED WITH SCREENING

A. Background and Studies Included

In conducting a systematic review to assess the impact of screening for diseases on the levels of psychological morbidity, Shaw et al. (1999) included finding the six studies presented in Table 1. Each considered the effect of testing positive or negative in a screening programme on levels of long-term anxiety. "Long term" was defined as being at least six months from the screening test and subsequent notification of the test result; but, as can be seen from Table 1, the follow-up interval varied between studies. The six studies considered screening for two specific diseases: studies 1 and 2 screened individuals for hypertension, while studies 3 to 6 tested individuals for HIV antibodies. In both cases, individuals were asymptomatic at testing; but, in studies 3 to 6, subjects were considered "at risk," while in studies 1 and 2 these were population screening programs. It could be argued that, since the two diseases are very different, and certainly testing positive for them has very different consequences, they should not be combined. However, the comparative nature of the studies enables their synthesis to allow an assessment of the *relative* effect of the screening process, and subsequent test results, on psychological morbidity.

Four different instruments, in the form of self-completed questionnaires, were used to assess long-term anxiety levels: HAD (Zigmond and

Table 1 Studies Included in Meta-analysis of Psychological Effects of Screening for Disease (● denotes statistics not reported in the original publication)

No.	Author	Year	Disease	Measure	Follow-up (Months)	Group	N	Baseline Mean	SD	Follow-up Mean	SD	Change Mean	SD
1	Ambrosio et al.	1984	Hypertension	Keller	24	Positive	81	3.6	3.6	2.8	2.7	●	●
						Negative	129	4.7	4.5	3.4	3.4	●	●
2	Rudd et al.	1986	Hypertension	STAI	6	Positive	325	34.5	8.9	●	●	−1.3	5.8
						Negative	362	37.1	8.9	●	●	−1.3	6.8
3	Ostrow et al.	1989	HIV	Hopkins	12	Positive	70	●	●	●	●	−2.3	16.7
						Negative	94	●	●	●	●	−3.8	14.5
4	Perry et al.	1993	HIV	STAI	12	Positive	106	42.6	12.3	37.8	11.6	●	●
						Negative	222	41.5	13.0	37.7	12.7	●	●
5	Jadresic et al.	1994	HIV	HAD	6	Positive	20	8.3	3.4	6.1	3.9	●	●
						Negative	68	8.2	4.3	7.7	4.2	●	●
6	Pugh et al.	1994	HIV	STAI	12	Positive	20	47.9	18.6	40.1	10.9	●	●
						Negative	41	45.3	12.9	38.7	13.2	●	●

Table 2 Maximum-likelihood (MLE) and Markov Chain Monte Carlo (MCMC) Estimates for Fixed and Random Meta-analysis Assuming Various Fixed Values of ρ

Model		ρ 0.00	0.25	0.50	0.75	0.95
Fixed						
μ MLE	Mean	0.056	0.058	0.062	0.070	0.106
	SD	0.053	0.053	0.053	0.053	0.053
	95% CI	(−0.050, 0.162)	(−0.048, 0.164)	(−0.044, 0.168)	(−0.036, 0.176)	(0.000, 0.212)
μ MCMC	Mean	0.057	0.059	0.063	0.072	0.113
	SD	0.053	0.053	0.053	0.053	0.053
	95% CrI	(−0.048, 0.161)	(−0.046, 0.163)	(−0.042, 0.167)	(−0.034, 0.175)	(+0.008, 0.217)
Random						
μ MLE	Mean	0.076	0.082	0.094	0.122	0.254
	SD	0.069	0.073	0.080	0.095	0.160
	95% CI	(−0.059, 0.211)	(−0.061, 0.225)	(−0.082, 0.251)	(−0.064, 0.308)	(−0.060, 0.568)
μ MCMC	Mean	0.079	0.084	0.093	0.119	0.275
	SD	0.090	0.093	0.100	0.123	0.300
	95% CrI	(−0.082, 0.263)	(−0.082, 0.277)	(−0.082, 0.301)	(−0.094, 0.374)	(−0.246, 0.845)
τ^2 MLE	Mean	0.008	0.011	0.015	0.029	0.121
τ^2 MCMC	Mean	0.026	0.029	0.036	0.063	0.452
	SD	0.066	0.074	0.095	0.170	1.830
	95% CrI	(0.001, 0.154)	(0.001, 0.170)	(0.001, 0.200)	(0.001, 0.322)	(0.058, 1.750)
χ^2_{HET}	Statistic	7.07	7.66	8.85	12.33	35.31
	P-value	0.21	0.18	0.12	0.03	<0.001

Snaith, 1983), STAI [(Spielberger et al., 1970), Kellner (Kellner and Sheffield, 1973), and HSCL (Derogatis et al., 1974). The actual reported results are presented in Table 1. There is considerable variation in not only the instruments used but also in what was actually reported. In combining the individual-study results, the standardized difference (between the two test groups) of mean change in long-term anxiety levels will be used (Fleiss, 1993).

B. Other Information

A number of other authors (Herrmann, 1997; Ford et al., 1990; Salkovskis et al., 1990; Hopwood et al., 1991; Jenkins et al., 1994; Visser et al., 1995) have considered the test–retest reliability of the HAD instrument for assessing anxiety levels over time intervals of between 1 and 6 months. The results of these studies, in terms of the within-subject correlation are reported later in Table 4. Visual inspection of the table reveals that the level of within-subject correlation of the HAD is clearly positive, and indeed relatively high, ranging from 0.55 to 0.89. Therefore, in any analysis of changes in anxiety levels over time, it is imperative to account for within-subject correlation.

III. CLASSICAL ANALYSES

This section considers Classical meta-analyses of the data identified in the systematic review and presented in Table 1.

A. Calculation of Standardized Differences

The different studies used a variety of outcome measures. In order for the results in Table 1 to be combined quantitatively, they first must be transformed into standardized group differences. We will use the following notation. Let x_{ij} denote the mean baseline anxiety level in the ith study and jth group, i.e., $j = 1$ are those subjects who ultimately tested negative, and $j = 2$ are those patients who ultimately tested positive. Similarly let y_{ij} denote the mean follow-up anxiety level in the jth group in the ith study. For each group, within each study, we can calculate the mean change between baseline and follow-up, d_{ij}, as $x_{ij} - y_{ij}$, and the mean difference between the two groups, d_i, as $d_{i1} - d_{i2}$. To standardize this group difference, d_i, so that studies using different instruments may be combined, it will be divided by the corresponding standard deviation.

If the variances of the mean change in the two groups in each study were reported, then an estimate of the corresponding variance of the difference in mean changes, between groups, $\hat{\sigma}^2_{di}$, is given by

$$\hat{\sigma}^2_{di} = \frac{(n_{i1} - 1)V(d_{i1}) + (n_{i2} - 1)V(d_{i2})}{n_{i1} + n_{i2} - 2}, \tag{1}$$

where n_{i1} and n_{i2} are the numbers of subjects in each group in the ith study. Consequently the standardized difference, z_i, for the ith study is given by

$$z_i = \frac{d_i}{\hat{\sigma}_{di}}. \tag{2}$$

However, as can be seen from Table 1, only studies 2 and 3 report the mean change and corresponding standard deviations for the two groups. The remaining studies report the anxiety levels, in terms of means and standard deviations, separately for baseline and follow-up. While this still ensures that the mean change can be calculated, the calculation of the corresponding variances, which are required in (1), is not as straightforward. Assuming, as is the case in Table 1, that the variances are reported at baseline and follow-up for the two groups, then the variance of the mean change is given by

$$V(d_{ij}) = V(x_{ij}) + V(y_{ij}) - 2\,\mathrm{cov}(x_{ij}, y_{ij}). \tag{3}$$

The covariance term in (3) may be expressed in terms of the individual variances and the within-subject correlation, ρ, as

$$V(d_{ij}) = V(x_{ij}) + V(y_{ij}) - 2\rho\sqrt{V(x_{ij})V(y_{ij})}. \tag{4}$$

Although the data in Table 1 do not provide any further information regarding ρ, for any of the studies based on the reported results, explicit calculation of $V(d_{ij})$ in terms of ρ does ensure that either sensitivity analyses, as considered in subsection III.C or analyses which use information from other studies regarding ρ, as in Sect. V, may be performed. It is useful to see how different values of ρ lead to different estimates of the standardized difference. From (4) it can be seen that higher values of ρ will lead to lower values of $V(d_{ij})$, which in turn will lead to smaller values of σ^2_{di}. A smaller value of σ^2_{di} in (2) will thus lead to a higher standardized difference. Finally, having calculated the standardized differences, z_i, the variance for the z_is is also required and is given by

$$\sigma_{zi}^2 = \frac{n_{i1} + n_{i2}}{n_{i1} n_{i2}} + \frac{z_i^2}{2(n_{i1} + n_{i2})}. \tag{5}$$

The second term of (5) includes z_i. This causes difficulty when adopting the Bayesian approach of Sect. V. A number of authors have suggested an approximation to σ_{zi}^2 based on only the first term of (5), and this will be used in all subsequent analyses (Fleiss, 1993).

B. Assuming $\rho = 0$

One possible assumption regarding the value of ρ in studies 1, 4, 5, and 6 is to assume that it is zero. While the studies shown later in Table 4 suggest that, even up to six months between assessments, ρ could be as great as 0.70, assuming ρ to be zero could be considered as a reference analysis, yielding wider confidence intervals than would be the case when $\rho > 0$. The individual standardized differences and associated 95% CIs (confidence intervals) are presented in Fig. 1, which shows that there is some variation between the studies with respect to both the point estimates and the width of the 95% CIs.

In combining the estimated standardized differences from the six studies, a number of possible models can be used (see Stangl and Berry, Chap. 1, for an overview). An initial analysis could be based upon a *fixed-effect* model, in which any between-study heterogeneity is ignored and a weighted average of the standardized differences is formed, each study being weighted by the reciprocal of its variance (Fleiss, 1993). Thus,

$$\hat{\mu} = \frac{\sum_{i=1}^{K}(z_i/\sigma_{zi}^2)}{\sum_{i=1}^{K}(1/\sigma_{zi}^2)}, \qquad V(\hat{\mu}) = \frac{1}{\sum_{i=1}^{K}(1/\sigma_{zi}^2)}.$$

A more appropriate analysis would take into account the between-study heterogeneity using a *random-effect* model assuming normality (Fleiss, 1993). Such a model can be described by

$$z_i \sim N[\theta_i, \sigma_i^2] \quad (i = 1, \ldots, 6), \qquad \theta_i \sim N[\mu, \tau^2], \tag{6}$$

where μ is the overall population standardized difference, and τ^2 is a measure of the between-study heterogeneity. Some authors have advocated methods for parameter estimation using random-effects models such as (6), including method of moments and maximum likelihood

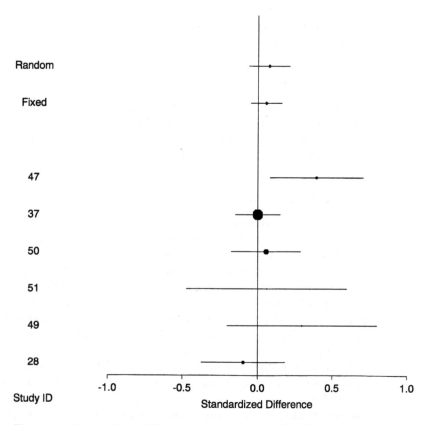

Figure 1 Standardized differences together with 95% CIs based on reported results from studies assessing differences in long-term anxiety between patients who tested positive and negative, and results from classical fixed and random effects meta-analyses (the size of the symbols is proportional to the total number of subjects in each study).

(DerSimonian and Laird, 1986; Harville, 1977). A disadvantage of these methods is that they have not fully accounted for uncertainty in the model parameters (Sutton et al., 1998, 1999). Namely, estimating μ, which is frequently the main parameter of interest, has not taken the uncertainty associated with τ^2 into account. Ignoring this uncertainty is problematic when there are a small number of studies. Though a number of empirical Bayes methods have been advocated as a solution (Morris, 1983; Louis, 1984; Carlin, 1992), more recently methods based on profile likelihood (Hardy and Thompson, 1996) and asymptotic likelihood methods (Biggerstaff and Tweedie, 1997) have been suggested. Sutton et al. (Chap. 14) discuss software to implement the various modelling approaches.

Discrimination between the fixed and random effects models from a classical perspective is frequently based on a χ^2 test for homogeneity, which has low statistical power and therefore should be interpreted cautiously (Fleiss, 1993; Thompson, 1994).

Table 2 shows the results of applying both fixed- and random-effect models to the long-term anxiety data of Table 1 using maximum likelihood estimation methods (MLE). For studies 2 and 3, the z_is, i.e., the standardized differences between the mean changes in subjects testing positive and negative, are calculated using the published statistics, while for studies 1, 4, 5, and 6, the $\hat{\sigma}_{d_i}^2$s are calculated using (1) and (4) assuming $\rho = 0$. Table 2 shows that the point estimates for the pooled standardized differences are not very different using either of the two methods, with the random-effect model producing slightly wider 95% CIs as a result of allowing for the between-study heterogeneity. There appears to be little difference in long-term anxiety levels between subjects who test positive and those who test negative. A χ^2 test for homogeneity yields a p-value of 0.21. Therefore there seems to be little evidence to reject the null hypothesis of homogeneity. However, this test has low power.

C. Sensitivity Analysis

Subsection III.B presents a reasonable initial analysis, but one that assumes that $\rho = 0$ in studies 1, 4, 5, and 6, while studies 2 and 3 automatically allow for $\rho \neq 0$ as they report mean changes from baseline. To assess the implications of not accounting for within-subject correlation in four of the six studies, a sensitivity analysis will be performed. Results of the meta-analysis will be compared across various values of ρ. While, in theory, ρ may take values between -1 and $+1$, it would seem reasonable to assume a non-negative within-subject correlation. Furthermore, the data presented in Table 4, from other studies that have examined the level of within-subject correlation when assessing anxiety levels longitudinally, suggest that ρ may be over 0.5 and as high as 0.9. Therefore, in addition to using $\rho = 0$, we consider values of 0.25, 0.5, 0.75, and 0.95.

In addition to performing a χ^2 test for homogeneity, the same fixed- and random-effect models as in Section 3.2 were applied. The results, in terms of maximum-likelihood (MLE) parameter estimates, are presented in Table 2, and Fig. 2 displays the results graphically. The sensitivity analysis shows that the results appear to be fairly consistent across the various values of ρ, except when ρ takes values near $+1$. Although as ρ increases the pooled standardized difference increases, the only model where such an effect approaches statistical significance at the 5% level is under a fixed-effect model with $\rho = 0.95$. Figure 2(a) also shows that the 95% CIs increase as ρ increases, but that the random-effect model yields wider CIs at all values of ρ compared to the fixed-effect model. Similarly, Figs 2(b, c) show that, in terms of the χ^2 statistic for homogeneity and its associated test, statistical significance is only approached at the 10% level for values of ρ of 0.75 and 0.95.

IV. BAYESIAN ESTIMATION AND MODEL AVERAGING

This section considers a Bayesian approach and contrasts it with the classical meta-analyses presented in Sect. 3.

A. Bayesian Model for Meta-analysis

A number of authors have considered Bayesian approaches to meta-analytic models; see Sutton et al. (1998, 1999) and Chap. 12. The approach adopted here utilizes a two-stage Gaussian hierarchical model similar to that used in the random-effects analysis of Section 3 (DuMouchel, 1990; Louis and Zelterman, 1994; Abrams and Sansó, 1998). The model is

$$z_i \sim \mathrm{N}[\theta_i, \sigma_i^2] \quad \sigma_i^2 \sim \mathrm{IG}[u_i, v_i], \quad \theta_i \sim \mathrm{N}[\mu, \tau^2] \qquad \text{for } i = 1, \ldots, K, \tag{7}$$

$$\mu \sim \mathrm{N}[a, b], \qquad \tau^2 \sim \mathrm{IG}[c, d],$$

where the hyper-parameters u_i, v_i, a, b, c, and d need to be specified by the analyst. Several authors have commented that replacing the σ_i^2s with the actual observed study-level variances in (7) makes little difference to the inferences made (Smith et al., 1995; Abrams and Sansó, 1998). Often *a priori* beliefs regarding μ, the overall pooled effect, are diffuse. In model (7), in which a normal prior distribution is assumed for μ, diffuse beliefs

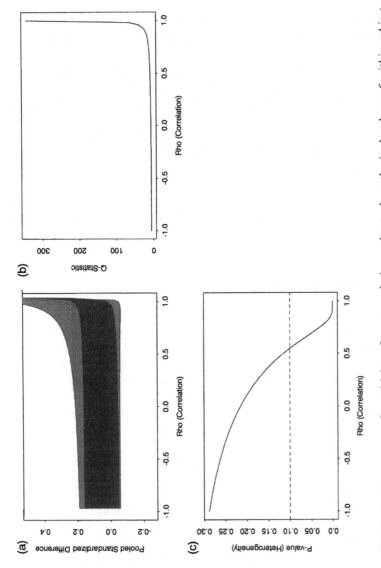

Figure 2 Assessment of sensitivity of meta-analysis results to hypothesized values of within-subject correlation (ρ); (a) pooled standardized difference (key: ■ fixed-effect model, ▨ random-effect model), (b) Heterogeneity Q-statistic, and (c) p-value associated with heterogeneity Q-statistic.

may be represented by specifying a suitably large variance, b. Alternatively a uniform prior distribution over the whole real line may be used.

The choice of an inverse-gamma prior distribution for τ^2 has the advantage of constraining τ^2 to be greater than zero, a feature that the classical approach does not possess, and yielding a prior distribution that can accommodate informative *a priori* beliefs. The latter aspect is particularly advantageous, since there is often evidence that cannot formally be incorporated into the meta-analysis, but which may provide information regarding the plausible level of variability observed in a population (Abrams and Jones, 1995; Sutton et al., 1999). In the analyses presented here, a prior distribution with $c = d = 0.001$ is used to represent diffuse *a priori* beliefs (Smith et al., 1995) for τ^2 and a N[0, 1000] prior distribution for μ.

B. Parameter Estimation

In a Bayesian analysis for which a conjugate analysis is not possible, there are a number of methods for parameter estimation available (Bernardo and Smith, 1993). While asymptotic methods have been considered in a meta-analysis setting (Abrams and Sansó, 1998), more recently Markov chain Monte Carlo (MCMC) methods, and Gibbs sampling in particular, have been advocated as a relatively efficient method for estimating parameters in models such as (7) (Smith et al., 1995; Gilks et al., 1996) and are used in the following analyses.

Implementation of Gibbs sampling can be accomplished in any programming environment that allows random sampling from known distributions. However, development of the BUGS software* (Thomas et al., 1992) has provided a unified and user-friendly method for implementing Gibbs sampling. In addition there is the S-Plus (Statistical Sciences Inc., 1990) suite of functions CODA* (Cowles et al., 1994) which implements many of the diagnostic methods that have been advocated (Cowles and Carlin, 1996; Gelman, 1996) and which works directly with the output from BUGS. The BUGS and CODA software is used to estimate parameters in the Bayesian models considered in this chapter. A discussion of

---

*Available at *http://www.mrc-bsu.cam.ac.uk/bugs/*

other software for explicitly implementing a Bayesian meta-analysis is given by Sutton et al. (Chap. 15).

C. Results

Table 2 displays the results of applying the above Bayesian model to the five values of ρ, denoted MCMC, for both a model in which τ^2 is constrained to be zero, corresponding to the classical fixed-effects model, and a model where it is left unspecified, i.e., as a parameter in the model. While the fixed-effect estimates obtained from the Bayesian model are in close agreement with those obtained in the classical analysis, there are slight differences with respect to the random-effect estimates. In general, the standard deviations are larger under the Bayesian analysis, reflecting the fact that the extra uncertainty regarding the estimation of τ^2 has been taken into account, and therefore the relative weighting of the six studies in the analysis has been changed, producing slightly different estimates of the overall pooled standardized difference. As with the classical analysis, only the fixed-effect model with $\rho = 0.95$ has a 95% CrI (credible interval) which excludes zero, indicating some evidence of an effect between groups. In terms of the estimation of τ^2, the estimates obtained by adopting a Bayesian approach are slightly larger than those obtained using maximum likelihood, reflecting the extra uncertainty induced by specification of the relatively vague prior distribution. Note though that the posterior distribution for τ^2 is highly skewed, and inference based solely on the posterior mean can be misleading.

Figure 3 shows trace plots of the samples obtained from applying Gibbs sampling to the Bayesian model with $\rho = 0$ for both μ and τ^2, together with kernel estimated posterior densities. Formal examination of the samples revealed neither lack of convergence nor auto-correlation to be present. In addition, the use of radically different starting values for μ and τ^2 produced quantitatively similar results, as did varying the length of both burn-in and sampling chain. The results presented were obtained using a burn-in of 1,000, together with a sampling of chain of length 5,000 iterations.

D. Model Comparison and Model Averaging

As with the classical models of Sect. III, we often wish to compare the various models that we have considered. In a meta-analysis context, one

Table 3 Model Comparison and Posterior Estimates for Fixed and Random Meta-analysis Assuming Various Values for ρ

ρ	Model	M_j	$\log_e[P(D|M_j)]$	$2\log_e(B_{10})$	B_{10}	$P(M_j|D)$	$\hat{\mu}_j$	$SE(\hat{\mu}_j)$
0.00	Fixed	M_0	−12.26			0.11	0.057	0.053
	Random	M_1	−11.21	2.10	8.17	0.89	0.079	0.090
	Averaged					1.00	0.077	0.087
0.25	Fixed	M_0	−12.00			0.08	0.059	0.053
	Random	M_1	−10.78	2.44	11.47	0.92	0.084	0.093
	Averaged					1.00	0.082	0.091
0.50	Fixed	M_0	−11.83			0.04	0.063	0.053
	Random	M_1	−10.22	3.22	25.03	0.96	0.093	0.100
	Averaged					1.00	0.092	0.099
0.75	Fixed	M_0	−12.32			0.03	0.072	0.053
	Random	M_1	−9.34	5.96	387.61	0.97	0.119	0.123
	Averaged					1.00	0.118	0.122
0.95	Fixed	M_0	−22.54			0.00	0.113	0.053
	Random	M_1	−6.13	32.82	1.79×10^{14}	1.00	0.275	0.300
	Averaged					1.00	0.275	0.300

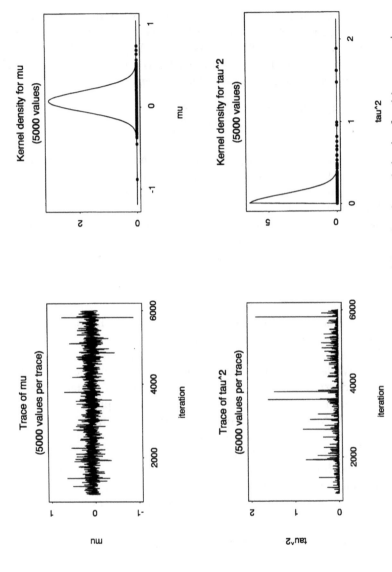

Figure 3 Trace of marginal posterior samples and kernel estimated density for (a) μ assuming an N[0, 1000] prior distribution, and (b) τ^2 assuming an IG[0.001, 0.001] prior distribution.

comparison of particular interest is that of fixed- and random-effect models. From a Bayesian perspective, one approach is to use what are termed *Bayes factors* (BFs) (Kass and Raftery, 1995). In the case when there are two models, say M_{FE} a fixed-effect model and M_{RE} a random-effect model, and that we have data D, then

$$\frac{P(M_{FE}|D)}{P(M_{RE}|D)} = \frac{P(D|M_{FE})}{P(D|M_{RE})} \times \frac{P(M_{FE})}{P(M_{RE})},$$

which—assuming that M_{FE} and M_{RE} are the only possible models under consideration—is equivalent to saying that the posterior odds of M_{FE} are equal to the prior odds of M_{FE} multiplied by the quantity $P(D|M_{FE})/P(D|M_{RE})$, which is termed the *Bayes factor*. Thus, the Bayes factor is the ratio of the posterior odds of M_{FE} to the prior odds of M_{FE}. Though a number of scales have been proposed on which to interpret Bayes factors, the key distinction with a classical hypothesis testing framework is that they convey a continuum of plausibility among the various models.

In addition to providing a means for model comparison, if the *a priori* model probabilities are specified, then Bayes factors enable calculation of the posterior model probabilities, and thus averaging of effects across all possible models, e.g., fixed- and random-effect models.

While estimation of Bayes factors is problematic when improper prior distributions are assumed for model parameters (Berger and Pericchi, 1995; O'Hagan, 1994), when proper prior distributions are used, MCMC methods may be used (Gelfand and Dey, 1994; Carlin and Chib, 1995; Gelfand, 1996). Assume that as a result of using a MCMC approach, a sample of L realizations $\{\hat{\theta}_{lk}; l = 1, \ldots, L\}$ are obtained for the kth model. In addition, if there exists an importance sampling density, $h(\theta)$, for the posterior density, then Gelfand (1996) suggested that an approximation to the marginal, $P(D|M_k)$, is given by

$$P(D|M_k) = \frac{1}{\dfrac{1}{L}\displaystyle\sum_{l=1}^{L} \dfrac{h(\hat{\theta}_{lk})}{P(D|\hat{\theta}_{lk}, M_k)P(\hat{\theta}_{lk})}}. \tag{8}$$

The better $h(\theta)$ approximates the posterior density, then the better the approximation in (8) will be. An alternative approach is to assume that $h(\theta)$ and $P(\theta)$ are equivalent, and therefore (8) reduces to

Table 4 Derivation of Prior Distribution for ρ Based on Studies which Considered the Test–retest Within-subject Correlation of the HAD for Assessing Anxiety Levels

Author	Year	Interval (Months)	N	$\hat{\rho}$	$S = \frac{1}{2}\log_e[(1+\rho)/(1-\rho)]$	SE(S)	95% CI for ρ
Ford et al.	1990	6	67	0.70	0.867	0.125	(0.55, 0.81)
Salkovskis et al.	1990	1	130	0.71	0.887	0.089	(0.61, 0.79)
Hopwood et al.	1991	3	155	0.55	0.618	0.081	(0.43, 0.65)
Jenkins et al.	1994	6	12	0.77	1.020	0.333	(0.35, 0.93)
Visser et al.	1995	1	42	0.89	1.422	0.160	(0.80, 0.94)
Shaw et al.	Unpublished	1	166	0.85	1.256	0.078	(0.80, 0.89)
Fixed				0.74	0.960	0.042	(0.71, 0.78)
Random				0.76	1.001	0.136	(0.63, 0.85)
Bayesian				0.75	1.001	0.165	(0.59, 0.87)

$$P(D|M_k) = \frac{1}{\frac{1}{L}\sum_{l=1}^{L}\frac{1}{P(D|\hat{\theta}_{lk}, M_k)}}.$$

E. Results

The log marginal densities, and the corresponding Bayes factors, were calculated using the output from the MCMC analysis for all ten models, i.e., $\rho = 0, 0.25, 0.5, 0.75, 0.95$ for both fixed and random effects, and the results are presented in Table 3.

Considering $2\log_e(B_{10})$ there is increasing evidence to favor a random-effect model as ρ increases, from slight evidence at $\rho = 0$ to overwhelming evidence when $\rho = 0.95$. In terms of the posterior model probabilities, assuming that *a priori* each model was equally plausible, i.e., $P(M_0) = P(M_1) = 0.5$, these range from 89% when $\rho = 0$ to 100% when $\rho = 0.95$. An advantage of the Bayesian approach, though, is that the posterior model probabilities may be used to average over both the fixed- and random-effect models at each of the values for ρ, in order to produce overall estimates that take into account both the uncertainty associated with the estimates obtained from each model and the model uncertainty. Table 3 shows that the average point estimates of the pooled standardized difference increase as ρ increases, but that in addition the pooled standard deviations also increase with ρ, so that for no value of ρ would a 95% CrI exclude zero, i.e indicating a difference in mean changes of anxiety levels for the two groups. This can be seen in Fig. 4 which displays the five model averaged posterior densities for μ.

V. INCORPORATING UNCERTAINTY REGARDING ρ

In their own particular ways, the methods of Sects III–IV have dealt with the question of uncertainty regarding ρ. However, in terms of a sensitivity analysis, the approach has not taken account of the evidence from the six studies in Table 4 regarding plausible values for ρ.

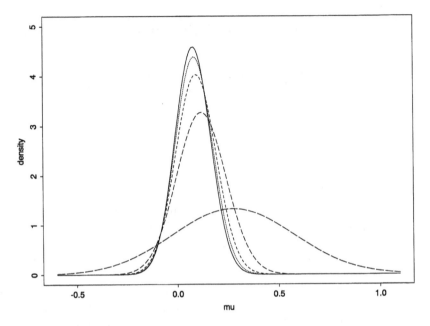

Figure 4 Posterior densities of μ for various values of ρ averaged over fixed- and random-effect models (Key: — $\rho = 0$, $\cdots\cdots$ $\rho = 0.25$, – – – $\rho = 0.5$, —— —— $\rho = 0.75$, — — $\rho = 0.95$).

A. Model Description

First, the Bayesian model of Sect. IV needs to be extended so that ρ is explicitly incorporated as a random parameter, so that a prior distribution, based on the other evidence, may be specified. ρ is used in calculating the variance of the change from baseline, for the two groups separately, i.e., positive and negative subjects, in the four studies that do not automatically adjust for it in their reporting. The variances for the two groups are then used to calculate the variance of the mean difference, which can then be used to standardize it. In order to estimate the model parameters in BUGS it is necessary to re-parameterize the model, so that each random quantity appears once on the left-hand side of the model, thus preventing each model entity from being both deterministic and stochastic. Thus, the difference in mean change is distributed with mean $\theta_i \sigma_{zi}$ and variance $\sigma_{zi}^2 \sigma_{di}^2$. Algebraically, the model is of the following form,

$$V(d_{ij}) = V(x_{ij}) + V(y_{ij}) - 2\rho\sqrt{V(x_{ij})V(y_{ij})}, \qquad i = 1, 4, 5, 6,$$

$$j = 1, 2,$$

$$\sigma_{di}^2 = \frac{(n_{i1} - 1)V(d_{i1}) + (n_{i2} - 1)V(d_{i2})}{n_{i1} + n_{i2} - 2}, \qquad \sigma_{zi}^2 = \frac{n_{i1} + n_{i2}}{n_{i1}n_{i2}},$$

$$d_i \sim N[\mu_i, \gamma_i], \qquad \mu_i = \theta_i\sigma_{zi}, \qquad \gamma_i = \sigma_{zi}^2\sigma_{di}^2, \qquad (9)$$

$$\theta_i \sim N[\mu, \tau^2],$$

$$\mu \sim N[a, b], \qquad \tau^2 \sim IG[c, d], \qquad \rho \sim \text{gamma}[e, f]I[, 1].$$

As with the model in Sect. IV, diffuse prior distributions are assumed for μ and τ^2, namely N[0, 1000] and IG[0.001, 0.001] respectively. In terms of the prior distribution for ρ, there are a number of possibilities. Assuming that the within-subject correlation is positive, then one possible diffuse prior distribution would be a uniform distribution on [0,1]. Alternatively a beta or truncated gamma distribution would enable more informative beliefs to be incorporated. Though a beta distribution would appear to be particularly suitable, the use of such a distribution produces conditional distributions that are particularly difficult to sample from. Therefore, the results that follow are based on a gamma distribution, which is truncated at 1, i.e., gamma[e, f]I[, 1]. Figure 5 shows the directed acyclic graph for the above model, and the relevant BUGS code is presented in Appendix A.

B. Derivation and Choice of Prior Distribution for ρ

In order to incorporate the information from the six studies presented in Table 4 into a prior distribution for ρ, it first needs to be quantitatively combined. However, in order to do so, we have to assume that ρ can be considered the same across all four instruments, and that differences in the study lengths, compared to the studies in Table 1, make little qualitative difference. As the correlation coefficient is unlikely to be normally distributed, the individual study results are first transformed to a new scale, S, such that $S = \frac{1}{2}\log_e[(1 + \rho)/(1 - \rho)]$, which is approximately normally distributed and for which an approximate standard error is $1/\sqrt{n - 3}$, where n is the number of subjects upon which the correlation coefficient is based (Gardner and Altman, 1989; Altman, 1992). Table 4 shows the results for the six studies that assessed the within-subject correlation for the HAD scale. Using S and its standard error, a formal

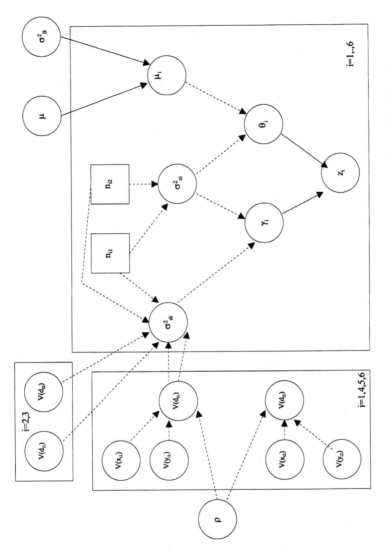

Figure 5 Directed acyclic graph (DAG) for fully Bayesian meta-analysis model taking into account prior information for ρ.

meta-analysis may be performed. Table 4 shows the results of using a classical fixed- and random-effects model, together with those obtained using a Bayesian analysis, with diffuse prior distributions on all parameters, i.e., $\mu \sim N[0, 1000]$ and $\tau^2 \sim IG[0.001, 0.001]$. The pooled point estimates for S are similar using all three methods, but the standard errors of these estimates are greater for the Bayesian analysis than for either the fixed- or random-effect classical analysis. Transforming back onto the correlation coefficient scale, the Bayesian analysis yields a point estimate of 0.75 with associated 95% CrI (0.59, 0.87). These results will be used with a "method of moments," to derive a gamma prior distribution for ρ.

If x has a gamma distribution with parameters α and β, then the mean of x, $E(x)$, and the variance of x, $V(x)$, are given by α/β and α/β^2 respectively. Therefore, equating the first and second moments of the gamma distribution with the results obtained from the meta-analysis in Table 4 yields

$$E(X) = \frac{\alpha}{\beta} = 0.754, \tag{10}$$

$$V(X) = \frac{\alpha}{\beta^2} = 0.073^2. \tag{11}$$

Solving (10) and (11) yields estimates for α and β of 103 and 140 respectively. Figure 6 displays this prior distribution, together with a uniform distribution on [0, 1] and a gamma distribution with parameters 3 and 4. This latter prior distribution was derived by assuming that values of ρ exceeding 0.3 were more likely than values ≤ 0.3, with values of ρ between 0.3 and 0.8 approximately equally likely. Figure 6 shows that the derived prior distribution is considerably more informative than either the uniform or gamma[3, 4] prior distributions, with a mode at 0.73.

C. Results

Table 5 shows the results obtained from (9) using the three different prior distributions for ρ outlined in Section 5.2 with parameters estimated using Gibbs sampling implemented via BUGS. In terms of the posterior estimates of μ, the pooled standardized difference, Table 5 shows that those estimates obtained using the three prior distributions are qualitatively similar, both in terms of the point estimates and the standard deviations. Figure 7 shows the posterior marginal densities for

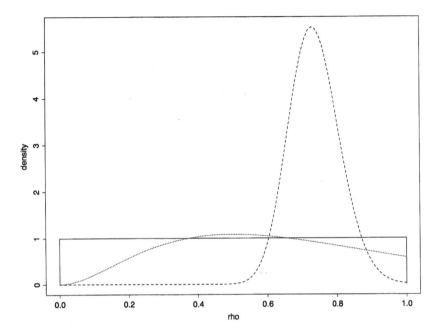

Figure 6   Prior distributions for ρ (key: — Uniform[0,1], ······ gamma[3, 4], − − − gamma[103, 140]).

μ obtained using the three prior distributions. The estimates obtained for τ^2 using the three prior distributions are also similar. However, as might be expected when the posterior estimates for ρ are considered, those corresponding to the derived informative prior distribution have a larger point estimate, similar in fact to the prior mean, and the standard deviation is approximately a third of those obtained using a vague prior distribution.

In terms of comparing the fully Bayesian approach adopted here with the classical and Bayesian sensitivity analyses of Sects 3 and 4, the overall pooled estimate of the standardized difference is consistent with values of ρ between 0.5 and 0.75, with similar posterior standard deviations.

Table 5 Posterior Estimates for Bayesian Model Using a Prior Distribution for ρ

Parameter		Prior distribution for ρ		
		Uniform [0, 1]	Gamma [3, 4]	Gamma [103, 140]
μ	Mean	0.111	0.105	0.117
	Median	0.100	0.094	0.110
	SD	0.120	0.113	0.118
	95% CrI	(−0.088, 0.394)	(−0.083, 0.375)	(−0.094, 0.377)
τ^2	Mean	0.054	0.049	0.060
	Median	0.019	0.018	0.027
	SD	0.118	0.106	0.128
	95% CrI	(0.001, 0.340)	(0.001, 0.283)	(0.001, 0.317)
ρ	Mean	0.528	0.564	0.732
	Median	0.559	0.579	0.732
	SD	0.248	0.202	0.066
	95% CrI	(0.049, 0.920)	(0.162, 0.902)	(0.605, 0.862)

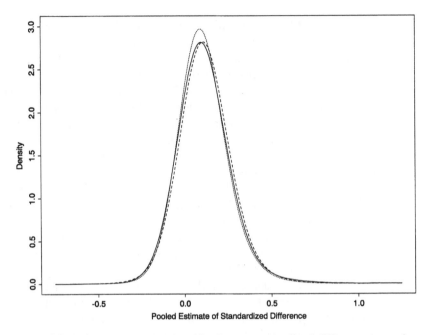

Figure 7 Marginal posterior densities for μ, standardized difference, assuming various prior distributions for ρ (key: — Uniform[0,1], ······ gamma[3, 4], – – –– gamma[103, 140]).

VI. DISCUSSION

The results of this chapter show that the application of either classical or Bayesian methods to the synthesis of incompletely reported study results may give rise to misleading conclusions, if such reporting is not taken into account. From a classical perspective, a sensitivity analysis may be conducted to examine the implications such reporting may have on the overall conclusions. However, although such an analysis may yield further insight into the potential for misleading conclusions, such uncertainty with respect to the overall pooled estimate cannot be formally incorporated. Though the Bayesian analysis of Sect. V yields consistently wider credibility intervals than either of the approaches of Sects III or IV, except when $\rho = 0.95$, whether such differences would have an impact upon screening policy can only be assessed within a decision theoretic framework (Petitti, 1994).

We have also seen in this chapter how a Bayesian sensitivity analysis combined with examining the *a posteriori* plausibility of the models considered enables model averaging to be used, which—although still not able to incorporate other sources of information—does take into account the uncertainty with respect to the different types of model in estimating the overall pooled estimate. The extension of the meta-analytic models to formally incorporate other information relating to the reporting of study results is a distinct advantage of the Bayesian approach.

A number of possible extensions to the methods described in this chapter are possible. Firstly, if individual-patient data had been available, then many of the difficulties induced by the incomplete reporting of study results would have been eliminated. However, the use of individual-patient data itself raises a number of problems (Stewart and Parmar, 1993), especially when there are some studies in a meta-analysis for which such data are available and others for which there is not (Higgins and Whitehead, 1997; Sutton et al., 1999).

In using the evidence from the other studies which had explored the within-subject correlation when using the HAD score, it has been assumed that ρ is the same for all four studies, even though different instruments were used to assess anxiety levels. It would be possible to extend the model (9) so that the prior distribution for ρ depended on which instrument had been used to measure anxiety levels. In addition, it would also be possible to extend the derivation of the prior distribution for ρ to take into account the fact that the studies in Table 4 did in fact have different follow-up times. In terms of the formulation of ρ in model (9), the assumption that a truncated gamma distribution is appropriate has been made. While such an assumption is not unreasonable, an alternative approach would have been to have transformed ρ onto the S scale, which is approximately normally distributed on the real line (Altman, 1992).

The parameters in the Bayesian models (7) and (9) have been estimated using the BUGS software, which (in conjunction with the CODA software) provides a user-friendly means of implementing Bayesian analyses in practice. In addition there is a suite of S-Plus functions which enables parameter estimation and inference in the standard Bayesian model (7) (DuMouchel, 1996; Sutton et al., Chap. 15).*

---

*Available at *ftp://ftp.research.att.com/dist/bayes-meta*

In conclusion, the quantitative synthesis of heterogeneously reported studies, using a variety of outcome measures, raises a number of important issues. By adopting a Bayesian approach, not only can allowance be made for all parameter uncertainty, but also other pertinent information may be formally included.

Acknowledgments

The authors would like to thank David R Jones and Alex J Sutton for their useful discussions. The first and third authors would also like to thank the British Council (Venezuela) for financial assistance.

Appendix A: BUGS CODE FOR FULL BAYESIAN ANALYSIS

```
model anxiety ;
const
    N=6;                    # number of studies
var
    study[N],               # study ID
    pos.n[N],               # no in +ve group
    pos.base.mean[N],       # baseline mean in +ve group
    pos.base.sd[N],         # baseline sd in +ve group
    neg.n[N],               # no in -ve group
    neg.base.mean[N],       # baseline mean in -ve group
    neg.base.sd[N],         # baseline sd in -ve group
    pos.post.mean[N],       # post trt mean in +ve group
    pos.post.sd[N],         # post trt sd in +ve group
    neg.post.mean[N],       # post trt mean in -ve group
    neg.post.sd[N],         # post trt sd in -ve group
    pos.change.mean[N],     # change mean in +ve group
    pos.change.var[N],      # change var in +ve group
    neg.change.mean[N],     # change mean in -ve group
    neg.change.var[N],      # change var in -ve group
    diff.mean[N],           # difference in means
    var.pool[N]             # pooled variance
    diff.stand[N],          # standardized diff
```

```
    var.stand[N],              # var standardized diff
    tau.stand[N],              # precision of standardized difference
    mu.study,                  # pooled estimate
    tau.study,                 # between study precision
    sigma.study,               # between study variance
    theta[N],                  #
    gamma[N],                  # Model Parameters
    mu[N],                     #
    rho;                       # within subject correlation

data  study, pos.n, pos.base.mean, pos.base.sd,
      neg.n, neg.base.mean, neg.base.sd, pos.post.mean,
      pos.post.sd, neg.post.mean, neg.post.sd, pos.change.mean,
      pos.change.var, neg.change.mean, neg.change.var,
      diff.mean in 'anxiety.dat';

inits in 'anxiety.in';

{
for (i in 1:4) {
   pos.change.var[i]  <- pow(pos.base.sd[i],2) + pow(pos.post.sd[i],2) -
                     2*rho*pos.base.sd[i]*pos.post.sd[i];
   neg.change.var[i]  <- pow(neg.base.sd[i],2) + pow(neg.post.sd[i],2) -
                     2*rho*neg.base.sd[i]*neg.post.sd[i];
}
for (i in 1:N) {
    var.pool[i]<- ((pos.n[i]-1)*pos.change.var[i] +
                 (neg.n[i]-1)*neg.change.var[i]) /
                 (pos.n[i] + neg.n[i] - 2);
    diff.mean[i] ~ dnorm(theta[i],gamma[i]);
    theta[i] <- mu[i] * sqrt(var.pool[i]);
    gamma[i] <- 1/(var.pool[i] * var.stand[i]);

    var.stand[i] <- (pos.n[i] + neg.n[i]) / (pos.n[i] * neg.n[i]);

    mu[i] ~ dnorm(mu.study,tau.study);
}

# Prior Distributions
mu.study ~ dnorm(0,0.0001);
```

```
tau.study ~ dgamma(0.001,0.001);
sigma.study <- 1/tau.study;
rho ~ dgamma(103,140)I(0,1);
}
```

References

Abrams, K. R. and Jones, D. R. (1995). Meta-analysis and the synthesis of evidence. *IMA Journal of Mathematics Applied in Medicine and Biology*, 12, 297–313

Abrams, K. R. and Sansó, B. (1995). Discrimination in meta-analysis—a Bayesian perspective. *Technical Report 95-03*. Department of Epidemiology and Public Health, University of Leicester, U.K. (http://www.prw.le.ac.uk/research/medstats/tech.html).

Abrams, K. R. and Sansó, B. (1998). Approximate Bayesian inference in random effects meta-analysis. *Statistics in Medicine*, 17, 201–218.

Altman, D. (1992). *Practical Methods in Medical Research*. Chapman and Hall, London.

Ambrosio, G. B., Dissegna, L., Zamboni, S., Santonastaso, P., Canton, G., and Dal Palu, C. (1984). Psychological effects of hypertension labelling during a community survey. A two year follow-up. *Journal of Hypertension*, 2(3), 171–173.

Berger, J. O. and Pericchi, L. R. (1995). The intrinsic Bayes factor for linear models. In: Bernardo, J., Berger, J., Dawid, A., and Smith, A., editors, *Bayesian Statistics 5*, Oxford. Oxford University Press.

Bernardo, J. M. and Smith, A. F. M. (1993). *Bayesian Theory*. John Wiley and Sons, Chichester, England.

Biggerstaff, B. J. and Tweedie, R. L. (1997). Incorporating variability in estimates of heterogeneity in the random effects model in meta-analysis. *Statistics in Medicine*, 16,753–768.

Carlin, B. P. and Chib, S. (1995). Bayesian model choice via Markov-chain Monte-Carlo methods. *Journal of The Royal Statistical Society, Series B*, 57(3), 473–484.

Carlin, J. B. (1992). Meta-analysis for 2 × 2 tables: a Bayesian approach. *Statistics in Medicine*, 11,141–158.

Cowles, M. K. and Carlin, B. P. (1996). Markov chain Monte Carlo convergence diagnostics: a comparative review. *Journal of the American Statistical Association*, 91,883–904.

Cowles, M. K., Best, N. G., and Vines, K. (1994). CODA—convergence diagnostics and output analysis software for Gibbs samples produced by the

BUGS language version 0.30. *Technical Report*. MRC Biostatistics Unit, Cambridge.

Derogatis, L. R., Lipman, R. S., Rickels, K., Uhlenhuth, E. H., and Covi, L. (1974). The Hopkins Symptom Checklist (HSCL): a self-report symptom inventory. *Behavioural Science*, 19(1),1–15.

DerSimonian, R. D. and Laird, N. (1986). Meta-analysis in clinical trials. *Controlled Clinical Trials*, 7,177–188.

DuMouchel, W. (1990). Bayesian metaanalysis. In: Berry, D., editor, *Statistical Methodology in the Pharmaceutical Sciences*, pages 509–529. Marcel-Dekker, New York.

DuMouchel, W. (1996). Predictive cross-validation for hierarchical Bayesian meta-analysis. In: Bernardo, J., Berger, J., Dawid, A., and Smith, A., editors, *Bayesian Statistics 5*, Oxford. Oxford University Press.

Fleiss, J. L. (1993). The statistical basis of meta-analysis. *Statistical Methods in Medical Research*, 2,121–145.

Ford, M. F., Jones, M., Scannell, T., Powell, A., Coombes, R. C., and Evans, C. (1990). Is group psychotherapy feasible for oncology outpatients attenders selected on the basis of psychological morbidity? *British Journal of Cancer*, 62,624–626.

Gardner, M. J. and Altman, D. G. (1989). *Statistics with confidence*. British Medical Journal.

Gelfand, A. E. and Dey, D. K. (1994). Bayesian model choice: asymptotics and exact calculations. *Journal of The Royal Statistical Society*, 56(3),501–514.

Gelfand, A. E. (1996). Model determination using sampling-based methods. In: Gilks, W. R., Richardson, S., and Spiegelhalter, D. J., editors, *Markov Chain Monte Carlo in Practice*. Chapman and Hall, London.

Gelman, A. (1996). Inference and monitoring convergence. In: Gilks, W. R., Richardson, S., and Spiegelhalter, D. J., editors, *Markov Chain Monte Carlo in Practice*, London. Chapman and Hall.

Gilks, W. R., Richardson, S., and Spiegelhalter, D. J. (1996). *Markov Chain Monte Carlo in Practice*, Chapman and Hall, London.

Greenland, S. (1987). Quantitative methods in the review of epidemiological literature. *Epidemiological Reviews*, 9,1–30.

Hardy, R. J. and Thompson, S. G. (1996). A likelihood approach to meta-analysis with random effects. *Statistics in Medicine*, 15,619–629.

Harville, D. A. (1997). Maximum likehood approaches to variance component estimation and relation problems. *Journal of the American Statistical Association*, 72,320–338.

Haynes, R. B., Sackett, D. L., Taylor, W., Gibson, E. S., and Johnson, A. L. (1978). Increased absenteeism from work after detection and labeling of hypertensive patients. *New England Journal of Medicine*, 299,741–744.

Herrmann, C. (1997). International experiences with the Hospital Anxiety and Depression scale—a review of validation data and clinical results. *Journal of Psychosomatic Research*, 42(1),17–41.

Higgins, J. P. and Whitehead, A. (1996). Borrowing strength from external trials in a meta-analysis. *Statistics in Medicine*, 15(24),2733–2750.

Higgins, J. P. and Whitehead, A. (1997). Inclusion of both patient-level and study-level covariates in a meta-analysis. *Controlled Clinical Trials*, 18(35):84.

Hopwood, P., Howell, A., and Maguire, P. (1991). Psychiatric morbidity in patients with advanced cancer of the breast: prevalence measured by two self-rating questionnaires. *British Journal of Cancer*, 64,349–352.

Jadresic, D., Riccio, M., Hawkins, D. A., Wilson, B., Shanson, D. C., and Thompson, C. (1994). Long-term impact of HIV diagnosis on moood and substance use—St Stephen's cohort study. *International Journal of STD and AIDS*, 5,248–252.

Jenkins, P. L., Lester, H., Alexander, J., and Whittaker, J. (1994). A prospective study of psychosocial morbidity in adult bone marrow transplant recipients. *Psychosomatics*, 35,361–367.

Jones, D. R. (1992). Meta-analysis of observational epidemiological studies: a review. *Journal of the Royal Society of Medicine*, 85,165–168.

Kass, R. E. and Raftery, A. E. (1995). Bayes factors. *Journal of The American Statistical Association*, 90,773–795.

Kellner, R. and Sheffield, B. F. (1973). A self-rating scale of distress. *Psychological Medicine*, 3,88–100.

Louis, T. A. (1984). Estimating a population of parameter values using Bayes and empirical Bayes methods. *Journal of The American Statistical Association*, 79,393–398.

Louis, T. A. and Zelterman, D. (1994). Bayesian approaches to research synthesis. In Cooper, H. and Hedges, L. V., editors, *The Handbook of Research Synthesis*, pages 411–421. Russell Sage Foundation, New York.

Marteau, T. M. (1993). Health related screening: Psychological predictors of uptake and impact. In Maes, S., Leventhal, H., and Johnston, M., editors, *International Review of Health Psychology* vol. 2, pages 149–174. John Wiley and Sons Ltd, Chichester.

Morris, C. N. (1983). Parametric empirical Bayes inference: theory and applications. *Journal of The American Statistical Association*, 78,47–65.

O'Hagan, A. (1994). Fractional Bayes factors for model comparison (with discussion). *Journal of The Royal Statistical Society*, 56,99–138.

Ostrow, D. G., Joseph, J. G., Kessler, R., Eller, M., Chmiel, J., and Phair, J. P. (1989). Disclosure of HIV antibody status: behavioral and mental health correlates. *AIDS Education and Prevention*, 1,1,1–11.

Perry, S., Jacobsberg, L., Card, C. A., Ashman, T., Frances, A., and Fishman, B. (1993). Severity of psychiatric symptoms after HIV testing. *American Journal of Psychiatry*, 150(5),775–779.

Petitti, D. (1994). *Meta-analysis, Decision Analysis and Cost-effectiveness Analysis—Methods for Quantitative Synthesis in Medicine.* Oxford University Press, New York.

Pugh, K., Riccio, M., Jadresic, D., Burgess, A. P., Baldeweg, T., Catalan, J., Lovett, E., Hawkins, D. A., Gruzelier, J., and Thompson, C. (1994). A longitudinal study of the neuropsychiatric consequences of HIV-1 infection in gay men. II psychological and health status at baseline and at 12 month follow-up. *Psychological Medicine*, 24,897–904.

Raftery, A. E. and Richardson, S. (1994). Model selection for generalized linear models via GLIB, with application to epidemiology. In: *Bayesian Biostatistics.* Marcel Dekker.

Rudd, P., Price, M. G., Graham, L. E., Beilstein, B. A., Tarbell, S. J., Bacchetti, P., and Fortmann, S. P. (1986). Consequences of worksite hypertension screening—differential changes in psychological function. *The American Journal of Medicine*, 80,853–860.

Salkovskis, P. M., Storer, D., Atha, C., and Warwick, H. M. (1990). Psychiatric morbidity in an accident and emergency department: characteristics of patients at presentation and one month follow-up. *British Journal of Psychiatry*, 156,483–487.

Shaw, C., Abrams, K. R., and Marteau, T. M. (1999). Psychological impact of predicting individuals' risks of illness: a systematic review. *Social Science in Medicine.* 49(12):1571–1598.

Smith, T. C., Spiegelhalter, D. J., and Thomas, A. (1995). Bayesian graphical modelling for random effects of meta-analysis: a comparative study. *Statistics in Medicine*, 14(24),2685–2699.

Spielberger, C. D., Gorsuch, R. L., and Lushene, R. (1970). *STAI Manual.* Consulting Psychologists Press, Palo Alto, California.

Statistical Sciences Inc. (1990). *Splus for Sun workstations.* Oxford, U.K. Version 2.3.

Stewart, L. A. and Parmar, M. K. B. (1993). Meta-analysis of the literature or of individual patient data: is there a difference? *The Lancet*, 341,418–422.

Sutton, A. J., Jones, D. R., Abrams, K. R., Sheldon, T. A., and Song, F. (1998). Systematic reviews of randomized trials. In: Black, N., Brazier, J., Fitzpatrick, R., and Reeves, B., editors, *Health Services Research Methods: A Guide to Best Practice*, pages 175–186, BMJ Books, London.

Sutton, A. J., Abrams, K. R., Jones, D. R., Sheldon, T. A., and Song, F. (1999). Systematic reviews of trials and other studies. *Health Technology Assessment*, 2,19. In Press (http://www.hta.nhsweb.nhs.uk/htapubs.htm).

Thomas, A., Spiegelhalter, D. J., and Gilks, W. R. (1992). BUGS: A program to perform Bayesian inference using Gibbs sampling. In: Bernardo, J., Berger,

J., Dawid, A., and Smith, A., editors, *Bayesian Statistics 4*. Oxford University Press, Oxford.

Thompson, S. G. (1994). Why sources of heterogeneity in meta-analyses should be investigated. *British Medical Journal*, 309,1351–1355.

Visser, M. C., Koudstaal, P. J., Erdman, R. A., Deckers, J. W., Passchier, J., van Gijn, J., and Grobbee, D. E. (1995). Measuring quality of life in patients with myocardial infarction or stroke: a feasibility study of four question-naires in The Netherlands. *Journal of Epidemiology and Community Health*, 49,513–517.

Zigmond, A. S. and Snaith, R. (1983). The hospital anxiety and depression scale. *Acta Psychiatry*, 67,361–370.

3
Meta-analysis versus Large Trials: Resolving the Controversy

Scott M. Berry
Texas A & M University, College Station, Texas

Abstract

There has been recent skepticism over the value of a meta-analysis. Some believe that meta-analysis is inferior to a large randomized trial. LeLorier et al. (1) compare the conclusions of 41 meta-analyses with those of 41 subsequent large trials. They conclude that the meta-analyses and large trials do not agree, and therefore the meta-analyses are faulty. In this paper, I argue that the large trials and meta-analyses disagree only because a key variance component is ignored.

I present hierarchical meta-analysis models which capture the heterogeneity in the treatment effect across trials. These models are applied to two examples. The first is a meta-analysis studying the efficacy of mammography in preventing breast-cancer deaths for women of 40–49 years of age. The second studies the effect of cholesterol-lowering treatment on coronary heart disease. These examples show that the hierarchical approach resolves—or at least sheds light on—the meta-analysis controversy.

I. INTRODUCTION

The development of hierarchical models may help ease recent skepticism in the literature about the usefulness of meta-analysis. LeLorier et al. (1)

study the relationship between meta-analyses and subsequent large randomized clinical trials that address similar issues. They summarize a meta-analysis by the mean of the population of trials, and a confidence interval. They conclude that some meta-analyses are faulty because the subsequent clinical trials don't "agree" with the meta-analysis interval. In this paper, I address their definition of "agree" and show that it ignores the possibility of an important form of heterogeneity. When variability across trials is explicitly modeled, and the distribution of trials is addressed, then the future clinical trials do in fact "agree" with the meta-analysis.

Two examples are considered. Section II presents a meta-analysis of eight trials addressing the efficacy of mammography screening for women aged 40–49. Three models are presented and discussed. Section III discusses the analysis of meta-analyses of LeLorier et al. Concluding remarks are given in Sect. IV.

See Stangl and Berry (Chap. 1) for an introduction to Bayesian hierarchical models. Calculations in this paper use Markov chain Monte Carlo (MCMC) techniques (2,3). Details of very similar algorithms can be found in Ref. 4. Pauler and Wakefield (Chap. 10) discuss general Bayesian hierarchical models and MCMC approaches.

II. MAMMOGRAPHY EXAMPLE

An NIH Consensus Development Conference was held on January 21–23, 1997, to answer several questions about the benefits and risks of mammography screening for women in their forties. One was: Is there a reduction in breast-cancer mortality due to screening women in this age group? The panel's answer was that each woman should make her own informed decision (5). This answer was controversial. In this paper, I examine evidence available from the randomized trials. Table 1 presents the data from a world-wide meta-analysis addressing the efficacy of mammograms for women in their forties (6). Figure 1 is a plot of the proportion of breast-cancer deaths per thousand life years for women assigned to mammography, against the corresponding proportion for control women.

In each trial, women in their forties were randomized to either the mammography or the control group. The women were followed, and the incidence of breast-cancer deaths was recorded. The number of life years

Table 1 The Data from the Meta-analysis on the Effectiveness of Mammography in Preventing Breast-Cancer Deaths*

Trial	Mammography			Control			
	B.C. Deaths	Years	$\hat{\lambda}$	B.C. Deaths	Years	$\hat{\lambda}$	$\hat{\rho}$
HIP (NY)	49	248	0.20	65	253	0.26	0.77
Edinburgh	46	170	0.27	52	155	0.34	0.79
Malmo	49	165	0.30	65	144	0.45	0.67
Kopparberg	23	144	0.16	18	75	0.24	0.67
Ostergotland	27	143	0.19	27	147	0.18	1.06
Canada	82	322	0.26	72	322	0.22	1.18
Stockholm	24	162	0.15	12	94	0.13	1.15
Gothenburg	18	138	0.13	39	168	0.23	0.57
Total	318	1492	0.21	350	1358	0.26	0.81

*The number of breast-cancer deaths, life years (in thousands), and proportion of breast-cancer deaths per thousand years ($\hat{\lambda}$) are presented for each trial. The observed $\hat{\rho}$ are presented for each trial.

for the women was also recorded. Follow-up in the trials ranged from 12 to 18 years.

This chapter presents three models, two of which are hierarchical. In each model, the results are presented and the relevant question of efficacy is addressed. In all three models, the number of trials is k. For trial i, let X_i and Y_i be the number of breast-cancer deaths in the control and mammography groups, respectively. Let n_i and m_i be the number of life years (in thousands) in trial i, for control and mammography respectively.

The number of breast-cancer deaths in each trial is modeled as a Poisson random variable. In the control group, the following model is adopted:

$$X_i \sim \text{Poisson}(n_i\theta_i) \quad (i = 1, \ldots, k).$$

Parameter θ_i is the mean number of breast cancer deaths per thousand life years at trial i. The Y_i are modeled as independent Poisson random variables:

$$Y_i \sim \text{Poisson}(m_i\theta_i\rho_i) \quad (i = 1, \ldots, k).$$

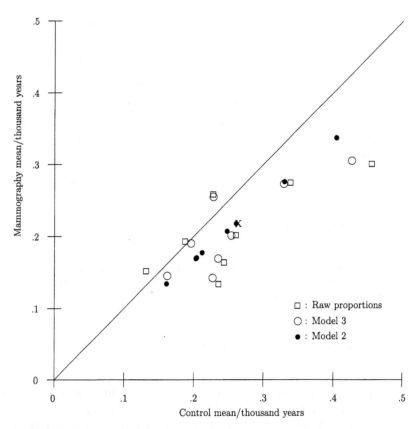

Figure 1 The observed mean numbers of deaths per thousand life years in the mammography example are shown as squares. X represents the posterior mean based on model 1. The shaded circles are the posterior means for the eight trials using model 2. The large unshaded circles are the posterior means for the eight trials using model 3.

The parameter ρ_i represents the relative risk between the control and the mammography group in trial i. If ρ_i is greater than 1, then the breast-cancer mortality rate is larger in the mammography group than in the control group in trial i.

In the simplest of the three models presented, model 1, the θ_i are equal and the ρ_i are equal, with common values θ_0 and ρ_0. The breast-cancer mortality rates within the respective treatment group are the same in all trials. Independent lognormal prior distributions are placed on θ_0 and ρ_0:

$$\theta_0 \sim \text{lognormal}(\mu_\theta, \sigma_\theta^2), \qquad \rho_0 \sim \text{lognormal}(\mu_\rho, \sigma_\rho^2). \tag{1}$$

I used $\mu_\theta = -1.5$, $\sigma_\theta^2 = 1$, $\mu_\rho = 0$, and $\sigma_\rho^2 = 1$. Due to the large number of life years and breast-cancer deaths, these priors have little effect on the results. The posterior mean for θ_0 is 0.259 with a standard deviation of 0.014. The posterior mean of ρ_0 is 0.824 with a standard deviation of 0.064. The posterior means of the control rate and the mammography rate are plotted in Fig. 1 with an "X."

The efficacy of mammography screening is easy to address in this model. Parameter ρ_0 indicates the relationship between the control-group rate and the mammography-group rate. Every trial has the same ρ and θ, and therefore a single conclusion applies for each trial and for the collection of trials. The posterior mean of ρ_0 is 0.83, and the posterior probability that ρ_0 is less than 1 is 0.992.

Model 1 includes assumptions that are seldom tenable. There is generally variation across trials due to differences in patient populations, differences in interventions in the various trials, differences in protocols, and differences in the way the protocols are applied.

In model 2 the variation across trials is captured using a hierarchical distribution for the trials. The θ_is are modeled with a lognormal distribution:

$$\theta_i \sim \text{lognormal}(\mu_\theta, \sigma_\theta^2) \quad (i = 1, \ldots, k). \tag{2}$$

The hyperparameters, μ_θ and σ_θ^2, are modeled with the following prior distributions:

$$\mu_\theta \sim \text{normal}(\delta_\theta, \tau_\theta^2), \qquad \sigma_\theta^2 \sim \text{inverse-gamma}(\alpha_\theta, \beta_\theta). \tag{3}$$

As in model 1, in model 2 the ρ_is are assumed to be equal. The prior distribution for the common value ρ_0 given in equation (1) is also used in model 2. The lognormal distribution in equation (2) represents the distribution of θs across trials. This allows the control rate to vary from trial to trial, and the mammography rate will also vary; but their ratios, and therefore the efficacy of screening, are assumed to be constant. The mean of the inverse-gamma with parameters α and β is $1/[\beta(\alpha - 1)]$ with a variance of $1/[\beta^2(\alpha - 1)^2(\alpha - 2)]$. This model is referred to as additively heterogeneous and interactively homogeneous. This model assumes the same form of heterogeneity as that assumed in the Mantel–Haenszel procedure (7,8).

For the following calculations, the hyperparameters are set as: $\delta_\theta = -1.5$, $\tau_\theta = 1.0$, $\alpha_\theta = 3$, and $\beta_\theta = 3$. The posterior means for each of the eight trials are shown as shaded circles in Fig. 1. The posterior mean of ρ is 0.834 with a standard deviation of 0.064. The posterior probability that ρ_0 is less than 1 is 0.990, about the same as in model 1. The variance of the θs is a critical parameter in model 2. When σ_θ^2 is small, model 2 is essentially model 1. The estimates for each of the eight trials will converge to the "X" on Fig. 1 as the prior distribution for σ_θ^2 becomes more concentrated on smaller values.

Model 2, as with model 1, provides a way to assess efficacy. While the control and mammography rates may vary, their relationship stays the same, and thus ρ represents the relative efficacy for patients in all trials. In this model, there is a 99% chance that mammography is better. This conclusion is the same for each of the trials.

Model 3 is still more general than model 2. It allows for the ρs to vary from trial to trial. The following hierarchical distribution for the ρs is used:

$$\rho_i \sim \text{lognormal}(\mu_\rho, \sigma_\rho^2) \quad (i = 1, \ldots, k). \tag{4}$$

The prior distributions for the hyperparameters are

$$\mu_\rho \sim \text{normal}(\delta_\rho, \tau_\rho^2), \qquad \sigma_\rho^2 \sim \text{inverse-gamma}(\alpha_\rho, \beta_\rho).$$

For model 3 the θs are assumed to have the distribution given by equations (2) and (3). This model allows the control and mammography group rates to vary from trial to trial. In this model, the ratios of the two rates are not necessarily equal. Equation (4) is the distribution for the ratios. This model is interactively heterogeneous (4). The adoption of this powerful and flexible model has important consequences in addressing the relative efficacy of the two treatments. There is no longer a common parameter ρ across trials. The question "which treatment is better?" is not well defined. The answer depends on the trial of interest. The parameter ρ_i represents the relative efficacy of mammography at trial i. Any general conclusion about the efficacy of mammography should take into account the distribution of ρs across trials. In some meta-analyses, the better treatment depends on the trial. In the mammography example, both screened and control groups differed in a number of ways across trials. These differences may give rise to different benefits across trials.

For the following results, the parameters for the priors are $\delta_\theta = -1.5$, $\tau_\theta = 1$, $\delta_\rho = 0$, $\tau_\rho = 1$, $\alpha_\rho = 3$, $\beta_\rho = 1$, $\alpha_\theta = 3$, and $\beta_\theta = 1$. The posterior means for the eight trials are plotted in Fig. 1 with large unshaded circles. The estimates vary from the line of constant ratio: the eight means are 0.80, 0.84, 0.75, 0.73, 0.90, 1.06, 0.84, and 0.66, given in the order of Table 1. The parameter σ_ρ^2 differentiates model 2 from model 3. As $\sigma_\rho^2 \to 0$, model 3 converges to model 2. The results from model 2 are sensitive to the choice of priors on σ_ρ^2 and σ_θ^2. However, they are not affected very much by the choices for other priors (as long as they are not concentrated in a subset of the parameter space).

The sensitivity of the results to the selection of the αs and βs is an advantage of this model. It enables the user, through the prior distribution, to capture the heterogeneity across trials. Figure 2 shows the difference in results when the prior distribution on σ_ρ^2 is changed. The data help to shape the posterior opinion of σ_ρ^2; but, with only eight trials, the prior is important. The small unshaded circles are the posterior means for each of the eight trials using the prior $\alpha_\rho = 3$ and $\beta_\rho = 0.1$, and the triangles are the posterior means based on the prior $\alpha_\rho = 3$ and $\beta_\rho = 10$. The first of these is diffuse, and the second puts all its probability on small values of σ_ρ^2 (I refer to it as a "strong" prior). Model 3 with a diffuse prior distribution borrows little strength across trials in the estimation of the individual ρs, while the model with a strong prior distribution borrows a great deal of strength (almost as much as does model 2). As the prior distribution ranges from diffuse to strong, the estimates range from the observed risk ratios to the model 2 estimates.

The sensitivity to the prior distribution of σ_θ^2 is similar to that of the prior distribution of σ_ρ^2. The estimates are regressed towards the "X" in Fig. 1 when the prior puts high probability on small values of σ_θ^2. This variance component differentiates model 2 from model 1.

The model estimates that, at seven of the eight trials, mammography is better—and control is better at the other trial. The idea that the efficacy of an intervention can vary by trial is not universally appreciated. Section III describes a debate about the usefulness of meta-analysis in which this idea plays a critical role.

Because the ρs are allowed to vary, a single conclusion about the relative risk of screening may be inappropriate. There is a distribution of ρs, and several features of this distribution are relevant for assessing efficacy. The mean of ρ and the probability of a ρ greater than 1 are examples. The posterior mean of the mean of the population of ρs is 0.85, and the standard deviation of the mean of the distribution of ρs is 0.12.

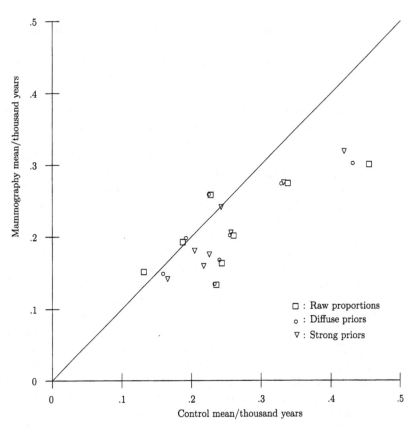

Figure 2 The observed mean number of deaths per thousand life years in the mammography example are shown as squares. The large unshaded circles are the posterior means for the eight trials using model 3 with the prior distributions given in the paper. The small unshaded circles and the triangles are the posterior means based on model 3 using alternative priors that are strong and diffuse. They are described in the text.

The probability that the mean of the ρs is less than 1 is 0.90. From the population of ρs, the probability that a future, hypothetical trial from the same population has a ρ that is less than 1 is 0.76. The predictive distribution of a future ρ is shown in Fig. 3. The posterior estimates for the ρs of the eight mammography trials are shown with shaded diamonds, while the ends of the 95% posterior intervals are marked with unshaded diamonds.

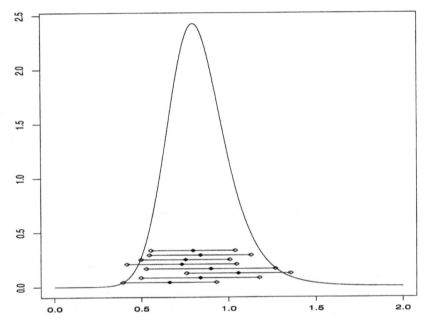

Figure 3 The predictive distribution of ρs from the mammography example based on model 3. The shaded diamonds are the posterior means of the eight trials. The unshaded diamonds are the ends of the 95% posterior intervals. The trials are in descending order from Gothenburg to HIP in Table 1.

These two probabilities, 0.90 and 0.76, are quite different as measures of efficacy, and their interpretations are likewise quite different. The probability that a new trial favors mammography is 0.76. This quantity represents the proportion of trials that favor mammography. The probability that the mean of the ρs is less than 1, i.e. 0.90, represents the average over many trials.

Goodman (9) criticizes using the mean of a distribution within random-effects models:

> If each trial is modeled as having its own intrinsic mean because of differences in protocols or participants, then the meta-analytic average becomes less relevant to any individual patient. For a clinician faced with a treatment decision, it would make the most sense to find the trial with patients or treatment protocol most like his own, not look at the random-effects grand mean. This may be why it has been

> claimed that the random-effects result does answer a question. But
> it's a very abstruse and uninteresting question.

I agree that, when one is interested in a future trial, an analysis of the
mean of the random-effects distribution is unimportant. Matching similar trials, as suggested by Goodman, is a good idea, but can be difficult to
do. In such cases, it is much more reasonable to use the predictive distribution of future trials. Brophy and Joseph (Chap. 4) discuss measuring
efficacy within Bayesian hierarchical models. In the next section, I discuss
how heterogeneity across trials can be misinterpreted.

III. META-ANALYSIS CONTROVERSY

LeLorier et al. (1) raise several concerns about the value of meta-analysis.
One concern is that the comparison of the trials for compatibility of the
entry criteria and protocols is not usually done very well. The availability
of the data and the possibility of publication bias are also difficulties in
preparing empirical information in a meta-analysis. These are legitimate
concerns for a meta-analyst. Smith et al. (Chap. 13) and Ref. 10 present a
Bayesian approach to meta-analysis in the presence of publication bias.
Simon (Chap. 12) discusses similar problems with meta-analyses.

To demonstrate the poor quality of meta-analyses, LeLorier et al.
compare the results of large (more than 1000 patients) clinical trials to
meta-analyses that addressed the same problem. They report on 12 large
randomized clinical trials in which a meta-analysis was previously done
on the same topic. From the 12 trials, 40 outcome variables were
addressed. LeLorier and his colleagues characterized whether the clinical
trial and the meta-analysis agreed. They make the claim that "agreement
between the meta-analyses and the large clinical trials was only fair."

In the meta-analyses, a 95% confidence interval for the relative risk
between two treatments is calculated assuming that the relative risk is
constant across trials (essentially model 2 from above). The relative risk
of the large clinical trial is compared to the relative risk from the meta-
analysis.

Of the 40 outcome variables studied, the meta-analysis and the large
trial were significantly different for 12 of them. In no case did the meta-
analysis and the clinical trial come to statistically significant conclusions
in the opposite directions (that is, favoring different treatments). They
also characterized a meta-analysis by a point value as to whether it

favored the control or the experimental treatment. They compared the conclusion of the meta-analysis with the conclusion of the clinical trial and found that the meta-analysis agreed with the clinical trial 65% of the time. A conclusion of the paper is that, because a large clinical trial is considered the gold standard in medicine, the meta-analyses were poor predictors of the "truth." They conclude: "The outcomes of the 12 large randomized, controlled trials that we studied were not predicted accurately 35 percent of the time by the meta-analysis published previously on the same topics." My first objection is to their assumption that the large clinical trial gives the correct answer. My second is that they are using statistical significance inappropriately. But perhaps most importantly, they are mistaken in summarizing the meta-analysis by a common relative risk (model 2). When relative risk between treatments can vary from trial to trial (model 3 above), the results observed by LeLorier et al. are expected. In trying to address whether a clinical trial agrees with a meta-analysis, the trial should be compared to a distribution rather than to the mean of the distribution.

As an illustration reconsider the mammography example of Sec. II. If a meta-analysis is done assuming the ρs equal, then trials that are in the meta-analysis can be considered incompatible with the meta-analysis. If the ρs are assumed to be equal, but the θs are allowed to vary, the 95% posterior interval for ρ_0, the common value of ρ, is $(0.71, 0.95)$. If the Canadian trial is analyzed separately from the other trials, the estimate of ρ_{Canada} is 1.16, with a 95% posterior interval of $(0.88, 1.54)$ (using Model 1 with $k = 1$, $X_1 = 72$, $n_1 = 322$, $Y_1 = 82$, and $m_1 = 322$). According to the attitudes of LeLorier et al., if the Canadian trial had been a separate large trial, it would be considered different from the meta-analysis and would be considered incompatible—yet the Canadian trial was part of the meta-analysis! If instead the meta-analysis assumes the ρs can vary (model 3) then the distribution of ρs (Fig. 3) shows that the Canadian trial is quite compatible with the meta-analysis as a whole.

Bailar (11), commenting on the meta-analysis skepticism raised by LeLorier et al., agrees with the notion that there is a distribution of trials, and that a one number summary is misleading. He writes:

> Although some meta-analyses stop with the presentation and discussion of the results of the individual trials, many others proceed further and combine the results into a single comprehensive "best" estimate, generally with statistical confidence bounds, that is meant to summarize what is known about the clinical problem. This last

step—preparing and presenting a single estimate as the distillation of all that is known—is the one that has drawn the most criticism. This is because there are often biological reasons, statistical evidence, or both, showing that the trials included in the meta-analysis have in fact measured somewhat different things, so that a combined estimate cannot be meaningful unless additional, doubtful assumptions are made.

Like Goodman, Bailar disagrees with the use of a single summary of a meta-analysis. Comparing future trials to a one number summary of a meta-analysis is sure to claim differences—despite the fact that the future trial may be quite likely under the predictive distribution derived in the meta-analysis.

While Bailar disagrees with using a single-number summary, he also raises concerns about using hierarchical models: "One . . . assumption is that the effects reported in the studies actually performed can be seen as a random sample of the effects observed in all possible studies that might have met the author's criteria. Unfortunately, there is little evidence to support an assumption such as this." There are few, if any, truly random samples. The question is whether the sampling procedure is biased in favor of studies of a particular type. Whether it is biased may not be accessible, but neither is bias accessible when selecting a single large trial. In comparing the two settings, a sample of studies offers more opportunity for accessing bias and for understanding heterogeneity among studies.

A. Example: Cholesterol and Coronary Death

I selected one of the examples from LeLorier et al. in which the results of the clinical trial and the meta-analysis are significantly different. A meta-analysis of six trials investigating the effect that cholesterol-lowering treatment has on coronary death for patients who have suffered a myocardial infarction is presented in Ref. 12. The Scandinavian Simvastatin Survival Trial Group (4S) (13) reported on a large clinical trial to investigate the same issue. The same cholesterol treatment and doses were not given at each trial. The type of cholesterol-lowering treatment and dose could be used as a covariate if these values were known. The data are presented in Table 2. Figure 4 shows the observed proportion of coronary deaths for each of the six trials in the meta-analysis (with squares) and the 4S trial (with an "4S").

Table 2 The First Six Trials Collected in a Meta-analysis (12) on the Number of Coronary Deaths for Patients who had Suffered a Myocardial Infarction*

Trial	Treatment			Control			
	Deaths	Patients	$\hat{\rho}$	Deaths	Patients	$\hat{\rho}$	$\hat{\rho}$
CDP	398	2224	0.18	535	2789	0.19	−0.09
Newcastle	25	244	0.10	44	253	0.17	−0.61
Edinburgh	34	350	0.10	35	367	0.10	0.02
Stockholm	47	279	0.17	73	276	0.26	−0.57
Oslo	37	206	0.18	50	206	0.24	−0.38
MRC	35	322	0.11	37	323	0.12	−0.06
4S	111	2221	0.05	189	2223	0.09	−0.57

*The treatment group received a cholesterol-lowering agent, while the control group did not. 4S was a clinical trial conducted after the meta-analysis and used by LeLorier et al. (1) to judge whether the clinical trial and meta-analysis agreed

For trial i, where i indexes the six trials in the meta-analysis, let X_i be the number of coronary deaths among the n_i patients in the control group. Likewise, let Y_i be the number of coronary deaths among the m_i patients in the treatment group. The following model is analogous to model 3 in the mammography example, except that the distributions are binomial. In particular, it exhibits both additive and interactive heterogeneity. Let

$$X_i \sim \text{binomial}(n_i, p_i) \quad (i = 1, \ldots, k),$$
$$Y_i \sim \text{binomial}(m_i, q_i) \quad (i = 1, \ldots, k).$$

The following log-linear model is used:

$$\log\left(\frac{p_i}{1 - p_i}\right) = \theta_i \quad (i = 1, \ldots, k),$$
$$\log\left(\frac{q_i}{1 - q_i}\right) = \theta_i + \rho_i \quad (i = 1, \ldots, k).$$

The θ_i, for $i = 1, \ldots, k$, are modeled as conditionally independent and having a normal distribution:

$$\theta_i \sim \text{normal}(\mu_\theta, \sigma_\theta^2).$$

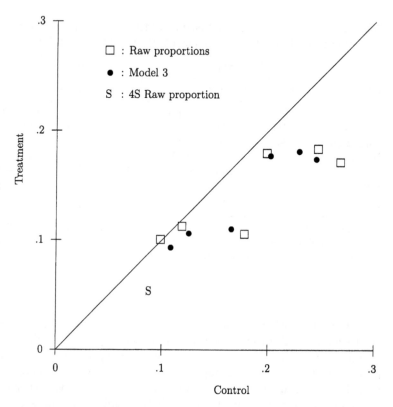

Figure 4   The observed proportion of deaths for the control and treatment are shown with squares. The shaded circles show the posterior mean values based on the hierarchical model presented in Sec. III. The "S" presents the mean estimate of the 4S trial.

The ρ_i, for $i = 1, ..., k$, are conditionally independent and have a normal distribution:

$$\rho_i \sim \text{normal}(\mu_\rho, \sigma_\rho^2).$$

The variability in the θ_i represents the control variation across trials. For example, the different diets of the people in the trials may partially explain the control variation across trials. The ρ_i represent this relationship between the control and the treatment at trial i. While diet may explain variation in the control response, the cholesterol treatment may affect people with different diets differently. This variation is captured by allowing the ρs to vary across trials. The following prior distributions are selected for the hyper-parameters:

$$\mu_\theta \sim \text{normal}(-1.5, 0.5^2), \qquad \mu_\rho \sim \text{normal}(0, 0.5^2),$$

$$\sigma_\theta^2 \sim \text{inverse-gamma}(3, 3), \qquad \sigma_\rho^2 \sim \text{inverse-gamma}(3, 3).$$

The posterior means based on this model are plotted in Fig. 4 with shaded circles. The predictive distribution of ρ for a future trial is given in Fig. 5.

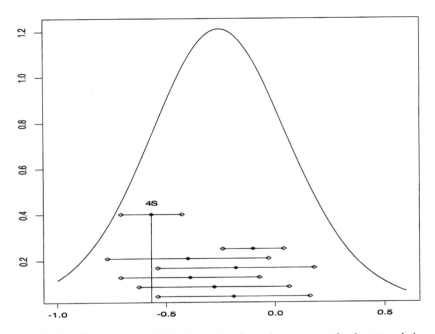

Figure 5 The predictive distribution of ρs from the coronary-deaths example is the density presented. The six horizontal lines represent the 95% posterior intervals for the six ρ_is. The shaded diamonds represent the posterior means, and the unshaded diamonds present the ends of the 95% posterior intervals. The trials are in descending order from MRC to CDP, as listed in Table 2. The 4S posterior mean and 95% posterior interval are presented. The predictive density and posterior means were calculated without the 4S trial included.

The posterior mean for the mean of the distribution of ρs is -0.26 with a standard deviation of 0.12. The posterior mean of the standard deviation of the distribution of future ρs is 0.22 with a standard deviation of 0.04. The probability that a future ρ is less than 0 is 0.85. Figure 5 shows the result of the 4S trial, which in relation to the meta-analysis is a future trial. LeLorier et al. point out that the 4S trial deviates (significantly) from the meta-analysis. But they assume that the ρs are constant across the trials in the meta-analysis. As seen in Fig. 5, the trial fits reasonably well in the population of trials, assuming that the ρ_i are allowed to vary across trials. In the meta-analysis, of the six trials, two had smaller observed ρs than that of 4S. The sample sizes were small in these two trials, and thus the model substantially shrunk their posterior means toward the overall mean. There is no reason to think that the 4S trial represents a different population from that of the meta-analysis. Only on assuming that the ρs of the six trials are equal does the 4S trial appear to be distinctive.

IV. DISCUSSION

Hierarchical models can handle trial heterogeneity that may be present. The conclusions from the model are more complicated, but also more accurate, when a hierarchical model is used. The question of "which treatment is more efficacious" is not uniquely defined. The answer depends on the trial and how efficacious is defined. When the trials are modeled as being exchangeable from some population, reporting the mean of that population is not an adequate description. Rather, one should report or describe the distribution of trials, including the possibility of giving characteristics of a future observation from this distribution.

LeLorier et al. (1) describe a collection of meta-analyses with future randomized clinical trials and explain that the agreement between the randomized clinical trials and the meta-analysis is only slightly better than chance. This is because their definition of agreement ignores the possibility of variability across trials. Their findings are no more surprising than the observation that the outcome of a roll of a die is not the mean outcome.

LeLorier and his colleagues, as well as Bailar, suggest that we accept the results of the 4S trial, and disregard the findings from the meta-analysis. There are no grounds for believing 4S and ignoring the other trials. Using hierarchical models not only explains the controversy, it resolves it. Someone interested in a particular trial, perhaps because it is large or well conducted or addresses the most relevant population, can borrow as much or as little strength as desired from the other trials by choosing a suitable prior.

References

1. J LeLorier, G Gregoire, A Benhaddad, J Lapierre, F Derderian. Discrepancies between meta-analysis and subsequent large randomized, controlled trials. The New England Journal of Medicine 337:536–542, 1997.
2. BP Carlin, TA Louis. Bayes and empirical Bayes methods for data analysis. Chapman & Hall, 1996.
3. WR Gilks, S Richardson, DJ Spiegelhalter (eds). Markov Chain Monte Carlo in Practice. Chapman & Hall, 1996.
4. SM Berry. Understanding and testing for heterogeneity across 2 × 2 Tables: Application to meta-analysis. Statistics in Medicine 17:2353–2369, 1998.
5. L Gordis, DA Berry, SY Chu, LL Fajardo, DG Hoel, LP Laufman, CA Ruferbarger, JR Scott, DC Sullivan, JH Wasson, CL Westhoff, RT Zern. Breast cancer screening for women ages 40–49. Journal of the National Cancer Institute 89:1015–1026, 1997.
6. DA Berry. Benefits and risks of screening mammography for women in their forties: A statistical appraisal. Journal of the National Cancer Institute 90:1431–1439, 1998.
7. N Mantel, and W Haenszel. Statistical aspects of the analysis of data from retrospective trials of disease. Journal of the National Cancer Institute 22:719–748, 1959.
8. LD Fisher, and G Van Belle. Biostatistics. Wiley, 1993.
9. SN Goodman. Meta-analysis and evidence. Controlled Clinical Trials 10:188–204, 1989.
10. NP Silliman. Hierarchical selection models with applications in meta-analysis. Journal of the American Statistical Association 92:926–936, 1997.
11. JC Bailar. The promise and problems of meta-analysis. The New England Journal of Medicine 337:559–561, 1997.
12. JE Rossouw, B Lewis, BM Rifkind. The value of lowering cholesterol after myocardial infarction. The New England Journal of Medicine 323:1112–1119, 1990.
13. Scandinavian Simvastatin Survival Trial Group. Randomised trial of cholesterol lowering in 4444 patients with coronary heart disease: The Scandinavian Simvastatin Survival Trial (4S). The Lancet 344:1383–1389, 1994.

4

A Bayesian Meta-analysis of Randomized Mega-trials for the Choice of Thrombolytic Agents in Acute Myocardial Infarction

James Brophy
University of Montreal, Montreal, Quebec, Canada

Lawrence Joseph
McGill University, Montreal, Quebec, Canada

Summary

Thrombolytic agents have been shown by several randomized trials to substantially reduce mortality following an acute myocardial infarction. Nevertheless, the choice of thrombolytic agents remains unclear. Three large randomized trials have reported results which directly compare two agents, streptokinase (SK) and tissue-plasminogen activator (t-PA). In this chapter, we evaluate the evidence from these trials via increasingly complex Bayesian meta-analytic models, including bias adjustments for possible differences between the trials. Our analyses suggest that the clinical superiority of t-PA over SK remains uncertain.

I. INTRODUCTION

Several randomized clinical trials (Gruppo Italiano per lo Studio della Streptochinase Nell'Infarto Miocardico (GISSI) (1), Wilcox et al. (2), and Fibrinolytic Therapy Trialists' (FTT) Collaborative Group (3)) have shown that thrombolytic agents improve 30-day survival by approximately 30 percentage points following an acute myocardial infarction (AMI). Thrombolytic agents are plasminogen activators which convert plasminogen, a proenzyme, to plasmin, an enzyme capable of cleaving fibrin and producing clot lysis. Streptokinase, SK, is the oldest identified plasminogen activator and was the first commercially available thrombolytic agent. SK not only acts on clots but also on circulating fibrinogen, giving rise to systemic fibrinogenolysis. Tissue plasminogen activator (t-PA) is a direct plasminogen activator which is produced endogenously by endothelial cells. Commercial production is available by means of recombinant DNA technology, which makes it approximately eight times more expensive than SK. This agent produces less systemic fibrinolysis than SK since it converts plasminogen to plasmin more efficiently in the presence of clot-bound fibrin.

While there is universal agreement about the clinical usefulness of thrombolytic agents, controversy remains about the choice of SK or t-PA, especially given the differences in the costs of these agents. Three large randomized clinical trials have directly compared SK to t-PA in AMI patients. The GISSI-2 trial (4) compared t-PA to SK both with and without subcutaneous heparin beginning 12 hours after the start of therapy. The ISIS-3 trial (5) compared t-PA and SK both with and without subcutaneous heparin in a similar factorial design, but began heparin 4 hours after the start of therapy. The 35-day mortality and stroke rates for these trials are shown in Table 1, where the different heparin groups have been combined since the survival rates in these groups were indistinguishably close. These two trials showed little difference in mortality rates between t-PA and SK, although the stroke rate was consistently elevated in the t-PA arm.

The next comparative trial was GUSTO-1 (6), which randomized 41,021 patients to four different thrombolytic strategies following an acute myocardial infarction. The treatment groups included two SK arms, one with intravenous and the other with subcutaneous administration of heparin, a t-PA arm, and a group given a combination of the two agents. This multi-center trial recruited patients from the United States

Table 1 Mortality, Stroke and Combined Endpoint for Three Mega-trials Comparing SK to t-PA

Trial	Agent	No. patients	Death	Stroke	Stroke or death
GISSI-2	SK	10396	958 (9.2%)	98 (0.9%)	1014 (9.8%)
	t-PA	10372	993 (9.6%)	136 (1.3%)	1067 (10.3%)
ISIS-3	SK	13780	1455 (10.6%)	141 (1.0%)	1530 (11.1%)
	t-PA	13746	1418 (10.3%)	188 (1.4%)	1513 (11.0%)
GUSTO-1	SK	21251	1475 (7.4%)	261 (1.5%)	1636 (8.1%)
	t-PA	10396	653 (6.3%)	161 (1.6%)	746 (7.2%)
	t-PA + SK	10374	723 (7.0%)	170 (1.6%)	817 (7.9%)

(17,796), Canada (2,898) and 13 other countries. Compared with the combined SK branches, the strategy of "front-loaded" or "accelerated" t-PA used throughout the trial showed a statistically significant lowered mortality (6.3% vs 7.3%, respectively; $p = 0.001$) and combined end point of 30-day mortality or disabling stroke (6.9% vs 7.8%, respectively; $p = 0.006$). See Table 1, where again the two SK groups have been combined, as in the original analysis of the GUSTO investigators. While GISSI-2 and ISIS-3 report 35-day mortality rates and GUSTO-1 only 30-day rates, the probability of a death occurring between 31 and 35 days is very low, in line with a 2% probability of death uniformly spread between 31 days and one year.

The GUSTO-1 trial was well executed, and the sample size was designed to have at least 80% power to detect a 15% reduction in mortality, equivalent to an absolute mortality difference of 1% between experimental groups. This value has been (somewhat arbitrarily) defined by the GUSTO-1 investigators as the minimal clinically important difference between the two agents. Here we will continue to accept a 1% decrease as a clinically meaningful difference, although pharmaco-economic analyses may be required to further investigate the cost-effectiveness ratio. Most clinicians would accept the frequentist analysis ($p = 0.001$) of this study as being conclusive (or almost conclusive) proof of the superiority of t-PA, that is, that the mortality rate for t-PA is less than that for SK, but in this chapter we will discuss whether this is an adequate summary of the available evidence. GUSTO-1 was a Herculean effort that was carefully carried out, but this does not exempt it from an equally careful examination of the conclusions, particularly in light of the cost differential and the contradictory data from the other trials. While many critiques of the GUSTO-1 trial have been published (7,8,9,10), these have mostly centered on design issues and the interpretation of the clinical relevance of the observed mortality differences. This chapter will raise further questions while performing increasingly complex meta-analyses of the data from the three major clinical trials comparing the two drugs.

It is important to note that, although all the trials were randomized with uniform entry criteria and drug dosages, reservations have been expressed about the relevance of any comparisons between these studies. The major sources of controversy are as follows:

The t-PA used in ISIS-3 was of a slightly different form, although the clinical difference is not believed to be materially important.

Adjunctive therapy accompanying t-PA in GUSTO-1 included more aggressive use of intravenous heparin.

In GUSTO-1, t-PA was administered in an accelerated fashion.

The t-PA arm in the GUSTO-1 trial experienced more revascularizations than the SK arm. Since the trial was not blinded and revascularization may improve survival, GUSTO-1 may have overestimated the benefits of t-PA compared to SK due to confounding with the revascularization rate.

Subgroup analysis of the GUSTO-1 data have revealed that there was a larger observed t-PA-to-SK difference in mortality in US patients compared to other patients. Therefore, the degree of benefit of t-PA over SK may in part be tied to an interaction with a particular health-care system.

Therefore, while there is an abundance of information comparing these two agents, there are varying opinions as to the extent to which the data from the trials could be combined. As we will show, conclusions about the advantages of t-PA can dissipate if data from the GISSI-2 and ISIS-3 trials are considered along with the GUSTO-1 data. Clinicians may vary in their weighting of the importance of the similarities and differences between the trials, leading to controversy about whether t-PA has been shown to be superior to SK.

The outline of this chapter is as follows. We begin in Section II with a very simple Bayesian analysis which considers that the three studies are sufficiently similar for a fixed-effects meta-analytic model to be appropriate, possibly with some downweighting of the GISSI-2 and ISIS-3 results compared to those from GUSTO-1. We also consider the conclusions that could be drawn from considering the GUSTO-1 results alone. Recognizing the differences in the settings of the three trials, we construct a hierarchical (random-effects) model for the difference between the death rates with t-PA and SK in Section III. Finally, the above discussion suggests that slightly different parameters may be operating in the studies, because of biases that may occur when studies use slightly different preparations, become unblinded, or use different rates of other treatments such as increased use of revascularization. Therefore, in Sec. IV, we attempt to adjust our hierarchical model to take into account some of these differences, using evidence in the literature concerning the possible magnitudes of these biases. Since the stroke rate was higher in the t-PA groups across all three studies, we will focus on mortality, where conclusions are less clear.

II. SIMPLE BAYESIAN META-ANALYSES

Perhaps the simplest form of meta-analysis is the pooling of data across trials as if all data arose from a single trial. From a Bayesian perspective, this is equivalent to using the data from previous trials to construct a prior distribution for the current trial, and updating via Bayes' theorem. However, as many authors have recently pointed out (11,12), no single prior distribution is likely to be sufficient to represent the diversity of clinical opinions that exists before a trial is carried out. Indeed, this diversity is usually a prerequisite for ethical randomization. Therefore, trial results should usually be reported starting from a range of prior distributions. The corresponding set of posterior distributions then summarizes the range of post-trial beliefs. If this latter set of distributions includes only a sufficiently narrow range of possible effects, conclusions could be drawn with which most clinicians should agree regardless of their initial opinions. Otherwise, the debate continues and further research is indicated.

Since our trial results are given in terms of binomial outcomes, a convenient conjugate family of distributions is the beta. A random variable θ follows a beta distribution with parameters α and β if

$$f(\theta) = \frac{1}{B(\alpha, \beta)} \theta^{(\alpha-1)}(1 - \theta)^{(\beta-1)} \quad \text{for } 0 < \theta < 1,$$

where θ represents the probability of a "success" (which in our case paradoxically is a death), and where $B(\alpha, \beta)$ is the Beta function which represents the required constant coefficient in a beta density. The mean of a beta(α, β) density is given by $\alpha/\alpha + \beta$, and the variance is $\alpha\beta/(\alpha + \beta)^2(\alpha + \beta + 1)$. The parameters α and β may be interpreted as the prior numbers of successes and failures, respectively. For example, according to Table 1, there were a total of $\alpha = 958 + 1455 = 2413$ deaths and $\beta = 9438 + 12325 = 21763$ survivors under SK in the GISSI-2 and ISIS-3 trials, so that a prior density for the probability of death from SK in the GUSTO-1 trial might be a beta$(2413, 21763)$ density.

This prior distribution considers that the information on each patient in the GISSI-2 and ISIS-3 trials is as important and relevant as in the GUSTO-1 trial. Alternatively, a researcher who believes that the difference in trial protocols cannot be ignored might elect to only partially consider the earlier results. For example, one could arbitrarily treat the value of each observation in the previous trials as worth only 50% or even 10% of each observation in the GUSTO-1 data, so that the above prior distribution would change to a beta(2413/2, 21763/2) or a beta(24 13/10, 21763/10) density, respectively. A more extreme position would be that the trials are too dissimilar to be combined, and that consequently all previous research should be ignored. This approach assumes that nothing is known about the mortality rate, implying that a diffuse prior distribution which spreads mass equally over the entire feasible region from 0 to 1 might be appropriate. Therefore, a uniform prior distribution, equivalent to a beta(1, 1) density, could be used as the prior density for each rate. Other prior distributions are also possible and are not necessarily derived by a weighting of previous data. Most of these would fall in between the above extremes. For example, an expert panel could be convened to elicit a prior distribution for the GUSTO-1 trial taking into account the similarities and differences between the three trials, and all other relevant information.

Similar prior densities can be constructed for the t-PA arm of the trial. We are interested in the difference in the mortality rates for patients given t-PA and SK, that is, the difference between two beta densities. Since no convenient closed-form solution exists to express this density, one can either use Monte Carlo simulation or normal approximations. With our large sample sizes, the normal approximation is virtually indistinguishable from the true posterior density, so this was used to create Figs 1 and 2, discussed below. We chose to use Monte Carlo simulations via a Gibbs sampler (13) for the analyses presented in Secs 3 and 4, which are more complex.

Figure 1 shows three prior probability densities for the difference in mortality between t-PA and SK derived from using 100%, 50%, and 10% weighting of the data from GISSI-2 and ISIS-3. The prior mean of these curves is close to zero (0.0002), suggesting that no important difference exists between the two agents. Fully accepting the results of these two trials would suggest almost no *a priori* possibility of t-PA being clinically superior to SK, since a decrease in the mortality rate with t-PA of greater than or equal to 1% is represented by the area under the curve to the left of −0.01, and this area is essentially zero in the case of using 100% of the prior data. This leads to a prior distribution representing skepticism as to the superiority of t-PA. Using weights of 50% or 10% of

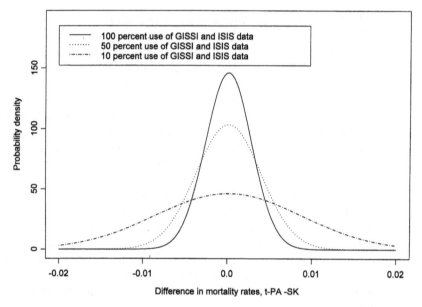

Figure 1 Plot of the prior distributions for the difference in mortality rates between tissue-type plasminogen activator (t-PA) and streptokinase (SK) using weights of 100 and ISIS-3 data, representing a range of prior beliefs in the relevance of these trials to the GUSTO-1 trial.

course widens the prior variance of the mortality difference, increasing the prior probability of clinical superiority. Using a 10% prior weight leads to a 12% prior probability of clinical superiority of t-PA, and this number increases to approximately 49% under uniform prior densities for the mortality rates of the two agents.

Combining a beta(α, β) prior density with binomial data of x successes in n trials, via Bayes theorem, leads to a beta($\alpha + x$, $\beta + n - x$) posterior density, so that the posterior density for the difference in death rates in using SK and t-PA is again a difference between two beta densities. Figure 2 displays the range of posterior densities from our four different prior distributions. If one assigns equal weight to each observation from GISSI-2, ISIS-3, and GUSTO-1, then the posterior mean difference in mortality between t-PA and SK is 0.13% (0.0013 in favor of SK), and the final (posterior) probability of t-PAs being superior to SK is only about 31% (area under the curve to the left of 0). The probability that t-PA is clinically superior to SK is almost zero, as indicated by the area under the curve to the left of –0.01. Similarly, if one considers observations from the previous randomized clinical trials to have 50% the value of each observation in GUSTO-1, then Fig. 2 shows that the probability that t-PA is superior to SK for mortality is about 53%. Further, accepting that a difference of 1% mortality is the minimum clinically significant value, the probability that t-PA is clinically superior remains negligible. If all prior data from GISSI-2 and ISIS-3 are considered irrelevant and are ignored, then t-PA is virtually certain to have a lower death rate than SK (99.96%), but the probability that t-PA exceeds the defined clinical superiority point is only 50.3%.

This preliminary analysis assumes a fixed-effects model, where the mortality difference between t-PA and SK is assumed constant between trials. Given the different settings and treatment regimens between the trials, this asumption may not be reasonable. Furthermore, neither the *p*-values reported by the GUSTO investigators nor the Bayesian analysis

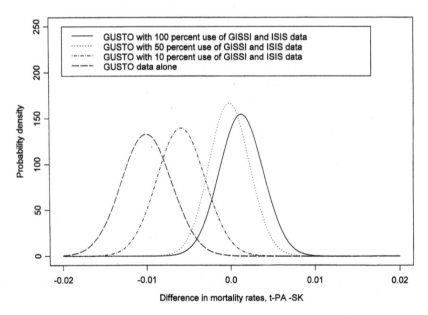

Figure 2   Plot of the posterior distributions for the difference in mortality rates between tissue-type plasminogen activator (t-PA) and streptokinase (SK), using data from the GUSTO-1 trial and a range of prior distributions.

presented above considers potential biases or confounding. The GUSTO-1 trial like GISSI-2 was unblinded, which may lead to some degree of confounding. For example, while not reported in the original article, it appears that 9.5% of the t-PA group underwent coronary-artery bypass surgery compared with 8.5% in the SK group. This difference may have contributed to the observed mortality differences between the trials. A Bayesian approach to adjustments for such biases is presented in Sec. IV. In the next section, we calculate a hierarchical model that predicts the difference in mortality for the next randomly selected study that is similar to the three trials for which we currently have data.

III. BAYESIAN HIERARCHICAL MODEL

Much of the controversy surrounding the interpretation of the three trials under discussion here arises from the heterogeneity of their observed mortality differences. It may therefore be reasonable to consider a Bayesian hierarchical random-effects model, wherein it is assumed that there is a distribution of possible "true effects" across different settings, and where each trial is assumed to be exchangeable (14). From this model, we can calculate a posterior density for the overall mean effect (difference in mortality rates between t-PA and SK) as well as predict what the effect might be for the "next" study.

Bayesian hierarchical models for meta-analysis of clinical trials with dichotomous outcomes are discussed by Carlin (15). Following this general approach, our model can be described in a hierarchy consisting of three levels. These levels describe the statistical relationships between data and parameters at the individual, study, and population strata.

Level I.   At the individual level, the number of deaths in each of the control and treatment groups follows a binomial distribution, so that

$$s_{ti} \sim \text{binomial}(p_{ti}, n_{ti}) \quad \text{and} \quad s_{ci} \sim \text{binomial}(p_{ci}, n_{ci}),$$

where s_{ti}, p_{ti}, and n_{ti} respectively denotes the number of successes, the probability of success, and the total sample size in the treatment group of the ith trial, for $i = 1, 2, 3$, and similarly s_{ci}, p_{ci}, and n_{ci} for the control group.

Level II.   At the study level, the logit of the control rate in study i, for $i = 1, 2, 3$, is defined as

$$\log\left(\frac{p_{ci}}{1 - p_{ci}}\right) = \mu_{ci}.$$

The difference between the logits of the control and treatment probabilities for the ith trial is denoted by δ_i. Therefore, we have

$$\delta_i = \log\left[\frac{p_{ti}}{1 - p_{ti}}\right] - \log\left[\frac{p_{ci}}{1 - p_{ci}}\right],$$

so that δ represents the log odds ratio. We assumed

$$\delta_i \sim N(\mu_\delta, \sigma_\delta^2).$$

This normal distribution represents the main hierarchical component of the model.

Level III.   The third stage sets the remaining prior distributions and prior parameters. We used:

$$\mu_{ci} \sim N(0, 1000^2) \quad \text{and} \quad \sigma_{ci}^2 \sim \text{gamma}(0.001, 0.001), \quad (i = 1, 2, 3),$$

$$\mu_\delta \sim N(0, 1000^2), \qquad \sigma_\delta^2 \sim \text{gamma}(0.001, 0.001).$$

Diffuse prior distributions were appropriate since, before these three trials were performed, little was known concerning the mortality differences between the two treatment agents. In addition, with the large numbers of subjects from the trials, unless they were very strong, prior densities would have little influence.

Since no closed-form solution is available, we followed Carlin (15) in using the Gibbs sampler (13) to approximate the marginal posterior densities. See Larose (Chap. 8), Dominici and Parmigianni (Chap. 5), and DuMouchel and Normand (Chap. 6) for other examples where the Gibbs sampler has been used in meta-analysis. Using the Raftery and Lewis Gibbsit algorithm (16), we decided that 10,000 iterations of the Gibbs sampler after a burn-in of 1000 iterations was more than sufficient for highly accurate estimation of all of our parameters.

The results are displayed in Fig. 3. Looking at the results from the individual trials, the mean t-PA–SK mortality difference in the GISSI-2 study is 0.002 (in favour of SK), with a 95% credible interval of (-0.005, 0.010). Similarly, the mean mortality difference from the ISIS-3 study was also close to zero at -0.003, with 95% credible interval of (-0.010 to 0.004), and the mean from GUSTO-1 was -0.009 (in favour of t-PA) with 95% credible interval (-0.015, -0.002). Comparing the values given by the hierarchical model with those from an analysis of the results from

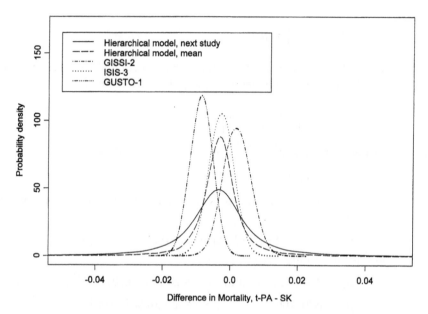

Figure 3 Plot of the posterior distributions for the difference in mortality rates between tissue-type plasminogen activator (t-PA) and streptokinase (SK) from a hierarchical model without bias adjustments.

each trial alone (no hierarchy) showed that the model had little influence on the individual trial estimates, with all differences of the order of 0.001.

Inference about the overall mean difference on the logit scale showed that there was a 0.74 probability that $\delta < 0$, meaning that there was a 74% posterior probability that the true mean across studies was less than 0 (less mortality in the t-PA group). This is calculated from the area to the left of 0 under the posterior density curve for δ (curve not shown). However, using the predictive distribution for the "next δ" gave a probability of only 66%, reflecting that the trial-to-trial variability still gives a 34% chance that the next trial would be in favor of SK. This analysis assumes exchangeability among the studies and among patients in the same study, but does not assume exchangeability of patients in different studies.

It is of course more satisfactory to draw conclusions about mean differences in mortality directly on the probability scale, rather than on the difference-in-logits scale. However, to do so requires setting a baseline rate for the logit of the probability of a death in the control group, since the hierarchical model for δ provides differences from this baseline rate. We compared the inferences obtained from using the mean of the posterior distributions from each of the three studies, and fortunately all were quite similar, with differences in means and 95% interval limits of at most 0.01 across all comparisons. Therefore, below we report the transformations from the logit scale back to the difference-in-probabilities scale using the mean of the posterior density from the SK group in the GUSTO study. This value on the logit scale was -2.55, which corresponds to a probability of death in the SK group of 0.072.

The posterior density for the mean difference and the predictive density for the mortality difference for the "next study" are given in Fig. 3. Inference about the overall mean across all studies gives a value very close to zero at -0.002, with 95% credible interval of $(-0.017, 0.016)$. Hence, overall, while the mean from this analysis is quite close to zero, a 1% advantage in either direction is not ruled out. Of course, the mean predicted difference in rates for the "next study" is also -0.003, but with a very wide credible interval of $(-0.030, 0.037)$. This wide interval results from the relatively wide variability in mortality differences observed among these trials, as illustrated in Fig. 3, and from the fact that only three trials are considered.

While the Bayesian hierarchical model presented here accounts for the trial-to-trial variability, it does not attempt to explain these differences. In the next section, we will examine whether protocol differences and possible biases in these trials can be explicitly modelled.

IV. BIAS MODELING

Some controversy has arisen because treatment assignment in GUSTO-1 was not blinded. The investigators have countered this criticism by pointing out that not all other thrombolytic trials have been blinded, that mortality trials do not need blinding since death is a "hard outcome" and multivariate logistic regression may account for any unbalancing introduced by the unblinded nature of the trial (17,18). However, blinding remains a key necessity even in a randomized controlled trial with hard outcomes, as unblinded randomized trials have been associated with exaggerated treatment effects (19).

The most important reason for blinding is to avoid intentional and unintentional bias in assessing outcomes across treatment arms. In this case, however, a death is a death, so does lack of blinding really matter? Despite better outcomes with accelerated t-PA (increased early reperfusion rates and decreased mortality), this strategy was associated with 1% more early revascularizations, an unbalancing well beyond what is expected by the play of chance. While reductions in mortality by appropriately employed revascularizations in stable angina may take a long time to be realized (20), this is not necessarily the case in the setting of acute ischemic syndromes. North American physicians are increasingly performing routine early post infarctus angiography and revascularization (21) with the widespread—albeit possibly unsupported—belief (22,23) that such interventions may reduce morbidity and mortality. The GUSTO-1 investigators (17) included a revascularization term in their logistic model and maintain this did not negate the "significant" mortality advantage for t-PA. Nevertheless, with a sample size of over 40,000, a mortality difference that had shrunk from 1% to perhaps 0.2% or 0.3% could remain statistically significant. Furthermore, mathematical modelling such as logistic regression can only adjust for confounding variables that are measured and included in the model. The difference in revascularization rates may not represent simple confounding (risk-factor control) but also suggests other unmeasured or unmeasurable bias in the selection of future treatments, which is of special concern in an unblinded trial.

In this light, the recent publication (24) of variations in patient management and outcomes between patients in the US and other countries participating in GUSTO is interesting. After controlling for baseline characteristics, overall prognosis was statistically improved in patients randomized in the United States. Randomization in the US may be a marker for increased revascularizations, since these procedures were three times more common in US patients. Further, there was a statistically insignificant trend for an interaction term between treatment arm/country ($p = 0.07$) and this reached statistical significance for the case of t-PA versus combination therapy ($p = 0.02$). Therefore, differences in patient management may have affected survival, although perhaps to a limited extent. While such post hoc subgroup analyses must be viewed very cautiously, the large sample size, the prospective planning, and consistency with previous European thrombolytic trials do suggest that part of the 30-day mortality advantage attributed to accelerated t-PA may be the result of an interaction with the significantly different healthcare system in the US.

While any bias corrections are necessarily speculative, we reasoned as follows. Of the several potential biases that may make GUSTO differ- ent from the other studies, the two most important are the differences in revascularization rate between the SK and t-PA groups, and the combina- tion of accelerated t-PA and more aggresive heparinization given in the GUSTO-1 trial but not in the others. We therefore decided to first create a prior distribution for the increase in the probability of death that might have been due to decreased revascularization in the SK arm of the GUSTO-1 trial. Using this prior distribution to impute the "true" prob- ability of death that might have occurred had the revascularization rate been balanced, we re-created the hierarchical analyses of Sec. III. This allows the examination of the impact this adjustment could have on the between-trial variability, and on predictions for the "next" trial. Second, we created prior densities for the change in the probabilities of death in the t-PA arms of the ISIS-3 and GISSI-2 trials, to impute the "true" rates had the possibly more effective accelerated t-PA protocol been used in these trials. Combining both of these biases, we again re-estimated the hierarch- ical model of Sec. III. The results of these analyses are presented below.

A. Adjusting for Possible Biases Arising from Differences in Revascularization Rates in GUSTO-1

The t-PA group benefited from a slightly higher rate of revascularization by approximately 1%, although the exact figures do not seem to be published. Overall, the 41021 GUSTO-1 patients received a total of 31% (22% angioplasty and 9% by-pass surgery) revascularization pro- cedures (24). Since undoubtedly some patients received both angioplasty and by-pass surgery, we will assume that 25% received one or both procedures. Suppose that the rates are such that 24% of the SK cohort and 25% of the t-PA cohort received a revascularization procedure. It seems reasonable to assume that the more severe cases were sent for early revascularizations. The most severe patients across all treatment groups included 2200 who experienced cardiogenic shock. Of these, 406 received a revascularization, with mortality rate of 38%, and 1794 did not receive the procedure, with a mortality rate of 62% (25). If 1% of the SK cohort (200 patients) were at high risk and did not receive revascularization, it is then conceivable that this may have lead to approximately 50 additional deaths $(200 \times (0.62 - 0.38))$, with a range of perhaps 0 to 100 deaths. This may have introduced a bias of 2.5 deaths per 1000 patients treated, with a range of 0 to 5 deaths per 1000 patients treated.

Using this range, we re-estimated the posterior density of the mortality rate difference in the GUSTO-1 study, and updated the hierarchical model discussed in Sec. 3. The resulting range of posterior predictive densities for the "next" study are displayed in Fig. 4. While the probability of mortality in the SK group decreased from 7.2% (95% CI = 6.9%, 7.6%) to 7.0% (95% CI = 6.5%, 7.5%) to 6.7% (95% CI = 6.4%, 7.1%) for the models without bias adjustments and with adjustments of 2.5 and 5 additional deaths per 1000 treated, respectively, Fig. 4 shows that this bias adjustment had a visible impact on the predictive distribution for the "next" study. Nevertheless, the overall message

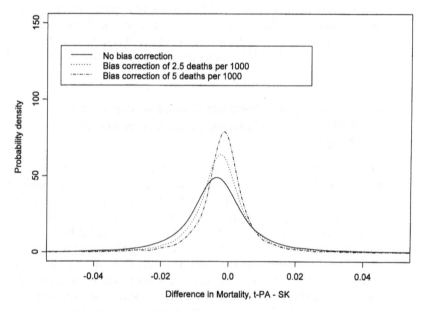

Figure 4 Plot of the posterior distributions for the difference in mortality rates between tissue-type plasminogen activator (t-PA) and streptokinase (SK) from the hierarchical model without bias adjustments (same curve as in Fig. 3), and with bias adjustments for revascularization of 2.5 and 5 lives saved per 1000 treated in the SK arm of the GUSTO-1 study.

remains the same. There is not much evidence of a mortality-rate difference between the trials, and although some narrowing of the posterior densities occurs due to the decreased between-study variability, the predictive intervals remain wide.

B. Adjusting for Possible Biases Arising from Differences in t-PA Administration in GISSI-2 and ISIS-3 Compared to GUSTO-1

The prevalent paradigm is that increased coronary patency as measured by the TIMI flow score is responsible for different survival rates between thrombolytic agents. In particular, the lower mortality with the accelerated t-PA strategy (drug administration over 90 minutes) employed in GUSTO-1 has been attributed to improved early 90-minute patency. There are nevertheless some observations that are difficult to reconcile with this theory. First, virtually all thrombolytic agents give similar patency rates at longer times such as 180 minutes (26). Also, the newer strategies for administering t-PA, for example as a double bolus injection, have shown marked improvement in TIMI flow but no improved clinical outcomes when compared to accelerated t-PA (27). Further, while a standard 3-hour infusion of t-PA has improved TIMI grade 3 flow compared to SK (50.2% vs 31.5%) (28), earlier comparative trials did not find any mortality reductions.

At 90 minutes, the rates of TIMI grade 3 flow for accelerated t-PA are improved by 13% (63.2% vs. 50.2%) compared with standard-dose t-PA (29). Accelerated t-PA had 54% TIMI 3 with a mortality rate of 4%, and 46% TIMI 0–2 with a mortality rate of 8% for an overall predicted mortality of approximately 5.9%. Standard t-PA might have had only 41% TIMI 3, with a mortality rate of 4%, and 59% TIMI 0–2 with a mortality rate of 8%, leading to an overall predicted mortality of 6.4%. The difference in rates (6.4% − 5.9%) suggests that accelerated t-PA could be responsible for 5 lives saves per 1000 treated. A reasonable range for the difference in the numbers of deaths might then be 0 (no advantage at all for accelerated t-PA) to 10 lives per 1000 treated (completely explaining the SK to t-PA difference seen in GUSTO).

Using this range to decrease the numbers of deaths in the t-PA arms of the GISSI-2 and ISIS-3 trials, both alone and combined with the revascularization adjustment of Section IV.A produced the range of curves in Fig. 5. As expected, assuming that the accelerated t-PA protocol

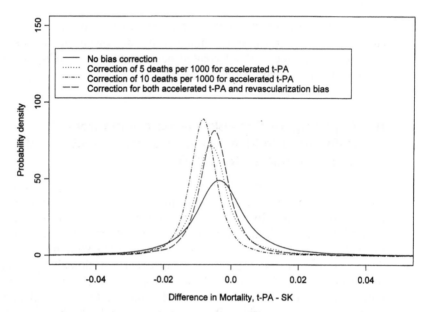

Figure 5 Plot of the posterior distributions for the difference in mortality rates between tissue-type plasminogen activator (t-PA) and streptokinase (SK) from the hierarchical model without bias adjustments (same curve as in Fig. 3), and with bias adjustments for accelerated t-PA in the GISSI-2 and ISIS-3 study, and for a combination of both revascularization and accelerated t-PA bias (see text).

saves 5 or 10 lives per 1000 patients treated shifts the predictions for the "next" study in favor of t-PA. Correcting for both revascularization (2.5 lives saved per 1000 treated in the SK arm of GUSTO-1) and accelerated t-PA (5 lives saved per 1000 treated in the t-PA arms in GISSI-2 and ISIS-3) does not substantially change the mean prediction for the "next" study compared to a model with no adjustments, although the prediction interval narrows somewhat.

Even in the scenario most favorable to t-PA, there is only a 32.7% chance of t-PA being clinically superior (mortality difference of at least 1%), compared to a 19.6% chance without bias adjustments, and a 15.9% chance with both adjustments included. We conclude that, while the biases discussed here may explain some of the differences in the results of the three trials, adjusting for them does not substantially change the conclusions of the simpler analyses presented in Secs II–III.

V. DISCUSSION

In this chapter, we have extended the analyses originally presented by Brophy and Joseph (30) to include random effects and bias adjustments. The range of different Bayesian analyses presented herein suggests that restraint in accepting t-PA into routine clinical practice would be appropriate, agreeing with our previous analyses. The same conclusion was reached by Diamond et al. (31), who used a Bayesian point null-hypothesis test. This is especially true when one remembers that the stroke rate was higher in the t-PA group across all three studies, as well as the cost differences.

In assessing the evidence relating to choosing a thrombolytic agent, the reporting of p-values from a single trial is a poor tool for formulating policy, even when there is a considerable amount of data from a well-designed randomized clinical trial. This is due to the shortcomings of standard significance tests in addressing clinically relevant questions— and to the problems in their interpretation, especially across different sample sizes. Furthermore, classical analysis of clinical trials does not easily permit the synthesis of trial results with the range of clinicians' prior beliefs, nor the adjustment for possible biases. This makes it difficult to evaluate the coherence of the conclusions and what clinical impact the conclusions should have. Following Eddy et al. (32), we have illustrated a method that can explicitly model possible biases, helping to explore the magnitude of the effects of various differences among a set of clinical trials with different protocols. Bayesian analyses along the lines presented herein may help to overcome these problems, thereby raising the level of debate following publication of a clinical trial, or the synthesis of a set of trials.

References

1. Gruppo Italiano per lo Studio della Streptochinase Nell'Infarto Miocardico (GISSI). Effectiveness of intravenous thrombolytic treatment in acute myocardial infarction. Lancet, number 8478:397–402, 1986.
2. RG Wilcox, G Von der Lippe, CG Olsson, et al. Trial of tissue plasminogen activator for mortality reduction in acute myocardial infarction. Anglo-Scandinavian Study of Early Thrombolysis (ASSET). Lancet (1988, number 8610):525–530.

3. Fibrinolytic Therapy Trialists' (FTT) Collaborative Group. Indications for fibrinolytic therapy in suspected acute myocardial infarction: collaborative overview of early mortality and major morbidity results from all randomised trials of more than 1000 patients. Lancet 343:311–322, 1994.

4. The International Study Group. In-hospital mortality and clinical course of 20,891 patients with suspected acute myocardial infarction randomised between alteplase and streptokinase with or without heparin. Lancet 336:71–75, 1990.

5. ISIS-3 (Third International Study of Infarct Survival) Collaborative Group. ISIS-3: a randomised comparison of streptokinase vs tissue plasminogen activator vs anistreplase and of aspirin plus heparin vs aspirin alone among 4,299 cases of suspected acute myocardial infarction. Lancet 339:753–770, 1993.

6. The GUSTO Investigators. An international randomized trial comparing four thrombolytic strategies for acute myocardial infarction. New England Journal of Medicine 329:673–682, 1993.

7. PM Ridker, C O'Donnell, VJ Marder, CH Hennekens. Large-scale trials of thrombolytic therapy for acute myocardial infarction: GISSI-2, ISIS-3, and GUSTO. Annals of Internal Medicine 119:530-532, 1993.

8. PM Ridker, C O'Donnell, VJ Marder, CH Hennekens. A response to 'holding GUSTO up to the light.' Annals of Internal Medicine 120:882–884, 1994.

9. P Sleight. Thrombolysis after GUSTO: a European perspective. J Myocard Ischemia. 5:25–30, 1993.

10. E Rapaport. GUSTO: assessment of the preliminary results. J Myocard Ischemia 5:15–24, 1993.

11. M Hughes. Reporting Bayesian analyses of clinical trials. Statistics in Medicine 12:1651–1663, 1993.

12. DJ Spiegelhalter, LS Freedman, MKB Parmar. Bayesian approaches to randomised trials. Journal of the Royal Statistical Society, Series A 157:357–416, 1994. Reprinted in: D. Berry and D. Stangl, eds. Bayesian Biostatistics. Marcel Dekker, 1996.

13. WR Gilks, S Richardson, DJ Spiegelhalter. Markov Chain Monte Carlo in Practice. New York: Chapman and Hall, 1996.

14. A Gelman, JB Carlin, H Stern, DB Rubin. Bayesian Data Analysis. New York: Chapman and Hall, 1995.

15. JB Carlin. Meta-analysis for 2 × 2 tables: a Bayesian approach. Statistics in Medicine 11(2):141–158, 1992.

16. A Raftery, S Lewis. How many iteration in the Gibbs sampler? In: J Bernardo et al. Bayesian Statistics 4. Oxford: University Press, 1992, pp 763–773.

17. KJ Lee, RM Califf, J Simes, F Van der Werf, EJ Topol. Holding GUSTO up to the light. Annals of Internal Medicine 120:882–885, 1994.

18. CD Naylor, PW Armstrong. From thrombolytic megatrials to clinical policy-making: Facing the facts. Canadian Journal of Cardiology 11:472–476, 1995.
19. K Schulz, I Chalmers, RJ Hayes, DG Altman. Empirical evidence of bias: Dimensions of methodological quality associated with estimates of treatment effects in controlled trials. JAMA 273:408–412, 1995.
20. S Yusuf, D Zucker, P Peduzzi, et al. Effect of coronary artery bypass graft surgery on survival: overview of 10-year results from randomised trials by the coronary artery bypass graft surgery trialists collaboration [see comments] [published erratum appears in Lancet 344 (no. 8934):1446, 1994 Nov 19]. Lancet 344:563–570, 1994.
21. DB Mark, CD Naylor, MA Hlatky, et al. Use of medical resources and quality of life after acute myocardial infarction in Canada and the United states. New England Journal of Medicine 331:1130-1135, 1994.
22. SWIFT (Should We Intervene Following Thrombolysis?) Trial Study Group. SWIFT trial of delayed elective intervention v conservative treatment after thrombolysis with anistreplase in acute myocardial infarction. BMJ 302:555–560, 1991.
23. The TIMI Study Group. Comparison of invasive and conservative strategies after treatment with intravenous tissue plasminogen activator in acute myocardial infarction. Results of the thrombolysis in myocardial infarction (TIMI) phase II trial. New England Journal of Medicine 320:618–627, 1989.
24. F Van de Werf, EJ Topol, KL Lee, et al. Variations in patient management and outcomes for acute myocardial infarction in the United States and other countries. Results from the GUSTO trial. Global utilization of streptokinase and tissue plasminogen activator for occluded coronary arteries. JAMA 273:1586–1591, 1995.
25. PB Berger, DR Holmes Jr, AL Stebbins, ER Bates, RM Califf, EJ Topol. Impact of an aggressive invasive catheterization and revascularization strategy on mortality in patients with cardiogenic shock in the global utilization of streptokinase and tissue plasminogen activator for occluded coronary arteries (GUSTO-1) trial. An observational study. Circulation 96:122–127, 1997.
26. The GUSTO Angiographic Investigators. The effects of tissue plasminogen activator, streptokinase, or both on coronary-artery patency, ventricular function, and survival after acute myocardial infarction. New England Journal of Medicine 329:1615–1622, 1993.
27. Anonymous. A clinical trial comparing primary coronary angioplasty with tissue plasminogen activator for acute myocardial infarction. The global use of strategies to open occluded coronary arteries in acute coronary syndromes (GUSTO-iib) angioplasty substudy investigators. New England Journal of Medicine 336:1621–1628, 1997.

28. RJ Simes, EJ Topol, DR Holmes Jr, et al. Link between the angiographic substudy and mortality outcomes in a large randomized trial of myocardial reperfusion. Importance of early and complete infarct artery reperfusion. Circulation 91:1923–1928, 1995.

29. NA Barbagelata, CB Granger, E Oqueli, LD Suarez, M Borruel, EJ Topol, RM Califf. TIMI grade 3 flow and reocclusion after intravenous thrombolytic therapy: a pooled analysis. American Heart Journal 133(3):273–282, 1997.

30. JM Brophy, L Joseph. Placing trials in context using Bayesian analysis: GUSTO revisited by Reverend Bayes. Journal of the American Medical Association 273(11):871–875, 1995.

31. GA Diamond, TA Denton, JS Forrester, PK Shah. Is tissue plasminogen really superior to streptokinase? Circulation 88:1-452 (Abstract), 1993.

32. DM Eddy, V Hasselblad, R Shachter. Meta-analysis by the Confidence Profile Method. New York: Academic Press, 1992.

5

Combining Studies with Continuous and Dichotomous Responses: A Latent-Variables Approach

Francesca Dominici
Johns Hopkins University, Baltimore, Maryland

Giovanni Parmigiani
Duke University, Durham, North Carolina

Abstract

A challenging problem in meta-analysis is combining studies in which similar medical outcomes are captured in some studies as continuous variables and in others as binary variables. A common approach is to dichotomize the continuous responses and proceed as in the simpler binary case. This approach is practical, but it has limitations. One is that there may be arbitrariness in the choice of the cutoff point. Another is that there is a loss of information. In this paper, we propose a strategy that overcomes both of these difficulties. It is based on assuming that the binary responses are the result of dichotomizing some underlying unobserved continuous variable. Bayesian reconstruction of the unobserved continuous variable preserves the full information from the studies reporting continuous variables and does not require the choice of arbitrary cutoff points.

We develop our method in the context of a simulation example. We then apply it to a meta-analysis of efficacy of calcium-blockers for preventing migraine headaches. In the headache example, we illustrate our approach in the context of hierarchical Bayesian models. We give details of the computations necessary to carry out posterior inference. In particular, both the hierarchical model and the latent-variable structure are handled conveniently using MCMC methods. We provide full conditional distributions to facilitate implementation.

I. INTRODUCTION

A difficulty in integrating results from different studies stems from the sometimes diverse nature of the studies, both in design and reporting methods employed. A common issue in combining evidence is that study responses, while similar, may not be directly comparable. In this paper, we propose a methodology for combining results when a particular medical outcome is captured in some studies as a continuous variable, and in others as a binary variable. When raw data are available, a simple approach to this problem is to dichotomize the continuous responses and proceed as in the simpler all-binary case. This approach is practical, but it has limitations. One is that the cutoff may be arbitrary. Another is that dichotomization may result in the loss of information.

In this chapter we propose a strategy that overcomes both limitations. It takes the view that the binary responses are the result of dichotomizing some underlying but unobserved continuous variable. This corresponds to what is frequently the case in practice. The variable is not necessarily the same one measured in the trials that report continuous variables, but it should measure the same clinical outcome. The idea is to generate, for each binary response, the results of a hypothetical continuous response which is consistent with the observed binary response. In this way, we preserve the full information from the studies reporting continuous variables, but without a loss of information from those reporting dichotomous variables. We do not require choosing a cutoff point, and we incorporate uncertainty arising from the heterogeneous nature of the responses across trials.

Various quantitative methods have been developed to synthesize information from multiple sources (see the various chapters in this volume and the reviews in Refs 1–3). Some approaches are frequentist and some are Bayesian. Here we take a Bayesian approach, for two reasons: the convenience of modeling and making inferences about latent variables, which lets us create a common underlying scale for all the studies; and the convenience of modeling study-to-study heterogeneity using hierarchical approaches. The use of Bayesian models in meta-analysis problems is well-established in the methodological literature (4–9) as well as in medical literature (10–15). A variety of Bayesian models can be handled using standard software, such as BUGS (16–17). An up-to-date review of software for meta-analysis is presented in Ref. 18. Bayesian methods are especially suitable for (a) developing probability distributions to be used as input in a subsequent decision analysis (19), or a probabilistic sensitivity analysis of decision models (20); (b) modeling unobserved parts of the data-generation process, such as overdispersion of study or treatment effects (21), missing covariates (22) and the process leading to publication bias (23–25); and (c) modeling heterogeneous outcomes (26).

In Sec. II., we introduce a simple simulation example, illustrating the latent-variable approach in a non-hierarchical case, and we give the details of the simple distributional theory necessary to implement a Gibbs sampler to draw from the posterior distributions of the parameters of interests. We provide a comparison with an alternative analysis based on discretizing the continuous variables, and show how the loss of information associated with discretization can be important. In Sec. III., we apply our method to the meta-analysis of the effectiveness of calcium-blocker treatments for prophylaxis of migraine headache. We present a hierarchical Bayesian model for synthesizing evidence and apply it to the nine studies. Eight of the studies report continuous responses, while one of the studies reports a two-by-two table. Estimation and ranking of treatments are all implemented through simulation-based methods.

II. SIMULATION EXAMPLE

To illustrate our approach and its properties, we first use a simple example with simulated data. Because the goal of this section is to illustrate the latent-variables approach, we consider a fixed-effect model, temporarily ignoring potential study heterogeneity. We consider four studies, indexed by s, each comparing a treatment arm ($t = 1$) with a placebo arm ($t = 0$). In each study, we have 20 observations per arm. Studies 1 and 2 record a continuous response, while studies 3 and 4 record a binary response x_{ti}^s. Write as y_{ti}^s the observed responses of the ith of the $n_t^s = 20$ individuals assigned respectively to arm t in studies $s = 1$ and $s = 2$.

Denoting by θ_0 and θ_1 the placebo and treatment effects respectively, we assume that:

$$y_{ti}^s \mid \theta_t, \sigma^2 \sim N(\theta_t, \sigma^2) \quad (t = 0, 1; \ s = 1, 2),$$

where $N(\mu, \sigma^2)$ denotes a normal distribution with mean μ and variance σ^2. The key idea of the latent-variables approach is to introduce, within studies 3 and 4, latent variables $y_t^s = (y_{t1}^s, \ldots, y_{t20}^s)$, where again

$$y_{ti}^s \mid \theta_t, \sigma^2 \sim N(\theta_t, \sigma^2) \quad (t = 0, 1; \ s = 3, 4),$$

and $x_{ti}^s = 1$ whenever $y_{ti}^s > 0$. The resulting likelihood for patient i in the binary studies is

$$x_{ti}^s \mid \theta_t, \sigma^2 \sim \text{Bernoulli}\left(\Phi\left(\frac{0 - \theta_t}{\sigma}\right)\right) \quad (t = 0, 1; \ s = 3, 4),$$

where Φ is the standard normal cdf. If the binary response can be thought of as a dichotomization of the continuous response, this model describes directly the data generation mechanism. Else, the latent variables may represent hypothetical continuous outcomes consistent with the observed discretized outcomes.

If continuous and discrete responses were thought to differ systematically, a random offset parameter α could be added to the model. The distribution of the binary variables in studies 3 and 4 could be specified as:

$$x_{ti}^s \mid \theta_t, \sigma^2 \sim \text{Bernoulli}\left(\Phi\left(\frac{\alpha - \theta_t}{\sigma}\right)\right) \quad (t = 0, 1; \ s = 3, 4),$$

where α can be interpreted as the overall difference in measured efficacy between the two types of response. Equivalently, this parametrization can be thought of as modeling the case where the cutoff for the discretization is not 0, but rather some unknown value that needs to be estimated from the data. We do not consider this case further. If the binary responses were dichotomization of the same underlying continuous variable according to different cutoff points in different studies, these differences would be erroneously attributed to differences in treatment effects. This difficulty can be overcome by using a hierarchical model, as done in Sec. III.

In our model, the marginal likelihood of the study effects and response variance is

$$L(\theta_0, \theta_1, \sigma^2) = \prod_{t=0,1}^{n_1} \prod_{i=1}^{n_1} \frac{1}{\sigma} \phi\left(\frac{y_{ti}^1 - \theta_t}{\sigma}\right) \prod_{i=1}^{n_2} \frac{1}{\sigma} \phi\left(\frac{y_{ti}^2 - \theta_t}{\sigma}\right)$$

$$\prod_{i=1}^{n_3} p_B\left(x_{ti}^3 \middle| \Phi\left(\frac{0 - \theta_t}{\sigma}\right)\right) \prod_{i=1}^{n_4} p_B\left(x_{ti}^4 \middle| \Phi\left(\frac{0 - \theta_t}{\sigma}\right)\right),$$

where ϕ is the standard normal probability density function and $p_B(x|\pi)$ is the probability distribution function of a Bernoulli random variable with success probability π. Using this expression, a latent-variable approach to combining continuous and dichotomous studies can be carried out from a likelihood perspective. However, deriving exact confidence intervals could be difficult. By using a Bayesian approach, it is relatively straightforward to obtain a probability distribution, and probability intervals, on all parameters of interest and their functions, without resorting to asymptotic approximations.

Our Bayesian formulation is completed by specifying prior distributions for θ_0, θ_1, and σ^2. In the absence of more specific information, we choose conjugate, very dispersed, priors: $\theta_t \overset{ind}{\sim} N(m, k^2)$, $t = 0, 1$ where $m = 0, k = 5$, and $\sigma^2 \sim IG(a, b)$, independent of θs. We use the parameterization of the inverse gamma with density function $b^a x^{-a-1} e^{-b/x}/\Gamma(a)$ for $x > 0$, with mean $b(a - 1)$ and variance $b^2[(a - 1)^2(a - 2)]$. We selected hyperparameters $a = 3$, the integer value giving the most diffuse finite-variance prior distribution to our variance component, and $b = 2$. Unless the number of studies with continuous outcomes is much larger than that of studies with binary outcomes, it is important to use a proper prior distribution on the response variance σ^2.

In the fixed-effect model considered here, all parameters are identified. In more general formulations, one may consider both treatment effects and variances to be study-specific. However, from the observed xs, we cannot make inferences on both the mean and the variance of the underlying latent-variables in each study. Therefore, identifying restrictions are necessary, such as requiring the variance of the latent-variable, σ^2, to be equal in all the studies.

Bayesian inferences are based on the posterior distribution of unknown quantities. We can draw inferences on any of the unknown parameters using the joint posterior distribution:

$$p(\theta_0, \theta_1, y_0^3, y_1^3, y_0^4, y_1^4 \mid \text{data}) \tag{1}$$

where data $= (y_0^1, y_1^1, y_0^2, y_1^2, x_0^3, x_1^3, x_0^4, x_1^4)$. Neither the posterior distribution (1) nor its marginals are available in closed form. However a practical choice for determining the marginal distribution, marginal probabilities and other summaries of interest is to draw a sample of values from the joint posterior distribution (1). This can be done using a Gibbs sampler (27) based on partitioning the unknown parameters and missing data into groups, and sampling each group in turn, given all the others. If the y_{ti}^ss for $s = 3, 4$ are known, then posterior inference and simulation can be obtained easily using normal theory (28). The y_{ti}^ss are of course unknown; however, given the data x_{ti}^ss, the conditional distribution of y_{ti}^s given $x_{ti}^s = 1$ (0) is truncated normal: $N(\theta_t, \sigma^2)$ truncated to the positive (negative) values. With this in mind, we can write the full conditional distributions as:

$$y_{ti}^s \mid \theta_t, \sigma^2, \text{data} \sim N(\theta_t, \sigma^2) I_{y_{ti}^s > 0}(y_{ti}^s)$$
$$\text{for } i : x_{ti}^s = 1 \text{ and } s = 3, 4,$$

$$y_{ti}^s \mid \theta_t, \sigma^2, \text{data} \sim N(\theta_t, \sigma^2) I_{y_{ti}^s \leq 0}(y_{ti}^s)$$
$$\text{for } i : x_{ti}^s = 0 \text{ and } s = 3, 4,$$

$$\theta_t \mid \sigma^2, \text{data}, \mathbf{y}_0^3, \mathbf{y}_1^3, \mathbf{y}_0^4, \mathbf{y}_1^4 \sim N\left(v \times \left[\sum_{s=1}^4 n_t^s \bar{y}^s / \sigma^2 + m/k^2\right], v\right),$$

$$\sigma^2 \mid \theta_0, \theta_1, \text{data}, \mathbf{y}_0^3, \mathbf{y}_1^3, \mathbf{y}_0^4, \mathbf{y}_1^4 \sim IG\left(a + \sum_{s=1}^4 \sum_{t=0}^1 n_t^s, b \right.$$
$$\left. + \frac{1}{2}\sum_{s=1}^4 \sum_{t=0}^1 \sum_{i=1}^{n_t^s}(y_{ti}^s - \theta_t)^2\right),$$

where $v = \left[\sum_{s=1}^4 n_t^s / \sigma^2 + 1/k^2\right]^{-1}$ and $t = 0, 1$. Simulation-based methods for handling uncertainty about missing data in Bayesian analysis are discussed in detail in Refs 29–30. Using latent normal variables to model binary observation is discussed by Refs 31–32.

We carried out a simulation to highlight the differences between the latent-variable approach and an alternative analysis that dichotomizes the first two studies, and then proceed as in the all-binary case. We generated the 500 datasets according to the distributions:

Study	Placebo	Treatment	
1	$y_{0i}^1 \sim N(0, 1)$	$y_{1i}^1 \sim N(0.4, 1)$	$(i = 1, \dots, 20),$
2	$y_{0i}^2 \sim N(0, 1)$	$y_{1i}^2 \sim N(0.4, 1)$	$(i = 1, \dots, 20).$

For the treatment arms of studies $s = 3$ and $s = 4$, we generated binary outcomes from Bernoulli distributions with probability $\Phi((0- 0.4)/1) \simeq 0.67$. Both the placebo arms were generated from a Bernoulli distribution with probability $\Phi((0 - 0)/1) = 0.50$.

Using the Gibbs sampler described above, we estimated the marginal densities of quantities of interests, such as the effect-size difference $\theta_1 - \theta_0$, and summaries such as the posterior probability of a positive effect size difference, that is $P(\theta_1 > \theta_0 |$ data). We focused on the probability of a positive treatment effect, because it can be defined in a fairly comparable fashion in the two analyses. To keep our task manageable, we used 250 iterations from chains of size 520. There are generally no convergence problems, even with such short chains. However, there remains some uncertainty about the estimated probabilities.

In our alternative analysis, we dichotomized the continuous variables at 0 and combined all studies. This leads to a two-sample comparison of the two proportions π_0 for the placebo group and π_1 for the control group. Using a Bayesian analysis with independent uniform priors, a posteriori π_0 and π_1 have two independent beta densities. Using a sample of 250 draws from these, we computed the probability that the proportion of successes is greater in the treatment group than in the placebo group.

In 90.8% of the simulated samples, the probability of a positive effect was greater in the latent-variable approach, or the two were equal. The mean difference between the probability of a positive effect computed in the latent-variable approach and the same probability in the discretized analysis was 4.3%. Another way of comparing the two approaches is to focus on one-sided testing of the hypothesis of negative or null treatment effect versus a positive effect. We proceed by looking at whether the probability of a positive effect is greater than 0.95. Table 1 cross-classifies the 500 simulated data sets depending on the conclusions reached using the latent-variable and the discretization approach. In 67% of the cases, the two approaches agree in reaching the correct conclusion that there is a significant difference between the treatment and the placebo effect. In 23% of the cases, the latent-variable approach leads to the correct conclusion, while the dichotomized approach does not; the converse occurs only in 1% of the cases.

While the prior distributions used in the two approaches are not equivalent, they are sufficiently similar for us to attribute much of the observed differences in tail probabilities to the loss of efficiency resulting from the dichotomization of the observed continuous data. The magni-

Table 1 Simulation Results. Cross-classification of 500 analyses according to whether one-sided testing at the 95% level would lead to accepting the hypothesis of positive treatment effect, in the latent-variable and discretized analyses. The datasets are simulated assuming that the true-effect difference is 0.4 standard deviations in the continuous scale.

	$P(\pi_1 > \pi_0 \vert \text{data}) \leq 0.95$	$P(\pi_1 > \pi_0 \vert \text{data}) > 0.95$
$P(\theta_1 > \theta_0 \vert \text{data}) \leq 0.95$	0.1	0.01
$P(\theta_1 > \theta_0 \vert \text{data}) > 0.95$	0.23	0.67

tude of these differences depends on the specific settings used in the simulation. When the treatment and the control group are either very similar (small $\vert \theta_0 - \theta_1 \vert$) or very different (large $\vert \theta_0 - \theta_1 \vert$) the dichotomized and latent-variable analysis are likely to give the same results. Differences between the two analyses will be crucial when $\vert \theta_0 - \theta_1 \vert$ is moderately large. However, the loss of information illustrated here is general. A measure of the loss of information due to the discretization can be developed using the Lindley information (33–34), but it is not pursued here.

III. RANKING BETA-BLOCKERS FOR HEADACHE TREATMENTS

We now discuss an application regarding efficacy of calcium-channel blocking agents for migraine headache. Migraine headache is an important public-health problem (35–36). A wide range of treatments are available, and there is disagreement about which treatments are most effective (37). We analyze data from nine studies, from a recent AHCPR evidence report (38). A more comprehensive analysis of headache treatments, including other groups of drugs, is presented in Ref. 39.

Each of the studies considered here includes two arms chosen among four drugs and a placebo. Eight of the nine studies report treatment effects and standard deviations derived from underlying quantitative or ordinal scales (38). The reported standardized treatment effects and standard deviations are summarized in Table 2. Study 3 reports a binary outcome: there were 14 successes and 15 failures in the placebo group, and 23 successes and 6 failures in the treatment group.

Table 2   Study/Treatment Table*

Study	n_s	Placebo	Verapamil	Nifedipine	Flunarizine	Nimodipine
			Calcium blockers			
1	42	0.230 (0.9)			0.449 (1.09)	
2	20	1.581 (1)		−1.423(1)		
3	29	*			*	
4	24	−1.364 (1.12)	−0.987 (0.87)			
5	30	0.325 (1.17)			0.949 (0.80)	
6	28	0.069 (0.64)			0.526 (1.05)	
7	78			0.408 (1.03)	0.556 (0.97)	
8	25				0.237 (1.25)	0.39 (0.63)
9	50	0.485 (1.2)				0.924 (0.82)

*The values in the table are the observed standardized treatment effects zs. In parentheses are the standardized standard deviations $\sqrt{q}$'s; ⋆ indicates that the study reports a two-by-two contingency table. Study 3 reports a binary outcome: there were 14 successes and 15 failures in the placebo group, and 23 successes and 6 failures in the treatment group.

Let $\mathcal{T} = \{1, 2, 3, 4\}$ be the indexes of the treatments under study and let $\mathcal{T}^s \subset \mathcal{T}$ be the indexes of the treatments included in study s. Write as y_{ti}^s the response of the ith of the n_t^s individuals assigned to treatment t in study s, and as $\bar{y}_t^s$ and v_t^s the sample mean and variance for treatment t in study s respectively. Finally, write the sample pooled estimate of the variance in study s as:

$$v^s = \frac{\sum_{t \in \mathcal{T}^s} (n_t^s - 1) v_t^s}{\sum_{t \in \mathcal{T}^s} (n_t^s - 1)}. \tag{2}$$

For all studies except study 3, the AHCPR report provides the standardized treatment effect $z_t^s = \bar{y}_t^s / v^s$ and their standardized variances $q_t^s = v_t^s / v^s$. Standardization was carried out by dividing the sample means $\bar{y}_t^s$ by the pooled estimate of the standard deviation in study s, i.e., $\sqrt{v^s}$, to give them a common dimensionless scale. Without this standardization, it would be difficult to compare heterogeneous responses such as ordinal measures of well-being; it is more meaningful to compare increments expressed in terms of population standard deviations. Combining standardized effect sizes is a common approach in meta-analysis (40, 1). For study 3—the study that reports a two-by-two contingency table—we use the notation x_t^s for the number of successes in arm t.

Our strategy is based on modeling the underlying responses y_{ti}^s. Because they are not directly available to us, we treat them as latent-variables. We thus create latent-variables for both the continuous and the discrete studies. In the continuous studies, the goal is to account for the uncertainty in the denominator of the standardized treatment effects. In the binary response study, we introduce normally distributed variables and regard the reported binary responses as dichotomizations of those, as done in Sec. 2. We model the latent variables with drug and study effects. Specifically, we assume that:

(i) the sample averages $\bar{y}_t^s$ are approximately normally distributed with means $\theta_t + \mu^s$, the sum of a treatment effect θ_t and a study effect μ^s, with variances σ_{ts}^2/n_t^s;

(ii) the sample variances v_t^s are approximately distributed as $\sigma_{ts}^2 \chi_{n_t^s-1}^2/(n_t^s - 1)$;

(iii) for study 3, the conditional distribution of y_{ti}^s given $x_{ti}^s = 1$ (0) is $N(\theta_t + \mu^s, \sigma_{ts}^2)$, truncated to the positive (negative) values.

Because only standardized quantities are observed, there is no information on the scales of the original responses. Define $\sigma_s^2 = \sum_{t \in T^s}(n_t^s - 1)\sigma_{ts}^2/\sum_{t \in T^s}(n_t^s - 1)$ and $\gamma_t^s = \sigma_{ts}^2/\sigma_s^2$. Then the distribution of the observed zs and qs is independent of σ_s^2 given the γs. We can therefore set $\sigma_s^2 \equiv 1$ and interpret θ_t, μ^s, and σ_{ts}^2 as parameters of a model for the correctly standardized sample averages $\bar{y}_t^s/\sigma_s$. By this convention, the parameters σ_{ts}^2 are constrained to have a weighted average of unity within each study. The structure of the model within a continuous study is shown in Fig. 1.

In our analysis, the parameter vectors $\theta = (\theta_1, \ldots, \theta_4)$ (treatment effects) and $\mu = (\mu^1, \ldots, \mu^9)$ (study effects) are modeled as random variables with a specified joint probability distribution, accounting for potential correlation of effects. The study effects μ^s, representing the differences across studies due to differing patient populations, protocol variations, and so on, are modeled as independent identically-distributed zero-mean random variables, independent of θ. In Ref. 39, we analyze a larger number of studies including three categories of headache treatments, and we model the distribution of study effects using a mixture of normals. Here the number of studies is not sufficient to justify such flexible distributional assumptions.

The origins of the original response scales may differ across studies. To obtain a meaningful comparison, we set the placebo effects to zero. For studies without a placebo arm, the study effect μ^s also embodies any difference in scale origins. The treatment effects θ are modeled as inde-

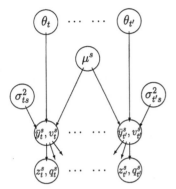

Figure 1 Summary of notation and conditional independence assumptions for the component model for continuous study s. The zs and qs are the observed standardized data; The $\bar{y}$s and vs are the unobserved sufficient statistics for the latent normal response; θs are treatment effects; μ^s is the study effect and σ^2_{ts}s are the arm-specific variances.

pendent identically-distributed normal random variables, with mean λ and variance τ^2. Our specification leads to a hierarchical model. General discussions of Bayesian hierarchical models are provided elsewhere in this volume by Refs 41–43.

In summary, the distributional assumptions of our model are as follows:

$$
\left.
\begin{aligned}
y_t^s | \theta_t, \mu^s, \sigma^2_{ts} &\overset{\text{ind}}{\sim} N(\theta_t + \mu^s, \sigma^2_{ts}/n_t^s) \\[2mm]
v_t^s | \sigma^2_{ts} &\overset{\text{ind}}{\sim} \sigma^2_{ts} \frac{\chi^2_{n_t^s} - 1}{n_t^s - 1} \\[2mm]
\theta_t | \lambda, \tau^2 &\overset{\text{ind}}{\sim} N(\lambda, \tau^2) \\[2mm]
\mu^s | \sigma^2_\mu &\overset{\text{iid}}{\sim} N(0, \sigma^2_\mu).
\end{aligned}
\right\} \text{ for } t \in \mathcal{T},
\tag{3}
$$

The hierarchical model (3) is specified in terms of unobserved variables y_t^s, which are related to observed quantities in different ways for different studies. In view of our Markov chain Monte Carlo implementation we give the conditional distribution of the latent Ys, given the reported summaries. In the studies with continuous data, the unobserved v^s are distributed as

$$v^s \sim \frac{\sum_{t \in T^s} \sigma_{ts}^2 \chi_{n_t^s-1}^2}{\sum_{t \in T^s}(n_t^s - 1)}.$$

Using v^s and the reported statistics, one can reconstruct all other unobserved statistics using $\bar{y}_t^s = z_t^s \sqrt{v^s}$ and $v_t^s = q_t^s v^s$. Similarly to the discussion in Sec. II, for the two-by-two table we treat the x_{ti}^s as indicators of $y_{ti}^s > 0$, so that

$$P(x_{ti}^s = 1 | \theta_t, \mu^s, \sigma_{ts}) = \Phi\left(\frac{0 - (\theta_t + \mu^s)}{\sigma_{ts}/\sqrt{n_t^s}}\right).$$

The constraint $\sigma_s^2 = 1$ still applies. The conditional distribution of y_{ti}^s given $x_{ti}^s = 1$ (0) is $N(\theta_t + \mu^s, \sigma_{ts}^2)$, truncated to the positive (negative) values.

For a Bayesian analysis of this model we specified prior distributions for all unknown parameters. A practical approach is to choose prior distributions that lead to conjugate full conditionals where possible. Here we use the following:

Overall effect mean	λ	$\sim N(\lambda^*, \gamma^2)$
Variance (treatment effects)	τ^2	$\sim IG(a_\tau, b_\tau)$
Variance (overall effect)	γ^2	$\sim IG(a_\gamma, b_\gamma)$
Variance (study effects)	σ_μ^2	$\sim IG(a_\mu, b_\mu)$
Variance (individual effects)	$p(\sigma_{ts}^2) \propto \sigma_{ts}^{-2}$	$(t \in T;\ s = 1, \dots S)$,

depending on hyperparameters v_α, λ^*, as, and bs. Unless otherwise implied by the hierarchical structure, we used independent prior distributions. For our analysis we selected hyperparameters $a_\tau = a_\gamma = a_\mu = 3$, the integer value giving the most diffuse finite-variance prior distributions to our variance components, and $b_\tau = b_\gamma = 1$, to ensure that the marginal expected treatment effect variance is $E(\tau^2 + \gamma^2) = 1$, appropriate for these standardized quantities. Also, $b_\mu = 2$.

Perfect homogeneity of our nine studies is unlikely a priori, in view of the fact that they are known to consider heterogeneous response measures, different sets of drugs, and different subpopulations. This is reflected in our choice of the inverse gamma density, which assigns a density of zero to the value $\sigma_\mu^2 = 0$. Also, the standardization of the response provides additional help for the elicitation of the prior hyperparameters. Similar considerations apply to the parameter τ^2, representing variation across treatment effects. Alternative prior specifications for high-level variance components in hierarchical models are discussed elsewhere in this volume. In Ref. 43, the authors suggest a default prior choice based on the half-normal distribution, a distribution also advocated in Ref. 44 in a similar context; a log-logistic distribution is discussed in Ref. 42. Both these distributions have positive density at zero.

We adopt the conservative approach of using prior mean $\lambda^* = 0$ for the overall treatment and placebo-effect mean λ, suggesting no prior information that any of the treatments are effective. This position ensures that any posterior difference between treatment arms must be attributable mostly to the data, and results in shrinking treatment effects for small studies towards 0.

An effective strategy for evaluating the posterior distribution of the parameters is again Markov chain Monte Carlo simulation. Most of the full conditional distributions, including those of the latent sufficient statistics, are available in closed form and lend themselves to straightforward sampling, making Gibbs sampling viable in most cases. An exception is the study-specific variances, whose distribution is complicated by the constraint $\sigma_s^2 = 1$. A closed-form expression is not available, and we use a Metropolis step (45). At each iteration, the latent-variables are simulated from the truncated normal distribution given earlier in (iii). We carried out graphical diagnostics for convergence of the chain, as implemented in CODA (46). A few hundred iterations typically suffice for convergence. We present results based on a subset of 1,000 equally spaced draws from the last 5,000 iterations of a single chain of 6,000 iterations.

Our objective is to make inferences about the treatment effectiveness, identifying which treatment seems best and how well each one works, by computing the posterior distribution of the vector of treatment effects θ, given the reported values z_t^s and q_t^s. The posterior distributions of the θs are shown in Fig. 2. All θs have positive posterior means, but there is significant mass to the left of zero.

An important contribution of meta-analysis to guideline development is a quantification of remaining uncertainty about which treatments are best. This can be expressed by a probability distribution over possible rankings. In addition to supporting treatment decisions, an accurate assessment of uncertainty may help guide the planning of new trials. In a simulation-based approach, the ranking problem can be solved easily even in presence of complex random effects and missing data that may induce significant dependencies in the joint distribution of the effects under comparison. Ranking is done simply by reporting the empirical

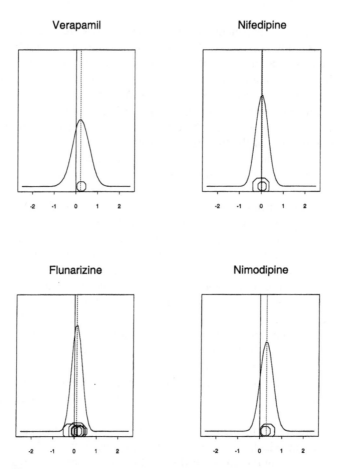

Figure 2 Marginal posterior distribution of treatment effects for the calcium channel blockers. Labels identify treatments; each circle corresponds to a study including the treatment in question, and is positioned at the posterior mean of $\bar{y}_t^s - \mu^s$; areas of circles are proportional to the sample sizes n_t^s. Posterior means and no-effect points are indicated by vertical dashed and solid lines, respectively.

frequency with which each treatment is better than its competitors within the same treatment group. In Fig. 3, for each treatment, we show the probability that its effect is the largest. Labeling the drugs in Fig. 3 from left to right as 1 to 4, the most probable ranking is 1, 4, 3, 2, with an estimated posterior probability of 20.6%, followed by 4, 1, 3, 2 and 4, 3, 2, 1, with estimated posterior probabilities of 15% and 11% respec-

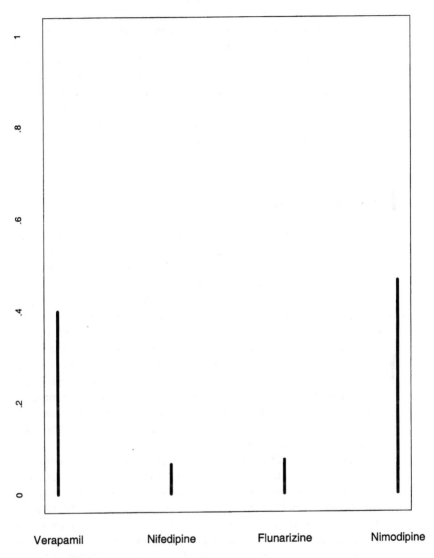

Figure 3 Probability that corresponding treatment effect is the largest.

tively. Drug 1 ranks first, second, and fourth in these three rankings, emphasizing the high uncertainty about which drug is most effective. Figure 4 illustrates the posterior distributions of the individual effects μ^s, supporting study heterogeneity.

While the headache studies considered here are all randomized clinical trials, not all drugs are utilized in all trials. Therefore, conclusions about ranking depend on indirect comparisons between arms. These comparisons rely on the assumption that the effect sizes are common to all studies, and that all study heterogeneity is adequately captured by an additive study-specific effect.

IV. DISCUSSION

We propose a Bayesian latent-variable approach to meta-analysis when some studies report continuous responses and other report binary responses. Bayesian methods are well suited to model unobserved parts of the data-generating process. In this paper we illustrated how this can be exploited to construct a common underlying scale for combining information from related, but different studies. Our model permits us to synthesize heterogeneous response information and to make inferences about treatment effects and the relative ranks of treatments, without losing information to discretization of continuous outcomes. Discretizing a continuous outcome can lead to loss of information, as illustrated in our simulated example. In the headache example, discretization would have to be carried out in all studies but one. The alternative of eliminating the binary study and confining attention to the continuous ones is also unattractive.

Standard analysis is based on discretizing continuous outcomes and proceeding as in the binary case. In Ref. 47, a maximum-likelihood approach is proposed, based on estimating the parameters of a normal distribution from the continuous studies, and the log odds ratios from the binary studies. By choosing a cutoff point for the continuous outcomes, the maximum-likelihood estimates and associated variances for the normal parameters can be transformed into log odds ratios and associated variances. Studies are then combined using standard weighting methods (1). This approach avoids direct dichotomization of the observed continuous outcomes and only requires sufficient statistics from the continuous studies (as does ours). The cutoff point(s) needs to be known. Reference

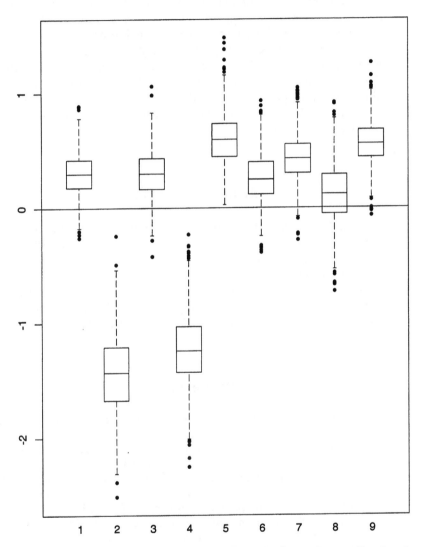

Figure 4 Posterior distribution of study effects. Study numbers are listed under each boxplot. Studies 1, 2, 4, 5, 6, 7, 8, and 9 report the estimated differences in effects sizes between the two arms. Study 3 reports a two-by-two contingency table.

48 presents an application to a problem where the binary variables are the dichotomization of an underlying continuous measure (blood loss) according to a cutoff well established in medical practice.

Alternatively, Ref. 49 proposes continuous scales based on the logits of the observed frequencies in the binary studies. These overcome the discretization problem in a nice way. Here we propose a Bayesian alternative, which has the added advantages of fitting conveniently within a hierarchical framework and leading to a simple solution of the problem of ranking treatments.

Acknowledgments

We thank Vic Hasselblad and Robert Wolpert, our co-authors on a larger meta-analysis of headache treatments, for many useful suggestions. We also thank Doug McCrory of the Duke University Center for Health Policy Research and Education for providing the data used in this article. These data were abstracted from published randomized controlled trials as part of the literature review in support of Clinical Practice Guidelines funded by the Agency for Health Care Policy Research.

References

1. LV Hedges, I Olkin. Statistical Methods for Meta-Analysis. New York: Academic Press, 1985.
2. V Hasselblad, DC McCrory. Meta-analytic tools for medical decision making: a practical guide. Med Decis Making 15:81–96, 1995.
3. AJ Sutton, KR Abrams, DR Jones, TA Sheldon, F Song. Systematic reviews of trials and other studies. Health Technology Assessment, 1998.
4. WH DuMouchel, JE Harris. Bayes methods for combining the results of cancer studies in humans and other species (c/r: pp 308–315). Journal of the American Statistical Association 78:293–308, 1983.
5. WH DuMouchel. Bayesian metanalysis. In DA Berry, ed. Statistical Methodology in the Pharmaceutical Sciences, vol. 104 of Statistics: Textbooks and Monographs. New York, Marcel Dekker, 1990, pp 509–529.
6. DA Berry. A Bayesian aproach to multicenter trials and meta-analysis. In: ASA Proceedings of Biopharmaceutical Section, 1–10, Alexandaria, 1990. American Statistical Association.

7. DM Eddy, V Hasselblad, R. Shachter. A Bayesian method for synthesizing evidence. The confidence profile method. International Journal of Technology Assessment in Health Care 6:31–55, 1990.

8. CN Morris, SL Normand. Hierarchical models for combining information and for meta-analysis. In: Bernardo et al. (50), pp 321–344.

9. DA Berry, DK Stangl, eds. Bayesian Biostatistics, vol. 151 of Statistics: Textbooks and Monographs. New York: Marcel Dekker, 1996.

10. RM Sorenson, NL Pace. Anesthetic techniques during surgical repair of femoral neck fractures: a meta-analysis. Anesthesiology 77:1095–1104, 1992.

11. LJ Baraff, SI Lee, DL Schriger. Outcomes of bacterial meningitis in children: a meta-analysis. Pediatr Infect Dis J 12:389–394, 1993.

12. JM Brophy, L Joseph. Placing trials in context using Bayesian analysis. GUSTO revisited by Reverend Bayes. JAMA 273:871–875, 1995.

13. BJ Biggerstaff, RL Tweedie, KL Mengersen. Passive smoking in the workplace: classical and Bayesian meta-analyses. Int Arch Occup Environ Health 66:269–277, 1994.

14. RL Tweedie, DJ Scott, BJ Biggerstaff, KL Mengersen. Bayesian meta-analysis, with application to studies of ETS and lung cancer. Lung Cancer 14(1S):171–194, 1996.

15. DA Berry. Benefits and risks of screening mammography for women in their forties: A statistical appraisal. J Natl Cancer Inst 90:1431–1439, 1998.

16. TC Smith, DJ Spiegelhatler, A Thomas. Bayesian approaches to random-effects meta-analysis: a comparative study. Stat Med 14:2685–2699, 1995.

17. TC Smith, DJ Spiegelhalter, MKB Parmar. Bayesian meta-analysis of randomized trials using graphical models and BUGS. In: DA Berry, D Stangl, eds. Bayesian Biostatistics. Marcel Dekker, 1996.

18. A Sutton, P Lambert, M Hellmich, K Abrams, D Jones. Meta-analysis in practice: A critical review of available software. In: D Stangl, DA Berry, eds. Meta-analysis in Medicine and Health Policy. New York: Marcel Dekker, 2000.

19. SM Berry. Proceeding to phase III. A Bayesian decision analysis. Technical report, Department of Statistics, Texas A&M University, 1997.

20. G Parmigiani, GP Samsa, M Ancukiewicz, J Lipscomb, V Hasselblad, DB Matchar. Assessing uncertainty in cost-effectiveness analyses: Application to a complex decision model. Medical Decision Making 17:390–401, 1997.

21. J Besag, D Higdon, K Mengersen. Meta-analysis via Markov chain Monte Carlo: Combining information through Bayesian random effects logistic regression. Discussion paper, Institute of Statistics and Decision Sciences, Duke University, 1997.

22. PC Lambert, KR Abrams, B Sansó, DR Jones. Synthesis of incomplete data using Bayesian hierarchical models: an illustration based on data describing survival from neuroblastoma. Technical report, University of Leicester, 1997.

23. GH Givens, DD Smith, RL Tweedie. Bayesian data-augmented meta-analysis that account for publication bias issues exemplified in the passive smoking debate. Statistical Science 12:221–250, 1997.

24. N Silliman. Hierarchical selection models with applications in meta-analysis. Journal of the American Statistical Association 92:926–936, 1997.

25. D Smith, G Givens, R. Tweedie. Adjustment for publication bias and quality bias in Bayesian meta-analysis. In: D Stangl, DA Berry, eds. Meta-analysis in Medicine and Health Policy. New York: Marcel Dekker, 2000.

26. K Abrams, P Lambert, B Sanso, C Shaw. Meta-analysis of heterogeneously reported study results—a Bayesian approach. In: D Stangl, DA Berry, eds. Meta-analysis in Medicine and Health Policy. New York: Marcel Dekker, 2000.

27. AE Gelfand, AFM Smith. Sampling-based approaches to calculating marginal densities. Journal of the American Statistical Association 85(410):398–409, 1990.

28. JM Bernardo, AFM Smith. Bayesian Theory. New York: Wiley, 1994.

29. MA Tanner. Tools for Statistical Inference—Observed Data and Data Augmentation Methods, vol. 67 of Lecture Notes in Statistics. New York: Springer-Verlag, 1991.

30. A Gelman, J Carlin, H Stern, D Rubin. Bayesian Data Analysis. London: Chapman and Hall, 1995.

31. B Carlin, NG Polson. Monte Carlo Bayesian methods for discrete regression models and categorical time series. In: Bernardo et al. (50), 577–586.

32. S Chib, E Greenberg. Analysis of multivariate probit models. Biometrika 85: 347–361, 1998.

33. DV Lindley. On a measure of the information provided by an experiment. Annals of Mathematical Statistics 27:986–1005, 1956.

34. G Parmigiani, DA Berry. Applications of Lindley information measure to the design of clinical experiments. In: AFM Smith and P Freeman, eds. Aspects of Uncertainty. A Tribute to D. V. Lindley. John Wiley & Sons, 1994, pp 351–362.

35. DK Ziegler. Headache. Public health problem. Neurol Clin 8:781–791, 1990.

36. RB Lipton, WF Stewart. Migraine in the United States: a review of epidemiology and health care use. Neurology 43:S6–S10, 1993.

37. WEM Pryse-Phyllips, DW Dodick, JG Edmeads, MJ Gawel, RF Nelson, RA Purdy, G Robinson, D Stirling, I Worthington. Guidelines for the diagnosis and management of migraine in clinical practice. Canadian Medical Association Journal 156:1273–1287, 1997.

38. RE Goslin, RN Gray, DC McCrory, J Tulsky, V Hasselblad. Drug treatments for prevention of migraine. Evidence report, Agency for Health Care Policy and Research, Bethesda, MD, 1998.

39. F Dominici, G Parmigiani, RL Wolpert, V Hasselblad. Meta-analysis of migraine headache treatments: Combining information from heterogeneous designs. Journal of the American Statistical Association 94, 1999.

40. R Rosenthal, DB Rubin. Comparing effect sizes of independent studies. Psychological Bulletin 92:500–504, 1982.

41. SL Berry. Understanding and testing for heterogeneity across 2 × 2 tables: Application to meta-analysis. Statistics in Medicine 1999.

42. WH DuMouchel, SL Normand. Computer modeling and graphical strategies for meta-analysis. In: D Stangl, DA Berry, eds. Meta-analysis in Medicine and Health Policy. New York: Marcel Dekker, 2000.

43. D Pauler, J Wakefield. Modeling and implementation issues in Bayesian meta-analysis. In: D Stangl, DA Berry, eds. Meta-analysis in Medicine and Health Policy. New York: Marcel Dekker, 2000.

44. P Gustafson. Large hierarchical Bayesian analysis of multivariate survival data. Biometrics 53:230–242, 1997.

45. L Tierney. Markov chains for exploring posterior distributions (with discussion). Annals of Statistics 22(4):1701–1762, Dec 1994.

46. NG Best, MK Cowles, K Vines. CODA: Convergence diagnostics and output analysis software for Gibbs sampling output, version 0.30. Technical report, MRC Biostatistics Unit, University of Cambridge, 1995.

47. S Suissa. Binary methods for continuous outcomes: a parametric alternative. Journal of Clinical Epidemiology 44:241–248, 1991.

48. A Whitehead, AJ Bailey, D Elbourne. Combining summaries of binary outcomes with those of continous outcomes in a meta-analysis. Journal of Biopharmaceutical Statistics 9:1–16, 1999.

49. V Hasselblad, LV Hedges. Meta-analysis of screening and diagnostic tests. Psychological Bulletin 117:167–178, 1995.

50. JM Bernardo, JO Berger, AP Dawid, AFM Smith, eds. Bayesian Statistics 4. Oxford: Oxford University Press, 1992.

6
Computer-modeling and Graphical Strategies for Meta-analysis*

William DuMouchel
AT&T Shannon Labs, Florham Park, New Jersey

Sharon-Lise Normand
Harvard Medical School, Boston, Massachusetts

Abstract

Over the past decade, meta-analyses have played an increasingly important role in summarizing clinical information and informing policy. A new generation of meta-analysts have come forth, and the demand for a streamlined modeling approach has emerged. Compared with linear model estimation, many aspects of selecting and fitting models for meta-analysis involve nonstandard data structures and statistical assumptions. For example, each data point in the analysis—that is, each study—is associated with its own measure of precision which must be accounted for in the estimation process. Other study variables, such as design covariates, within-study predictors, and other aspects of the studies, also play an important role

*The authors are grateful to Richard Tweedie, Colorado State University; Sue Taylor (University of Colorado Health Sciences Center); Zhezhen Jin (Columbia University); and David Fram, Barbara Snow, & Sally Cassells (Belmont Research, Cambridge, MA), with whom they have collaborated in the development of MetaGraphs,© a software package designed to perform meta-analysis using Hierarchical Bayes Linear Models. The authors thank Mariko Golden (Harvard Medical School) for word processing assistance.

when making inferences. Model selection therefore can be considerably facilitated by exploratory graphical analyses. In this chapter, we describe exploratory graphical methods. Also, we present a unified modeling approach to meta-analysis, one that integrates fixed-, random- and mixed-effect models, as well as Bayesian hierarchical models, into a single framework. Our approach places emphasis on model selection, estimation, fitting, diagnostics, and interpretation of results.

I. INTRODUCTION

Meta-analysis involves combining summary information across related but independent studies. The objectives of this type of analysis include increasing power to detect an overall treatment effect, estimation of the degree of benefit associated with a particular study, assessment of the amount of variability between studies, or identification of study characteristics associated with a particularly effective treatment. Compared with linear model estimation, many aspects of model fit for meta-analysis involve nonstandard data structures and nonstandard statistical assumptions.

Consider, for example, the data displayed in Table 1. This table summarizes fifty-nine studies evaluating the effect of nicotine-replacement therapy (henceforth denoted NRT) on smoking cessation. These data are taken from the Cochran Database of Systematic Reviews (1). The question of interest is whether the use of NRT enables someone to stop smoking and whether the effect is influenced by the form of NRT (chewing gum or transdermal patches) or by the intensity of the support offered to the smoker. Low-intensity support was defined as part of the provision of routine care; however, if the duration of time spent with the smoker exceeded 30 minutes during the initial consultation, or if the number of further assessment and reinforcement visits exceeded two, then the level of support was defined as high. Table 1 lists the study author, the year, the number of smokers who quit smoking, and the number of smokers assigned to control and treatment groups, respectively. The studies are stratified by the form of NRT and level of support, and within each, are ordered by year of study publication.

In this chapter, computer-modeling strategies for meta-analysis using a hierarchical Bayes linear model are presented. Data requirements are described and exploratory graphical methods are demonstrated using the NRT example. Using the Bayes linear model, we present a unified method for modeling within and between-study variation as well as discuss model diagnostics. In what follows, studies comprising the meta-analysis are referred to as primary studies. We focus on the methodology and defer many of the formulae to an Appendix.

II. DATA REQUIREMENTS

Assume that there are K studies available, and that it is desired to combine the results regarding treatment benefit. To conduct the meta-analysis, it is first assumed that the studies summarize information that is common across studies. The goal facing the meta-analyst is to transform each of the summary measures to a common scale. Often there is little control over the choice of the summary measure. For example, if risk differences are reported, rather than survival times, then the meta-analyst has few options but to employ risk differences as the summary estimate. However, in many situations, different summary measures may be reported across the K studies, and the meta-analyst will need to decide upon a (single) meaningful measure and subsequently transform the study-specific summaries to the chosen measure. There may be many choices for a summary effect, although the ultimate decision will largely rely upon what summary data are available in the primary studies. In the case of binary-valued responses measured in the primary studies, risk differences can be used to summarize the results, are easy to interpret, are defined for boundary values (proportions of zero or one), and are approximately normally distributed for modest sample sizes. Rate ratios and odds ratios, on the other hand, are typically analyzed on the logarithmic scale but, unlike risk differences, are not defined for boundary values. Alternatively, when the primary studies measure a continuous response, an *effect* estimate, defined as the difference in response means divided by the standard deviation of the response within the control population, may be employed. The discussion contained in Ref. 2 provides a useful summary of different measures. Our recommendation is to utilize estimates of quantities of direct scientific or decision-making relevance and to avoid ambiguous quantities that are difficult to interpret, such as p-values.

Let Y_i represent the summary statistic of the chosen type provided by the ith study and let s_i^2 denote the variance of Y_i. For $i = 1, \ldots, K$, we

Table 1 Fifty-nine Nicotine Replacement Therapy (NRT) Trials Examining the Efficacy of NRT on Smoking Cessation*

ID	Study Name	Year	Treated y_T/n_T	Control y_C/n_C	ID	Study Name	Year	Treated y_T/n_T	Control y_C/n_C
	Gum: high-intensity support					Patch: high-intensity support (contd.)			
1	Puska	1979	29/116	21/113	31	Richmond	1994	40/160	19/157
2	Malcolm	1980	6/73	3/121	32	Kornitzer	1995	19/150	10/75
3	Fagerstrom	1982	30/50	23/50	33	Stapleton	1995	77/800	19/400
4	Fee	1982	23/180	15/172	34	Campbell	1996	24/115	17/119
5	Jarvis	1982	22/58	9/58		Gum: low-intensity support			
6	Hjalmarson	1984	31/106	16/100	35	BR SOCIETY	1983	39/410	111/1208
7	Killen	1984	16/44	6/20	36	Russell	1983	81/729	78/1377
8	Schneider	1985	9/30	6/30	37	Fagerstrom	1984	28/106	5/49
9	Hall	1987	30/71	14/68	38	Jamrozik	1984	10/101	8/99
10	Tonnesen	1988	23/60	12/53	39	Jarvik	1984	7/25	4/23
11	Blondal	1989	37/92	24/90	40	Clavel-Chapel	1985	24/205	6/222
12	Garcia	1989	21/68	5/38	41	Schneidera	1985	2/13	2/23
13	Killen	1990	129/600	112/617	42	Page	1986	9/93	13/182
14	Nakamura	1990	13/30	5/30	43	Campbell	1987	13/424	9/412
15	Campbell	1991	21/107	21/105	44	Sutton	1987	21/270	1/64
16	Jensen	1991	90/211	28/82	45	Areechon	1988	56/99	37/101
17	McGovern	1992	51/146	40/127	46	Harackiewicz	1988	12/99	7/52
18	Pirie	1992	75/206	50/211	47	Llivina	1988	61/113	28/103
19	Zelman	1992	23/58	18/58	48	Sutton	1988	5/79	2/82
20	Herrera-1	1995	37/76	17/78	49	Gilbert	1989	11/112	9/111

Table 1 (Cont'd)

ID	Study Name	Year	Treated y_T/n_T	Control y_C/n_C
	Patch: high-intensity support			
21	Buchkremer	1981	11/42	16/89
22	Hurt	1990	8/31	6/31
23	Ehrsam	1991	7/56	2/56
24	Tnsg	1991	111/537	31/271
25	Sachs	1993	28/113	10/107
26	Westman	1993	16/78	2/80
27	Fiore-1	1994	15/44	9/43
28	Fiore-2	1994	10/57	4/55
29	Hurt	1994	33/120	17/120
30	ICRF	1994	76/842	53/844

ID	Study Name	Year	Treated y_T/n_T	Control y_C/n_C
	Gum: low-intensity support (contd.)			
50	Hughes	1989	23/210	6/105
51	Hughes	1990	15/59	5/19
52	Mori	1992	30/178	22/186
53	Nebot	1992	5/106	13/319
54	Fortmann	1995	44/262	42/261
	Patch trials: low-intensity support			
55	Abelin	1989	17/100	11/99
56	Daughton	1991	28/106	4/52
57	Tonneson	1991	17/145	2/144
58	Burton	1992	29/115	22/119
59	Paoletti	1996	15/60	4/60

*The numbers quitting and enrolled in the treatment (T) and control (C) groups are denoted y_T, n_T, y_C, and n_C, respectively.

assume that the variance of the study estimate is known. This assumption will be effectively met if the individual estimates are based on moderately large samples. Characteristics of the ith primary study are denoted by $\mathbf{x}_i = \{x_{ij}; j = 1, \ldots, J\}$ and the set of observed data by $\mathbf{Y} = \{Y_i; i = 1, \ldots, K\}$.

A. Strategies for Estimating Standard Errors When Sample Sizes are Small

Although our basic assumption is that sample sizes within each study are large enough for us to assume that the standard errors of the outcome measures are known, most reviews include some studies having small sample sizes. Therefore, a strategy is needed for integrating such data smoothly into the meta-analytic calculations. The question of how to handle small-sample estimates of variance arises in two common situations: small cell sizes for binary responses and unequal group variances for a continuous response.

1. Binary Response

The usual estimate of the variance of a sample proportion is $\hat{p}(1 - \hat{p})/n$, where $\hat{p} = x/n$ is the observed proportion and x is the observed number of "successes" in n trials. The usual estimate is obviously wrong if $x = 0$ $(\hat{p} = 0)$ or if $x = n$ $(\hat{p} = 1)$, because then s is estimated to be 0, even if n is small. What is often not appreciated is that meta-analytic estimates can be seriously biased by this variance-estimation problem (3). When the true proportion is near zero or one, the sample value $\hat{p}$ is highly correlated with the variance estimate, $\hat{p}(1 - \hat{p})/n$. The overall estimate of an effect in a meta-analysis is similar to weighted averages of study-effect estimates with weights inversely proportional to their estimated variances. If $\hat{p}$ is near 0, the smaller observed proportions get more weight than larger observed proportions, resulting in biasing towards effect sizes from the studies with the smallest proportions. This biasing effect is almost entirely due to the studies having $x = 0, 1, 2, n - 2, n - 1$, or n. Our recommended strategy for reducing this bias is as follows:

> If any of the proportions $\hat{p} = x/n$ used in the meta-analysis have $x = 0, 1, 2, n - 2, n - 1$, or n, then, throughout the meta-analysis, use the modified measures of outcome and standard error shown in Table 2, where $p' = (x + \frac{1}{2})/(n + 1)$ and $p'' = (x + 1)/(n + 2)$.

Table 2 repeats the common recommendation of adding $\frac{1}{2}$ to each cell when logarithmic effects are being computed, but what is new is our further recommendation to add 1 to each cell for the purpose of computing variance estimates, even if linear effects are being used. Unpublished simulations of our own show that this seemingly minor substitution of p'' $(1 - p'')$ for $\hat{p}(1 - \hat{p})$ reduces the correlation between the effect estimate and the standard error estimate. A Bayesian rationale for this variance correction is the belief that $p(1 - p)$ is not very near 0.

2. Continuous Response

Estimation of standard errors in the presence of small samples from a continuous response is also problematic. The question is whether to use variance estimates separately for each treatment group, or to use a pooled estimate of variance. This is not a major problem if there are only two treatment groups with approximately equal sample sizes, since then the estimate of the standard error of the difference in means is about the same for both choices. However, if there are three or more groups, it can make a substantial difference. Suppose group 1 has a much larger sample variance than either group 2 or 3. It seems inappropriate that group 1's variance should inflate the standard error of the contrast between group 2 and group 3. On the other hand, when group sample sizes are very small, the pooled estimate of variance is much more reliable than the individual variance estimates. Our recommendation is ad hoc (see Table 3), namely:

Table 2 Binary-response Effect Estimates and Standard Errors If Small Counts Are Present*

Summary measure	Effect estimate	Standard error
Rate difference	$p_1 - p_2$	$\left[\dfrac{p_1''(1 - p_1'')}{n_1} + \dfrac{p_2''(1 - p_2'')}{n_2}\right]^{\frac{1}{2}}$
(log) Relative risk	$\log\left(\dfrac{p_1'}{p_2'}\right)$	$\left[\dfrac{(1 - p_1'')}{n_1 p_1''} + \dfrac{(1 - p_2'')}{n_2 p_2''}\right]^{\frac{1}{2}}$
(log) Odds ratio	$\log\left(\dfrac{p_1'(1 - p_2')}{p_2'(1 - p_1')}\right)$	$[n_1 p_1''(1 - p_1'') + n_2 p_2''(1 - p_2'')]^{-\frac{1}{2}}$

*Recommended strategy if $x = 0, 1, 2, n - 2, n - 1$, or n. Note that $\hat{p} = x/n$, $p' = (x + \frac{1}{2})/(n + 1)$, $p'' = (x + 1)/(n + 2)$, and subscripts denote treatment group.

Table 3 Continuous-response Standard Errors of Group Differences*

Statistic	Definition
t_0^2	$\Sigma_j(n_j - 1)\, t_j^2/\Sigma_j(n_j - 1)$
t_j'	t_j if $n_j > n_0$
	$[(n_j - 1)\, t_j + (n_0 + 1 - n_j)t_0]/n_0$ if $nj \leq n_0$
standard error $(x_j - x_k)$	$\sqrt{(t_j'^2/n_j + t_k'^2/n_k)}$

*The sample size, sample mean, and sample standard deviation corresponding to the jth group $(j = 1, 2, \ldots, J)$ are denoted, respectively, n_j, x_j, and t_j. Choose an arbitrary sample size n_0 (we suggest $n_0 = 10$).

Choose a sample size n_0 (e.g., $n_0 = 10$). The choice is somewhat arbitrary. Use the sample-specific standard deviation for all groups having $n > n_0$. Use the pooled standard deviation for $n = 1$ (no other possible choice!). Interpolate between these values for $1 \leq n \leq n_0$.

B. Example: Effect of NRT On Smoking Cessation

Of the fifty-nine studies collated in Table 1, 20 report quit rates using chewing gum accompanied by low support, 14 report quit rates using transdermal patches accompanied by high support, 20 with chewing gum and high support, and 5 with transdermal patch followed by low support. Two study-design covariates can be used to characterize the fifty-nine studies. Let $x_{i1} = 1$ if the form of NRT in the ith study was transdermal patch (32% of the studies), and 0 otherwise. Similarly, let $x_{i2} = 1$ if high-intensity support was provided in the ith study (58% of the studies), and 0 otherwise. Letting T and C denote the treatment and control groups respectively, we use the difference in quit rates, defined as $Y_i = p_{Ti} - p_{Ci}$ as the study summary estimate. The corresponding measure of precision is calculated as $s_i^2 = p_{Ti}q_{Ti}/n_{Ti} + p_{Ci}q_{Ci}/n_{Ci}$, where p_i is the fraction of smokers quitting in the group, $q_i = 1 - p_i$ is the fraction not quitting, and the numbers of smokers in the treatment and control groups are respectively n_{Ti} and n_{Ci}. For example, the quit difference in study 34, a patch intervention accompanied by high-intensity support is $0.066 = 24/115 - 17/119$ with standard error equal to

0.0496. Thus an excess of 6.6% of treated smokers quit compared to the control group, with 95% confidence interval ($Y_i \pm 1.96^* s_i$) for the underlying (unknown) quit difference ranging from -3% to 16%. Because the raw numbers of quitters in both the treated and control groups are reported, odds ratios or rate ratios could also be utilized.

III. EXPLORATORY GRAPHICS

An important first step in any data analysis is to investigate the raw data graphically to identify any patterns, anomalies, or data-entry errors. In this section, we describe several different graphs that are useful in meta-analysis.

A. Preliminary Graphs

Figure 1(a), denoted a *baseline plot*, describes the relationship between the effect estimates and baseline risk or mean. In the figure, each primary study is reported by its quit difference on the y-axis (raw effect size) with its corresponding 95% confidence interval and plotted against the baseline risk of quitting in the control group on the x-axis. The quit difference for each study is represented by a square box, the area of which is proportional to the number of subjects in the study. The average baseline risk of quitting varies substantially, from less than 0.10 to approximately 0.45, across the fifty-nine studies. It does not appear that the quit difference associates with the baseline risk of quitting.

The second graph type (Fig. 1(b)), called a *L'Abbe plot* (4), is a scatterplot of the baseline risk (y-axis) against the treatment risk (x-axis) in each study. For most trials, the quit rate is higher in the treated arm than in the control arm.

The L'Abbe and the baseline plots can help measure the effect of treatment when responses are dichotomous. The question addressed is "what measure of effect is reasonably constant from study to study?" There are at least three types of homogeneity that these plots can be used to address. If the risk difference is constant, then the points in the L'Abbe plot should follow a straight line with slope one, parallel to the 45-degree line in Fig. 1(b). If the risk ratio is constant, then the points should follow a straight line going through the origin, with slope equal to the risk ratio. Finally, if the odds ratio is constant, the points should follow a curve that

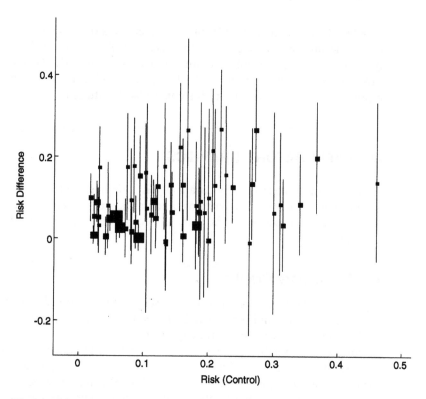

Figure 1(a)   Baseline graph displays the risk of smoking cessation as a function of the *baseline* risk of quitting for fifty-nine NRT trials.

goes through both the points (0,0) and (1,1), with maximum deviation from the 45-degree line near the middle of the graph.

Correspondingly for the baseline plot, if the risk differences (x-axis) are approximately constant, then the risk difference may be a suitable effect to analyze, while if the risk ratio is constant, then a linear trend will show up on the baseline plot. If the odds ratio is constant, then a curved trend will be expected, tied to a risk difference of zero at the values of baseline risk equal to zero or one. Of course, most meta-analyses do not have such a wide range of baseline risks to make such fine distinctions. When all of the baseline risks are less than $\frac{1}{2}$, as in the NRT data, the risk ratio and the odds ratio are practically equivalent. The baseline plot has one advantage over the L'Abbe plot, however, in that confidence inter-

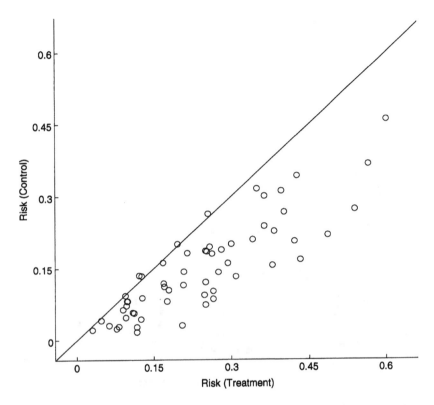

Figure 1(b)　L'Abbe graph depicts the relationship between the risk of smoking cessation in each of the two groups.

vals are available and provide a measure of sampling error to help judge whether any perceived pattern could be due to chance.

In the case of Figs 1(a, b), the plots do not disagree radically with any of the three measures of homogeneity. For the baseline plot of Fig. 1(a), if we draw a horizontal line at risk difference = 0.1, it would intersect all but two of the fifty-nine 95% confidence intervals, which is consistent with the constant risk difference hypothesis. On the other hand, one can perceive a slight upward trend in the points as the control risk increases from 0 to 0.5, consistent with both the hypothesis of the constant risk ratio and that of constant odds ratio. In Fig. 1(b), the L'Abbe plot seems to show points drawing away from the 45-degree

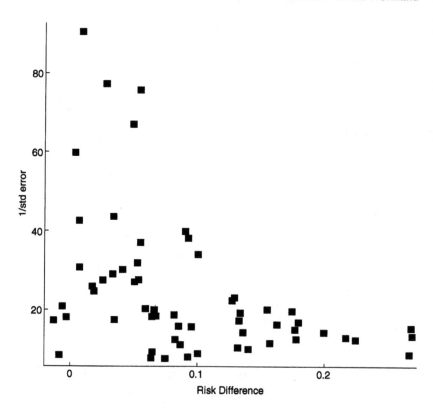

Figure 1(c)   Funnel plot depicts the relationship between study effect and study size.

line. This is inconsistent with the assumption of constant risk difference. However, the standard errors from studies with larger observed proportions are generally larger, so no conclusive statement can be made. In what follows, we will continue to use the risk difference as our summary measure.

A third graph type, called a *Funnel plot* (Fig. 1(c)), is a scatterplot of the study effect on the x-axis and a measure of study size (here taken to be the reciprocal of the effect's standard error) on the y-axis. We expect there to be no trend in this plot, but the points should be more spread out left-to-right at the bottom of the plot than at the top, like an inverted funnel. There is a "tail" of studies in the lower right corner of the plot, but the cloud of points drops off sharply at the lower left of the plot, with no studies extending much beyond the "chimney" of large (low-standard-error) studies. In other words, the average effect of treatment is larger for the small studies than for the large studies. This could be because smaller studies tend to be more poorly controlled and to be subject to misleading biases with overly optimistic effect measures, or it could be due to *publication bias* (5), in which studies that show no effect of NRT, or a negative effect of NRT, are less likely to be published and show up in the meta-analysis. That is, there may be a missing group of unpublished studies corresponding to the missing left tail Fig. 1(c). We will return to this feature of these data when discussing modeling for the example.

B. Effect Sizes Comprising the Meta-analysis

Figure 2(a), called a *ladder plot*, displays a graph of 95% confidence intervals for the 59 NRT studies, stratified by form of nicotine replacement (columns) and intensity of follow-up (rows). The plots display the quit differences (x-axis) and corresponding 95% confidence intervals for each of the individual studies. The study-specific estimates (Y_i) are indicated by squares, the areas of which are proportional to the total number of patients. Thirty-five of the 59 confidence intervals cover the line $\mu = 0$, although 55 of the 59 point estimates of μ are greater than zero.

Figure 2(b) displays a ladder plot for the same data, except that the studies are ordered by year of publication. Ladder plots ordered by a study-design covariate provide a way to check for systematic trends in study results. In this case there appears to be no time trend. Some researchers recommend cumulative ladder plots in which each confidence bar includes all previous data, as a primary method for describing systematic effects (6). We prefer the regression-modeling approach discussed in Secs IV–VI.

C. Study-design Covariates

To explore whether the effect of interest varies depending on characteristics of the studies, a graphical display of the relationship between one or more covariates is worth constructing. Figures 3(a, b) present two exploratory graphs relating risk differences to two covariates in the NRT example. Figure 3(a) depicts the quit difference (y-axis) and corresponding 95% confidence interval against the route of NRT (gum vs.

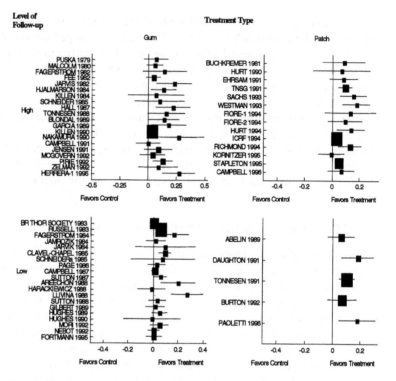

Figure 2(a) Ladder plots display quit differences for the fifty-nine NRT trials, stratified by form of nicotine replacement (columns) and by intensity of follow-up offered to the smoker (rows).

patch) (x-axis) for each of the 59 NRT trials. There does not appear to be a strong relationship between efficacy and route of nicotine, although the patch trials seem to have a slight advantage.

Figure 3(b) displays the risk difference against its standard error. The purpose of this plot is similar to that of the funnel plot (Fig. 1(c)). The larger trials have a lower average quit difference. One advantage of Fig. 3(b) over Fig. 1(c) is the presence of confidence bars to help evaluate the significance of any trend.

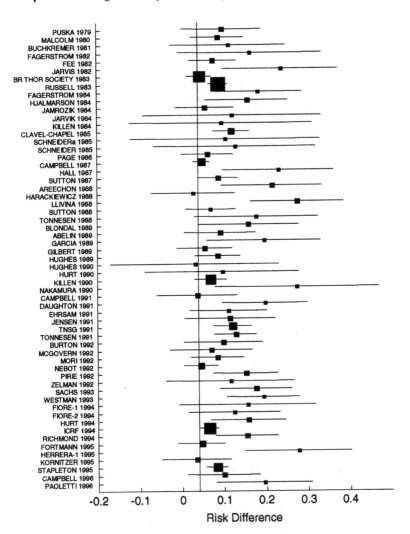

Figure 2(b) Quit differences ordered by year of study publication.

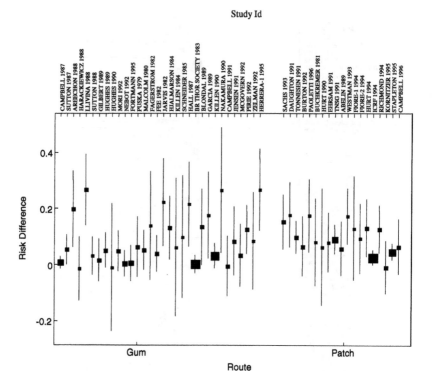

Figure 3(a)   Covariate plot depicts the relationship of the quit difference as a function of form of NRT.

IV. MODELING VARIATION IN META-ANALYSIS

There are two distinct approaches to combining information across studies that can be utilized when undertaking a meta-analysis. The type of approach adopted depends on whether the investigator believes that each study's summary statistic provides an estimate of the *same* underlying parameter, or that each study provides an estimate of a *different but related* parameter. The latter belief will be shown to be equivalent to assuming there is inter-study variation. The approach often recommended when modeling variation in meta-analysis is to first *test* for inter-study variation. In most meta-analyses, these tests have very low power, and other chapters comment on this. If the investigator fails to

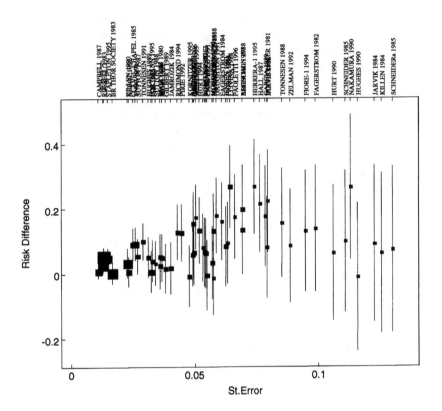

Figure 3(b) Covariate plot indicates the association of the quit difference to its standard error.

reject the hypothesis of no inter-study variability, then the meta-analyst combines the data to estimate the common parameter *assuming that inter-study variation is zero*. This approach is commonly referred to as a fixed-effect approach. Conversely, a random-effect approach explicitly allows for an inter-study variance component in the estimation of the common parameter.

A. Sources of Variation

There are at least three sources of variation to consider when combining study summaries. First, sampling error may vary across the K studies. This component of error arises as a result of the variability in the response variable *within* a study. Second, treatment efficacy may vary *across* subgroups defined by study-design covariates. For example, the efficacy of nicotine-replacement therapy may depend on the form of the therapy. Third, the underlying treatment effect may vary *across* studies. This component of error arises as a result of inter-study variation due to differences in conducting each study. The size of the inter-study variance component measures how much artifactual study-specific factors such as population variations, investigator differences, uncontrolled conditions, etc. contribute to unpredictable deviations within the subgroups.

The sources of variation are not mutually exclusive. Sampling error will almost always contribute to variation across studies. Moreover, treatment efficacy may vary across subgroups *and* within each subgroup, so that each study provides an estimate of a different but related parameter. Two approaches to modeling the variance components are described next. We conclude with a description of the hierarchical Bayes linear model (HBLM), a model that integrates both approaches.

B. Fixed-effect Model

Let μ represent the *common* parameter of interest, so that, for moderately large study sample sizes,

$$Y_i \sim \text{normal}(\mu, s_i^2).$$

This model asserts that each study summary is independently distributed according to a normal distribution with expectation μ and within-study variance s_i^2. The parameter μ is the *overall* or *combined* treatment effect. If there are study-design covariates, denoted by the row vector $\mathbf{x}_i$, then these may also be incorporated into the model by assuming that the expectation of Y_i is a linear function of the covariates: $Y_i = \mathbf{x}_i \boldsymbol{\beta}$. In this case, between-study variation is permitted only through fixed characteristics of the studies, $\mathbf{x}_i$. For example, if the underlying quit difference for the ith study depends additively on the form and intensity of support, then we would expect the average value of Y_i to be $\mu + \beta_1 x_{i1}^* + \beta_2 x_{i2}^*$. For convenience, we have centered the covariates: $x_{ij}^* = x_{ij} - \bar{x}_j$, where $\bar{x}_j$ is the unweighted mean of the jth covariate. Thus, regardless of the inclusion of covariates, μ is the overall efficacy of NRT. The parameter β_1 is defined as the average increase (or decrease) in NRT efficacy for transdermal patch trials compared to chewing-gum trials. Similarly, the regression coefficient β_2 is interpreted as the increase (or decrease) in efficacy for high intensity support trials compared to low intensity support trials.

C. Random-effect Model

In contrast to the fixed-effect formulation, the random-effect formulation asserts that each study has its own effect, μ_i, such that

$$Y_i \mid \mu_i, s_i^2 \sim \text{normal}(\mu_i, s_i^2).$$

Given μ_i and s_i^2, therefore, each study summary is independent and normally distributed. Furthermore, each study-specific mean, μ_i, is assumed to arise from a distribution of effects with mean μ and variance (inter-study variation) τ^2 such that

$$\mu_i \mid \mu, \tau^2 \sim \text{normal}(\mu, \tau^2).$$

If the study-specific effects are identical across all K studies, then the inter-study variance component, τ^2, is 0. A common recommendation is to formally test the hypothesis $\tau = 0$ using a statistic, denoted Q, based on the residual sum of squares (formula (9) in the Appendix). If Q is greater than the $100(1 - \alpha)$ percentile of the χ^2_{K-1} distribution, then the hypothesis is rejected, the test statistic is used to estimate τ, and the analysis proceeds as if τ were known (7). However, there are two fundamental and related problems with this approach. In both, the issue is that a particular value of τ is used. First, it is not correct to assume $\tau = 0$ just because the hypothesis cannot be rejected. This is the difference between rejecting the alternative hypothesis and proving that the null hypothesis is true. Second, if the hypothesis is rejected and τ is estimated, the uncertainty associated with the estimate is not taken into account in the subsequent analyses. Neglecting this source of uncertainty may result in the identification of a significant treatment benefit when there may be insufficient evidence to support this claim. Another commonly used estimate of τ is the restricted estimate of maximum likelihood (REML) which is obtained by maximizing the marginal likelihood for the data (8). Although the REML captures the correct features of the underlying model, the uncertainty associated with the REML of τ still needs to be accounted for in the subsequent analyses.

Similar to the fixed-effect model, study-level covariates may also be included by assuming that the average value of the study-specific effect, μ_i, is a linear function of the covariates, $\mathbf{x}_i\boldsymbol{\beta}$. In contrast to the fixed-effect model, although studies characterized by $\mathbf{x}_i$ are more similar than studies without characteristic $\mathbf{x}_i$, each study provides an estimate of a *different* underlying parameter. Moreover, τ is the standard deviation of any extra variation in μ_i not explained by the study covariates. In the NRT example, if on average NRT efficacy depends on both the form of replacement as well as on the intensity of support offered to the smoker, say $E(\mu_i \mid \beta_1, \beta_2, \tau^2) = \mu + \beta_1 x_{i1}^* + \beta_2 x_{i2}^*$, a random-effect model postulates that the efficacy within trials characterized by the same covariates will differ due to inter-study variation.

D.   A Unified Approach: Hierarchical Bayes Linear Model

The hierarchical Bayes linear model (HBLM) integrates the fixed-effect and random-effect models into one framework and provides a unified modeling approach to meta-analysis (9–10). The three sources of variation are modeled using the following formulation:

$$Y_i \mid \mu_i, s_i^2 = \mu_i + \epsilon_i, \epsilon_i \sim \text{normal}(0, s_i^2),$$

where

$$\mu_i \mid \beta, \tau^2 = \mathbf{x}_i\boldsymbol{\beta} + \delta_i \quad \text{and} \quad \delta_i \sim \text{normal}(0, \tau^2),$$
$$\boldsymbol{\beta} \sim \text{normal}(\mathbf{b}, D), \qquad \tau \sim \pi(\tau).$$

This is a simple re-expression of the random-effect model into sampling error (s_i^2), inter-study variation explained by fixed characteristics of the study ($\mathbf{x}_i\boldsymbol{\beta}$), and unexplained study-to-study variation (τ^2). Because the values of the s_i are assumed known, any excess variation in the study effects, Y_i, is fitted by choosing non-zero values of $\boldsymbol{\beta}$ and τ^2. The HBLM includes two new specifications, namely the prior distributions of $\boldsymbol{\beta}$ and τ, which are assumed to be independent.

Several models are special cases of the HBLM. If the prior for τ concentrates its probability near the value $\tau = 0$, then the HBLM is equivalent to the fixed-effect model. On the other hand, if the number of studies, K, is very large, or if the prior for τ is concentrated near the estimate of τ, then the HBLM is equivalent to the random-effect model. This is because the estimates of μ, τ, and the study-specific effects μ_i will then be equivalent to their maximum-likelihood estimates. The HBLM permits a prior for the regression coefficients, β, under either a fixed-effect or a random-effect model. In particular, our model designations (i.e., "fixed", "random", "Bayes") reflect how to handle τ, but not how to handle β.

V. MODELING STRATEGIES

A. Prior Distributions

The Bayesian specification is completed by assigning *prior* distributions to β and τ^2. A prior distribution is the unconditional distribution before information regarding $\mathbf{Y} = \{Y_i; i = 1, \ldots, K\}$ is observed. In the absence of specific prior information on β, a diffuse prior (formula (24) in Appendix) may be used. We assume that β and τ are a priori independent.

Our modeling strategy depends critically on the use of a prior distribution for the inter-study standard deviation, generically denoted $\pi(\tau)$. All inferences regarding other parameters are averaged with respect to the posterior distribution for τ. We employ a *proper* prior distribution, that is, one that satisfies the requirements $0 \leq \pi(\tau)$ and $\int_\tau \pi(\tau)d\tau = 1$. Discussions describing informative proper priors can be found in Refs 9, 11, and 12. Because the posterior distribution is proportional to the product of the prior distribution and the likelihood, if the number of studies, K, is large and the prior for τ is highly dispersed, then the likelihood will determine the shape of the posterior. However, in many meta-analyses, K is small, and therefore it is important to determine how inferences are affected by changes in the prior for τ. The fixed-effect model, which assumes that every μ_i equals μ, is equivalent to assuming that $\pi(\tau)$ is concentrated near 0. Researchers opposed to meta-analysis, who believe that it is hopeless to combine study summary statistics, can be described as assuming that τ is very large. We adopt a diffuse, but proper, prior distribution that accommodates prior probabilities across the parameter space. In the examples considered in this chapter, a log-logistic prior distribution for τ is assumed:

$$\pi(\tau) = \frac{s_0}{(s_0 + \tau)^2} \text{ for } \tau \geq 0, \quad \text{where } s_0^2 = \frac{K}{\sum s_i^{-2}}.$$

Here, s_0^2 is the harmonic mean of the K sampling variances, s_i^2, and this choice for s_0 implies that the density $\pi(\tau)$ has median equal to s_0. This distribution is highly dispersed, since the expectations of τ and $1/\tau$ are both infinite. Other choices for $\pi(\tau)$ are available, but this particular selection offers several advantages. First, it has a maximum at 0 and is a decreasing function of τ. This conforms with our belief that τ near 0 is a definite possibility. Second, the quartiles of the distribution are $s_0/3$, s_0, and $3s_0$ so that the distribution is automatically scaled to a sensible range in the correct units. If the sample standard deviations are not equal, then s_0 will be weighted towards the smaller s_i whose associated values of Y_i are more informative about τ. This feature prevents the results from being unduly influenced by the addition of one or two very imprecise studies having large values of s_i. Third, because the posterior means for the study-specific effects are convex combinations with weights determined by shrinkage factors, the choice of $\pi(\tau)$ above as a prior permits an a priori range of shrinkage. Let $B_0(\tau) = s_0^2/(s_0^2 + \tau^2)$. This is a "typical" shrinkage factor with s_0^2 defined as the harmonic mean of the K sampling variances. The prior quartiles of $1 - B_0(\tau)$ assuming the log-logistic prior are 0.10, 0.50, and 0.90. So there is a 25% probability of virtually no shrinkage ($B_0 \leq 0.10$) while there is a 25% probability of essentially complete shrinkage ($B_0 \geq 0.90$).*

There are many choices of highly dispersed distributions for τ that would retain the properties above. However, we emphasize that the improper prior often used by default in the Bayesian literature when estimating a scale parameter, $\pi(\tau) \propto 1/\tau$, leads to a non-integrable posterior distribution for τ, which collapses to a one-point distribution at $\tau = 0$. The other commonly suggested noninformative prior, the uniform diffuse prior $\pi(\tau) = 1$, $\tau \geq 0$ produces an integrable posterior but tends to skew the posterior distribution towards very large values of τ. See the Appendix for discussion of a 2-parameter family of distributions extending $\pi(\tau)$.

B. Posterior Distributions

Once the data have been observed, the Bayesian analysis proceeds by computing the posterior distribution for the parameters of interest

---

*In the MetaGraphs software, the prior for τ is set by using a scale-invariant transformation of τ^2 to $1 - B$. The quantity $1 - B$ is represented by R^2 in the MetaGraphs' documentation.

through an application of Bayes' theorem. The mean and standard deviation of μ, β, and μ_i are typically reported, as well as the probability that $\mu > 0$, $\beta_j > 0$, and $\mu_i > 0$. The posterior distribution for τ is calculated in the usual prior-to-posterior manner. Our approach to computing the posterior distributions for μ, β, and μ_i is to first calculate the posterior for τ and then to average the remaining distributions conditional on τ with respect to the posterior for τ (formula (20) in the appendix). Visually, graphs of $\pi(\tau \mid \mathbf{Y})$ can be constructed to display the dependence of the meta-analysis on the values of τ.

1. Exchangeable Study Effects: No Study-design Covariates

In the case of no study-design covariates, the mean for μ conditional on the data and on τ, denoted $\mu(\tau)$, is a weighted average of the study observations, in which the weights are the precision of the study estimates (formula (27)).

The estimator for the unknown treatment effect in the ith study, denoted $\mu_i(\tau)$, is a compromise between the observed value, Y_i, and the prior mean, μ (formula (29)). The relative weight of each component in the estimator is determined by the fraction of total variance that is due to sampling error in the ith study (the shrinkage factor). This weight, $B_i(\tau) = s_i^2/(s_i^2 + \tau^2)$, is such that $0 \leq B_i(\tau) \leq 1$. Very large values of τ imply that $B_i(\tau)$ is close to 0, so that not much is gained in combining the studies. Alternatively, studies *borrow strength* from each other when smaller values of τ prevail. For these reasons, the estimator for μ_i presented in formula (29) is referred to as a *shrinkage formula*.

The (unconditional) posterior mean of the combined treatment effect, μ^*, is then obtained by weighting the conditional mean corresponding to different values of τ by the posterior probability of τ (formula (31)). It is straightforward to calculate the (unconditional) posterior variance of the combined effect, μ^{**} (formula (32)), so that an approximate 95% confidence interval for μ is given by $\mu^* \pm 1.96\{\mu^{**}\}^{1/2}$. The (unconditional) posterior mean and variance for each study-specific effect, μ_i, are calculated in a similar manner (formulas (33)–(34)). The posterior probability that the combined effect or any particular treatment effect is greater than 0 can be easily computed (formulas (35)–(36)).

a. NRT Example Revisited

The main question of interest is whether nicotine-replacement therapy increases the probability of smoking cessation. To answer this question we employed an HBLM assuming the quit differences across the 59 studies are exchangeable. Figures 4(a, b) are *trace plots* that depict the impact of the inter-study variance component on the meta-analysis. In the trace plots, a histogram displays the approximate posterior distribution of τ, scaled on the left axis.

This histogram is overlaid with a curve that shows the estimated posterior expectation of μ given τ (formula (27)), which is scaled on the

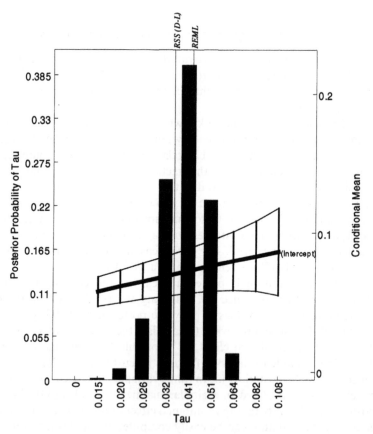

Figure 4(a)   Trace plot demonstrates the impact of the variance component on the meta-analysis. The histogram presents the approximate posterior probability of the inter-study variance component. The solid lines trace the conditional posterior expectation of the overall effect.

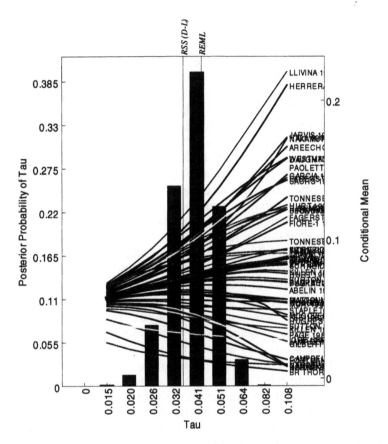

Figure 4(b)   Trace plot similar to (a) but with solid lines that trace the change in the shrinkage estimates as a function of the inter-study variation.

right axis. Also included are point-wise confidence bands for each posterior expectation. The posterior expectation is obtained by averaging the curve with respect to the posterior distribution of τ. In the NRT example, the modal value for τ is 0.04 and the overall effect, μ, ranges from 0.06 to 0.09 depending on the value of τ. In addition to the modal value of τ as depicted in the histogram, two other estimates of inter-study variation are indicated on the trace plots: the DerSimonian and Laird estimate, denoted RSS(D–L) (formula (10)), and the REML estimate (maximizing the function given in formula (11)). If study-level covariates are present, the RSS(D–L) estimator is a generalization of the usual DerSimonian and Laird estimator. The value of the overall effect at $\tau = 0$ is the fixed-effect model estimate, while the rightmost values of τ correspond to models in which the studies are not combinable. The value on the curve at the modal value is close to the HBLM estimate, which is actually computed as a weighted average of the values near each bar of the histogram, with weights equal to the heights of the corresponding bars.

Figure 4(b) is a trace plot for the NRT data, but rather than a single curve for the combined estimate, 59 curves representing the individual study estimates, $(\mu_i^*(\tau) :$ Formula (29)), are superimposed on the histogram. When τ is large, the heights of the curves are close to the observed values of the primary study estimates, Y_i, while as τ decreases to 0, the curves shrink to a common value, namely $\mu^*(\tau = 0)$. The rate of shrinkage among the studies varies, with studies having more sampling variability (larger s_i) shrinking faster.

Figure 5, denoted a *shrinkage* plot, displays a summary graph comparing the Bayesian results with the individual primary study results. The shrinkage plot provides a visualization of how the Bayesian computations convert the set $\{Y_i \pm 2s_i\}$ into the set $\{\mu_i^* \pm 2\{\mu_i^{**}\}^{1/2}\}$. The advantage of borrowing strength is demonstrated by the reduction in the posterior standard deviations $\{\mu_i^{**}\}^{1/2}$ as compared to the within-study errors s_i. The y-axis lists each of the primary studies, and the x-axis displays the quit differences. The estimates for the observed data, Y_i, and corresponding 95% confidence intervals are indicated by squares and solid horizontal lines; the shrinkage estimates, μ_i^* and corresponding 95% intervals, by circles; and the overall estimate, μ^*, by an elongated diamond. The location of μ^* is repeated for each study; because there are no covariates, it is the same value for all studies.

Table 4 presents the parameter estimates as a result of fitting the exchangeable model. The harmonic mean of the study sampling variances was used to construct the prior for τ. The posterior mean and standard deviation for τ are also reported, as well as the results of two tests of inter-study variation: one based on a Bayes factor (formula (19)) and one based on the Q-statistic (formula (9)). Both tests strongly favor the random-effect model (Sec. VI.A describes model choice in more detail). The posterior mean and standard deviation of the overall combined effect is reported together with the probability that the overall effect is positive.

In the case of no study-design covariates, a 95% credible interval for the overall effect is $\mu^* \pm 1.96\{\mu^{**}\}^{1/2} = 0.0720 \pm 0.0344$ and $P(\mu > 0 \mid \mathbf{Y}) = 1.00$ indicating overwhelming evidence that the use of nicotine-replacement therapy increases the probability of smoking cessation. A

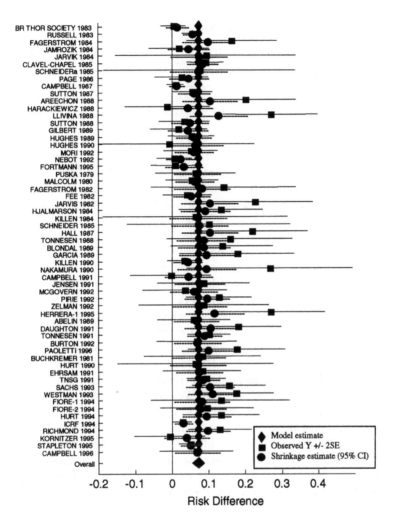

Figure 5 Shrinkage plot compares the observed data with the results of the model-fitting process. The overall model estimated is depicted by an elongated diamond on the bottom of the graph and stretched to indicate an approximate 68% confidence interval of the estimate. The shrinkage estimates are displayed as circles and the primary study data by squares.

Table 4   Results of Exchangeable Model*

	Mean	SD	Prob > 0
Constant term	0.0720	0.0176	1.000

*Results for estimating NRT efficacy using data from 59 trials. Coefficients for model with no covariates; $\tau = 0.0401$ (0.0090), BF = 100,000, Q statistic = 132, df = 58, p-value = 0.0000.

similar summary can be generated at the level of the individual studies, although we do not present one here. For example, we found that the posterior mean (standard deviation) for study 34 is 0.069 (SD = 0.0314) yielding a 95% confidence interval for μ_{34} as (1%, 13%). The 95% interval for the primary data yielded μ_{34} of (−3%, 16%) before conditioning on all the observed primary study data. Moreover, $P(\mu_{34} > 0 \mid Y) = 0.986$ implying that there is strong evidence that there was a beneficial treatment effect within this particular trial, in spite of the lack of statistical significance restricted to this trial's data.

Figure 6 is a scatterplot in which a cross-validation residual is plotted against the study standard errors. The cross-validation residual is a measure of how much each study effect differs from its prediction based on the remaining $K - 1$ studies. It is normalized so that, if the model is true, the cross-validation residuals will each have a standard normal distribution (formula (6)). We elaborate further on this in Sec. VI. In Fig. 6, there is a distinct trend in these residuals as a function of the study standard errors: there are "holes" in the upper left and lower right sections of the plot. This is another reflection of the problems identified by the funnel plot: for whatever reason, the smaller studies show larger effects of NRT, on average. We next explore an analysis using study covariates to help explain this deviation from the model assumptions.

2. Study-design Covariates

When study-design covariates are included in the model, the size of the inter-study variance component, τ, measures how much artifactual study-specific factors contribute to unpredictable deviations from the regression prediction, $\mathbf{x}_i\boldsymbol{\beta}$. As in the case with no covariates, the mean for $\boldsymbol{\beta}$ conditional on the data and on τ, denoted $\boldsymbol{\beta}(\tau)$, is the usual weighted regression estimator with weights equal to the reciprocal of the sum of the precision of the study-specific estimates and the prior precision. All our calcula-

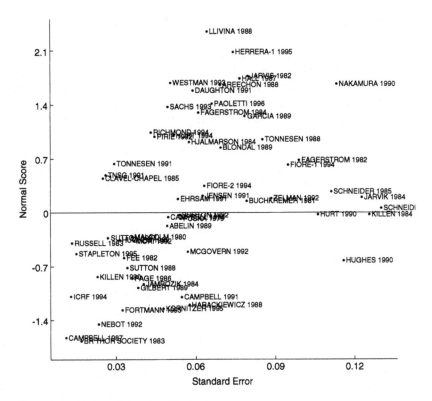

Figure 6 Cross-validated residual plot.

tions are analogous to the case without study covariates, that is, we obtain the unconditional mean of β and μ_i by averaging with respect to the posterior distribution of τ. The formulae are presented in the Appendix.

a. NRT Example Revisited

Having determined that NRT is efficacious in smoking cessation, the next question of interest is whether NRT efficacy is influenced by the route of therapy or by the intensity of support offered to the smoker. Let $x_i\beta = \mu + \beta_1 x_{i1}^* + \beta_2 x_{i2}^*$, where the covariates have been centered at their respective unweighted averages so that the parameter μ represents the overall effect of NRT on smoking cessation across all studies. In this case, x_{i1}^* has the value of 0.68 for transdermal-patch trials and -0.32 for chewing-gum trials, while x_{i2}^* has a value of 0.42 for trials followed by high support, and otherwise $x_{i2}^* = -0.58$. Although it is not necessary to center the study-level characteristics, doing so facilitates the interpretation of the intercept as well as the other parameters. The parameter β_1 represents the effectiveness of patch compared to gum, and β_2 represents the effectiveness of low-intensity support compared to high support. If present, the interaction term $(x_{i1}x_{i2})$ has a value of 1 for patch studies having high support and is also centered at its mean: $(x_{i1}x_{i2})^* = 0.76$ for the patch-high-support studies and -0.24 for all other studies. We assume that $\mu_i \mid \boldsymbol{\beta}, \tau^2 \sim \text{normal}(\mathbf{x}_i\boldsymbol{\beta}, \tau^2)$ so that, while μ_i is the mean of Y_i, we also assume that $\mu_i \sim \text{normal}(\mathbf{x}_i\boldsymbol{\beta}, \tau^2)$, where τ is the standard deviation of any extra variation in μ_i not explained by $\mathbf{x}_i\boldsymbol{\beta}$. We also posit that the distribution of μ_i given $\mathbf{x}_i$, is the same for all studies with the same value of $\mathbf{x}_i$. We utilize a diffuse prior for $\boldsymbol{\beta}$ (formula (24) with $d_j \to \infty$) and a log-logistic prior for the inter-study variance component (formula (15)).

Table 5 displays the results from fitting the hierarchical Bayesian model to three sets of covariates. Table 5(a) tabulates the results when all available covariates are used in the model. In addition to Support, Route, and their interaction, the variables Year of study and St.Error of quit difference are also included in the model, for a total of 6 degrees of freedom for coefficients $\boldsymbol{\beta}$. The numeric columns of Table 5 display the posterior mean of the regression coefficients, followed by the posterior standard deviations, and the posterior probability that the coefficient is positive. Because each covariate has been centered by subtracting its mean (the value of the last column of Table 5) from each x-value before performing the regression, the posterior mean of the constant term can be interpreted as the overall estimate of mean quit differences across all studies, $\mu = 9.5\%$, which is very significant since the posterior standard deviation is less than 1%. Table 5(b) shows the simplified model resulting when the terms Year, Support, and Support*Patch, whose means are each less than their respective standard deviations in absolute value, are removed from the model. The most striking covariate effect is the very significant coefficient of St.Error whose mean is $1.4854/0.2608 = 5.7$ standard deviations from 0. This is a quantitative and more formal confirmation of the trend that we previously noticed in the funnel plot (Fig. 1(c)) and in Figs 3(b) and 6. The interpretation of the fit of the model described by Table 5(b) is quite problematic, since the main causal model for the large observed effect of St.Error that comes easily to mind is the possibility of publication bias, which calls into question the entire meta-

Table 5 Regression Results Using Study-level Covariates
(a) Coefficients for model using study design variables, plus Year and St.Error;
$\tau = 0.0161$ (0.0083), BF = 1.41.

	Mean	SD	Prob > 0	Center
Constant Term	0.0951	0.0085	1.0000	
Route_Patch	0.0492	0.0254	0.9728	0.3220
Support_High	0.0135	0.0171	0.7864	0.5763
Support_High*Route_Patch	−0.0294	0.0316	0.1764	0.2373
Year	−0.0013	0.0017	0.2205	1988.9
St.Error	1.3787	0.2912	1.0000	0.0577

(b) Coefficients for model after removing non-significant covariates; $\tau = 0.0161$
(0.0076), BF = 1.92.

	Mean	SD	Prob > 0	Center
Constant Term	0.0953	0.0085	1.0000	
Route_Patch	0.0243	0.0120	0.9793	0.3220
St.Error	1.4854	0.2608	1.0000	0.0577

(c) Coefficients for model after removing first St.Error and then non-significant
covariates; $\tau = 0.0339$ (0.0090), BF = 1400.

	Mean	SD	Prob > 0	Center
Constant Term	0.0744	0.0085	1.0000	
Route_Patch	0.0642	0.0292	0.9846	0.3220
Support_High	0.0494	0.0194	0.9944	0.5763
Support_High*Route_Patch	−0.0784	0.0363	0.0166	0.2373

analysis. If we view these results as providing a method of adjusting for publication bias by using model predictions at the value of St.Error = 0 (i.e., predicting what would happen if only very large studies were used), then we arrive at the following estimates of mean quit difference, based on the numbers in Table 5(b):

$$0.0953 + (0 - 0.0577)1.4854 + (1 - 0.3220)0.0243 = 0.0261,$$
$$0.0953 + (0 - 0.0577)1.4854 + (0 - 0.3220)0.0243 = 0.0018.$$

These calculations suggest that the patch works better than gum (posterior probability 97.9%), but that even the patch only increases the quit rate by 2.6 percentage points more than in control groups, after attempting to adjust for publication bias by extrapolating the model to St.Error $= 0$. Because, as shown in Table 5(b), the posterior mean of $\tau = 1.61\%$ (standard deviation of $\tau = 0.76\%$, Bayes factor against the fixed effects model $= 1.92$), the model further predicts that individual-study effects of using patch will vary in the range 0–5 percentage points improvement over control.

Another modeling approach that many analysts prefer is to avoid using the St.Error variable in predictions, since the effect of this variable has an uncertain interpretation. The significant effect of St.Error in Table 5(a) might be viewed as a sign of the need for caution in subsequent interpretation, but we may prefer to avoid using it as a formal study-design covariate. When it is dropped from the model, the coefficient of Year remains insignificant, and one is led to the results displayed in Table 5(c), in which the coefficients for Route, Support, and their interaction are more than two standard errors away from 0. The large negative interaction term implies that, although high-intensity Support works better for gum studies, low-intensity Support works better for patch studies. If the patch requires less support, that would seem to be a further advantage of patch over gum. Table 5(c) indicates that the posterior mean of $\tau = 3.39\%$ (standard deviation of $\tau = 0.90\%$, Bayes factor 1400) for the model without St.Error as a covariate. The value of τ more than doubles in order to compensate for the decreased variance explained by the covariates. This is still less than the estimate $\tau = 4.01\%$ shown in Table 4 for the model without any covariates.

Figures 7(a, b) display the trace and shrinkage plots for the meta-analysis using the covariates in Table 5(c). Compared to Fig. 4, the histogram of the trace plot in Fig. 7(a) is shifted slightly to the left, and the estimates $E[\mu_i \mid \mathbf{Y}, \tau]$ do not shrink to a common value as τ approaches zero. Rather, they shrink toward the estimates of $\mathbf{x}_i\boldsymbol{\beta}$, which is just one estimate for each of the four study designs. This can be observed in the trace plot as well as in the shrinkage plot (4 distinct diamond values, one for each group of studies). The four model estimates are, in the order given in Fig. 7(b): (Gum, Low) $= 4.4\%$; (Gum, High) $= 9.3\%$;   (Patch,   Low) $= 10.8\%$;   (Patch,   High) $= 7.9\%$. However, because the value of τ is about 3.4%, we know that there will be overlap from the last three groups. In general, studies whose regression-fitted values are far apart will remain far apart after shrinking. The shrinkage plot displays the shrinkage estimates (formula (45), shown as circles, with error bars equal to 2 times the square root of formula (46)), the primary study values, shown as squares with error bars equal to

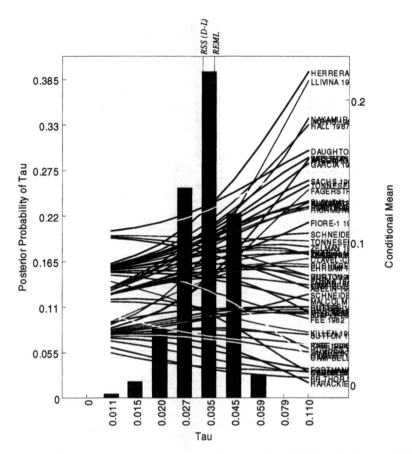

Figure 7(a) Trace plot after incorporating covariates displays the impact of the estimate for τ on the meta-analysis in Table 5(c).

$2s_i$, and the values of $\mathbf{x}_i\boldsymbol{\beta}$ (filled diamonds, formula (43)) toward which each set of studies is being pulled. The "Overall" diamond at the bottom of Fig. 7b is extended to show an approximate 68% confidence interval for μ (the constant term in the regression), but the other diamonds showing the individual model estimates do not display their corresponding uncertainties.

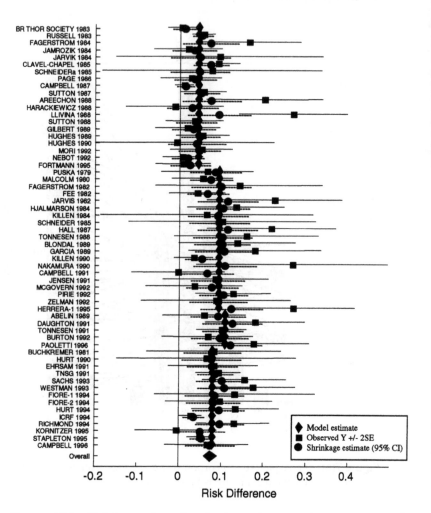

Figure 7(b) Shrinkage plot after incorporating covariates compares the observed data with the results from fitting the model of Table 5(c).

VI. MODEL DIAGNOSTICS

The HBLM depends on assumptions similar to any regression model, each of which needs to be assessed. In particular, the meta-analyst needs to check the appropriateness of the normality and equal-variance assumption of the random-effects model, the choice of prior distribution for the inter-study variance component, τ, and the appropriateness of the functional form $\mathbf{x}_i\boldsymbol{\beta}$ in the regression model. The appropriateness of the normality and unbiasedness of Y_i are assumptions of the primary studies and will not be discussed here. However, if these assumptions are not met, the integrity of the inferences made after combining studies will be questioned. In this section, we discuss how to use Bayesian ideas to choose between competing models and how to apply regression-based diagnostic techniques to assess model adequacy.

A. Fixed or Random Effects: Bayes Factors

The choice between a fixed-effect or random-effect meta-analysis relies upon the meta-analyst's belief as to whether each study summary is an estimate of the same underlying parameter or whether each study provides an estimate of a different but related parameter. Choosing a fixed-effect model implies that the inter-study variance component is zero, which is a rather strong assumption. Because the fixed-effect model for meta-analysis is used so widely, it is appropriate to test the corresponding hypothesis that $\tau = 0$ to possibly prevent an unjustified analysis. Within the framework of classical statistics, the Q-statistic described in formula (9) can be used. The most common Bayesian tool for testing a point null hypothesis is the Bayes factor (13), denoted BF (formula (19)), which describes how the data modify the prior odds against the null hypothesis:

Posterior Odds = Prior Odds × BF.

Thus, if Prior Odds = 1, indicating 50% prior probability that $\tau > 0$, and BF = 10, then Posterior Odds = 10, indicating $10/(10 + 1) = 91\%$ posterior probability that $\tau > 0$. If the Bayes factor is large, then the random-effect model is relatively more likely, and if BF is small then the random-effect model is relatively unlikely. An obvious modeling strategy is to set a threshold for BF, say BF_0 and to use the fixed-effect model if $BF < BF_0$. The choice of BF_0 should depend upon the prior belief in the plausibility of $\tau = 0$ and also on the relative seriousness of choosing the wrong model. In the absence of more specific information, our strategy is to set $BF_0 = 1$.

B. Residual Analysis

A useful tool for assessing model assumptions is that of cross-validated residuals. This involves calculating residuals formed from the predictive distributions of each data point obtained by analyzing the sample with that data point deleted (formula 49). Denote by u_k the value of the cumulative predictive distribution for the kth study effect Y_k, obtained using the model fit after removing the kth study from the analysis. Then if the model is specified correctly, each u_k will have a uniform distribution over the unit interval and $z_k = \Phi^{-1}(u_k)$ will have a standard normal distribution. These normalized residuals can be examined using standard tools of residual analysis such as construction of: histograms, normal probability plots (Q–Q plots), scatterplots, etc. By transforming the residuals, outliers may be identified, and then further examined. A numerical measure of whether the largest absolute residual is surprisingly large can be easily calculated (formula (51)). See Ref. 14 for a detailed discussion of the use of cross-validation in the context of meta-analysis.

C. Sensitivity Analysis for $\pi(\tau)$

More than one probability model will provide an adequate fit to any set of data. The real question is how much do inferences change when other *reasonable* models are used in place of a given model. In the HBLM, several models are specified: one for the primary study data, one for the random effects, and then models describing the uncertainty in the parameters of the random-effect distribution. Assuming that the model for the primary study data is correctly specified, the meta-analyst needs to examine the sensitivity of inferences with regard to changes in the remaining probability models. We discuss assessing sensitivity to the prior for the inter-study variance component; analogous methods can be utilized for the other assumed probability models.

The prior distribution we assume for τ is $\pi(\tau) \mid s_0 = s_0/(s_0 + \tau)^2$, where s_0 is the harmonic mean of the study sampling variances. The Appendix provides details of a two-parameter family $\pi(\tau \mid \tau_0, \gamma)$ in which $\tau_0 = s_0$ and $\gamma = 1$ is the default prior. The hyperparameters

(τ_0, γ) correspond one-to-one to (q_1, q_3) where $P(1 - B < q_1) = P(1 - B > q_3) = 25\%$, i.e., q_i are the quartiles of $1 - B(\tau) = \tau^2/s_0^2 + \tau^2$.

Table 6 summarizes how various analysis results depend upon the prior distribution for τ. In the first column, the row numbers 1–7 correspond to various choices of prior distribution (see below). The second and third columns display the first and third quartiles of the prior distribution of $1 - B_0(\tau) = \tau^2/(0.0330 + \tau^2)$, and the fourth column shows τ_0, the corresponding prior median of τ. The remaining columns display the posterior mean and standard deviation of τ, the shrinkage B_0 evaluated at the posterior mean of τ, the posterior mean and standard deviation of the overall effect μ, and the posterior probability that the interaction coefficient, β_3, is greater than 0. The first row yields results from the prior used to create Table 5(c), while the remaining rows display the results from a variety of other choices of q_1 and q_3. Row 2 is based on a tighter prior distribution around the same prior median of τ; rows 3 and 4 indicate what happens when the prior distributions of τ, and hence $B_0(\tau)$, are moved moderately in either direction; rows 5 and 6 show what happens when the prior distributions are changed more extremely in either direction. The results in rows 2–4 are very close to those in row 1. The 59 studies provide enough information about τ to almost completely erase differences in the prior distributions. Even the extreme prior distributions in rows 5 and 6 do not produce great changes in the posterior distributions, since most results remain within 5 to 10 percent of the corresponding values in row 1. When 75% prior probability is attached to $1 - B_0 < 0.1$ $(B_0 > 0.9)$, as in row 5, the posterior estimate of B_0 is 0.53, because the family of prior distributions we are using has very high tails and can be overruled by a likelihood with a quite different mode. In contrast, row 7 shows what happens when the fixed-effect prior, assuming $\tau = 0$ exactly, is used. The inferences about μ and β_3 differ substantially from those in all other rows.

D. Other Strategies

In addition to the suggestions above, we recommend some supplementary sensitivity analyses. Because our assumption is that there may be study-to-study variation, we have allocated the extra source of uncertainty to characteristics of the study, as measured by a non-zero β, and to between-study variation, as measured by a non-zero τ. However, it is entirely possible that all or some study-specific variances, s_i^2, are understated.

Table 6 Sensitivity of Inferences in the NRT Meta-analysis to the Prior for τ^*

Row	q_1	q_3	τ_0	$E[\tau\mid Y]$	$V[\tau\mid Y]^{\frac{1}{2}}$	$B_0 = (E[\tau\mid Y])$	$E[\mu\mid Y]$	$V[\mu\mid Y]^{\frac{1}{2}}$	$P(\beta_3 > 0\mid Y)$
1	0.1	0.9	0.0330	0.0339	0.0090	0.4933	0.0744	0.0085	0.0166
2	0.4	0.6	0.0330	0.0340	0.0067	0.4925	0.0745	0.0082	0.0160
3	0.1	0.5	0.0190	0.0318	0.0086	0.5093	0.0736	0.0083	0.0142
4	0.5	0.9	0.0571	0.0361	0.0089	0.4776	0.0752	0.0086	0.0189
5	0.01	0.1	0.0060	0.0296	0.0089	0.5272	0.0726	0.0082	0.0122
6	0.9	0.99	0.1802	0.0379	0.0093	0.4654	0.0758	0.0088	0.0211
7	0	0	0	0	0	1	0.0600	0.0052	0.0001

*The covariates are those defined in Table 5(c).

For example, suppose that we are less confident in the estimates obtained from studies accompanied by high-intensity support. In this case, we could increase the s_is for those studies suspected to be of poor quality, say by a factor of 2 (9). Alternatively, if study-quality scores are available, then these could be incorporated as a study design covariate in the analysis; see Ref. 15 and Smith et al. (Chap. 13). This is similar in spirit to using the study's standard error as a covariate. Finally, we also recommend removing one or two studies from the meta-analysis and re-analyzing. If inferences do not change substantially, then the scientific conclusions are on a firmer ground.

E. NRT Example With Covariates Revisited

In the NRT example described in section B.2, all evidence points toward the random-effect model. Both the Q-Statistic and the Bayes factor firmly favored the random-effect model (see Tables 4–5). The harmonic mean of the study sampling variances is 0.0330, which was utilized as the prior median for τ; the posterior mean for τ increased from its prior value to 0.0378 when using these covariates.

Table 7 displays the results of the cross-validation calculations for the NRT example when the covariates of Table 5(c) are included in the model. The first two columns list the primary study data (Y_i and s_i); the next two columns ("Pred Mn" and "Pred SD") present the mean and standard deviations of the posterior distribution for each μ_i given all the data except Y_i. The column labeled "Pred Prob" lists u_i, the value of this cumulative distribution evaluated at Y_i. The remaining three columns yield the posterior mean for τ; the mean for the study-specific efficacy rate and corresponding standard deviation; and the posterior probability that the study-specific treatment efficacy measure is positive. For example, study 35 had an observed quit difference of 0.3%. Based on all other studies, the predicted quit difference in study 35 is 4.8% (SD = 3.7%). The study 35 result is in the 10.6% tail of its predictive distribution given the other 58 studies. On the other hand, the study 47 result is in the 0.06% upper tail of its predictive distribution given the other 58 studies (observed = 26.8%, predicted = 3.8%).

A Q–Q plot of the normalized residuals is presented in Fig. 8. The solid line in the figure depicts the 45-degree line, and is used as a visual aid in diagnosing large deviations from normality. Is the normalized residual corresponding to study 47 too large? If u_{47} is viewed as the

Table 7 Cross-validation Results*

Study	Year	Y	SE	Pred. mean	Pred. SD	Pred. prob.	Tau	Post mean	Post SD	Prob > 0
...										
35.	1983	0.0032	0.0167	0.0479	0.0373	0.1062	0.0340	0.0119	0.0153	0.7835
36.	1983	0.0545	0.0132	0.0431	0.0388	0.6364	0.0354	0.0526	0.0123	1.0000
37.	1984	0.1621	0.0609	0.0403	0.0355	0.6916	0.0324	0.0717	0.0334	0.9910
38.	1984	0.0182	0.0404	0.0454	0.0382	0.3027	0.0349	0.0332	0.0266	0.8962
39.	1984	0.0925	0.1226	0.0435	0.0372	0.6504	0.0340	0.0474	0.0353	0.9197
40.	1985	0.0900	0.0250	0.0398	0.0359	0.8955	0.0325	0.0724	0.0213	0.9998
41.	1985	0.0744	0.1300	0.0437	0.0372	0.5907	0.0341	0.0458	0.0354	0.9117
42.	1986	0.0253	0.0361	0.0452	0.0383	0.3443	0.0351	0.0351	0.0251	0.9197
43.	1987	0.0088	0.0110	0.0480	0.0375	0.1204	0.0341	0.0172	0.0106	0.8830
44.	1987	0.0547	0.0269	0.0434	0.0384	0.6093	0.0351	0.0499	0.0213	0.9910
45.	1988	0.1993	0.0691	0.0400	0.0352	0.9817	0.0321	0.0742	0.0351	0.9911
46.	1988	−0.0134	0.0576	0.0457	0.0377	0.1871	0.0345	0.0288	0.0306	0.8349
47.	1988	0.2680	0.0642	0.0375	0.0322	0.9994	0.0292	0.0929	0.0380	0.9986
...										

*A portion of the results from a leave-one-out cross-validation of the analysis reported in Table 5(c).

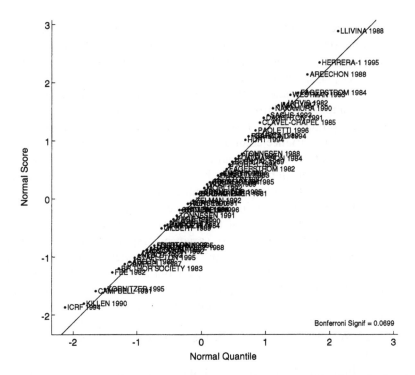

Figure 8 Cross-validated Q–Q plot presents the cross-validated residuals using the NRT example with the study covariates defined in Table 5(c).

most extreme deviation of the fifty-nine studies, this is not such an unusual result, because the Bonferroni bound for the probability of this event is given by 0.0699 (formula (51)). Thus, although there is strong evidence that $\mu_{47} - x_{47}\beta$ is more extreme than most other $\mu_i - x_i\beta$, we do not find strong evidence against the assumption of normality of the random effects.

VII. CONCLUDING REMARKS

In this chapter, we have presented a unified modeling approach to meta-analysis using a hierarchical Bayes linear model. This modeling approach permits incorporation of fixed-, random-, and mixed-effect models into one framework. Our focus, however, was on an implementation strategy for undertaking the meta-analysis, including the construction of a set of exploratory graph types, model selection techniques, model fitting, summary displays, and regression diagnostics.

An appropriate computational approach to performing meta-analysis is important. Throughout this chapter we have viewed the variance of each primary study result as the sum of two quantities: a known within-study variance component and an unknown inter-study variance component. Consequently, our computational approach focuses on making inferences on all parameters by averaging with respect to the posterior distribution of the inter-study variance component.

Informative prior distributions can be quite useful. Prior assumptions on the upper levels of the hierarchy, such as the choice of prior distribution for the inter-study variance component as well as the functional form of the impact of study-level covariates, are less risky. In fact, the use of a prior distribution helps avoid even more restrictive assumptions, such as assuming the prior for the inter-study variance component is concentrated at zero. Indeed, diffuse priors have strong implications in that they tend to skew the distribution towards large values of the variance component, even though the possibility of τ near zero is likely a priori.

In conclusion, a modern interactive computing environment for performing meta-analyses is essential. Clearly, hierarchical Bayes linear modeling requires special computations and graphics. Having quick access to a full range of exploratory regression tools that accommodates these needs is highly desirable.

Availability of Software. A set of S-PLUS functions, denoted hblm, designed to provide graphs and parameter estimates for the fitted model may be obtained at:

```
ftp:ftp.research.att.com/dist/bayes-meta/
```

User documentation and a list of references are available in a postscript file (hblm_doc.ps) and can be downloaded from the same site. A more advanced version that does not depend on S-PLUS, namely MetaGraphs©, is available from Belmont Research Incorporated, 84 Sherman Street, Cambridge MA 02140 (Telephone: 617-868-6878; Fax: 617-868-2654; http://www.belmont.com).

Appendix: Formulas For Meta-Analysis Modeling

A. Notation

$$\mathbf{Y} = (Y_1, \ldots, Y_K)^{\mathrm{T}}, \tag{1}$$

$$V = \mathrm{diag}(s_1^2, \ldots, s_K^2), \tag{2}$$

$$X = K \times (J+1): \text{Design matrix of study-level covariates}, \tag{3}$$

$$\boldsymbol{\beta} = (\mu, \beta_1, \ldots, \beta_J)^{\mathrm{T}}, \tag{4}$$

$$\mathbf{x}_i = i\text{th row of } X, \tag{5}$$

$$\mathbf{b} = (b_0, \ldots, b_J)^{\mathrm{T}}: \text{prior mean of } \boldsymbol{\beta}, \tag{6}$$

$$D = \mathrm{diag}(d_0^2, \ldots, d_J^2): \text{prior covariance of } \boldsymbol{\beta}, \tag{7}$$

$$\mu_i = \mathbf{x}_i\boldsymbol{\beta} + \delta_i: \text{the study-specific effect}. \tag{8}$$

B. Tests and Measures of Interstudy Variation

Q-Statistic without Covariates. Used to test the hypothesis: $H_0 : \mu_1 = \cdots = \mu_K = \mu$ versus $H_A :$ at least one $\mu_i \neq \mu_j$. Define

$$Q = \sum_i s_i^{-2}(Y_i - Y^*)^2, \tag{9}$$

where $Y^* = \sum_i s_i^{-2} Y_i / \sum_i s_i^{-2}$. If Q is greater than the $100(1-\alpha)$ percentile of the χ^2_{K-1} distribution, then the hypothesis of equal means would be rejected at the 100α level. Note that Q is the residual sum of squares (RSS).

Q-Statistic with Study-Level Covariates. Define this as in Formula (9), but with

$$Y^* = X(X^{\mathrm{T}}W_0X)^{-1}X^{\mathrm{T}}W_0Y,$$

where $W_0 = \mathrm{diag}(s_1^{-2}, \ldots, s_K^{-2})$, and compare Q to a χ^2_{K-J-1} distribution.

RSS Estimate of τ^2. This estimate is due to H. Chernoff, as cited in Ref. 9. In the absence of covariates, where $J = 0$ and X consists of a single column of 1s, it reduces to the estimate of DerSimonian and Laird (7).

$$\tau_{\text{RSS}}^2 = \max\left\{0, \frac{Q - K + J + 1}{\text{tr}[(X^T W_0 X)^{-1} X^T W_0^2 X]}\right\} \tag{10}$$

where "tr" denotes the trace of its matrix argument, i.e., the sum of the diagonal elements.

REML Likelihood Function for τ. The REML likelihood is the result of integrating the joint likelihood $L(\tau, \beta)$ with respect to β. This is equivalent to assuming a diffuse (constant) prior density for the coefficient vector β, in which case the REML likelihood is $L(\tau)$, where

$$-2\log L(\tau) = \sum_i \log(\tau^2 + s_i^2) + \log[\det(X^T W_\tau X)] + S_\tau^2 + \text{constant}, \tag{11}$$

$$W_\tau = \text{diag}((\tau^2 + s_1^2)^{-1}, \ldots, (\tau^2 + s_K^2)^{-1}), \tag{12}$$

$$S_\tau^2 = \sum_i \frac{(Y_i \mathbf{x}_i \beta(\tau))^2}{(\tau^2 + s_i^2)}, \tag{13}$$

$$\beta(\tau) = (X^T W_\tau X)^{-1} X^T W_\tau Y. \tag{14}$$

REML Estimate of τ. This estimate, denoted τ_{REML}, is the value that maximizes $L(\tau)$ for $\tau \geq 0$.

Prior Distribution for τ.

$$\pi(\tau) = \frac{s_0}{(s_0 + \tau)^2}, \quad \text{where } s_0^2 = \frac{K}{\sum s_i^{-2}}. \tag{15}$$

This prior distribution is a special case of an extended two-parameter family of prior distributions, formed as a location-scale family based on $\log(\tau)$. For $\tau > 0$ and $\gamma > 0$, let

$$\pi(\tau; \tau_0, \gamma) = \frac{\gamma \tau_0 \tau^{\gamma - 1}}{(\tau_0 + \tau^\gamma)^2}. \tag{16}$$

Our default prior distribution corresponds to choosing $\tau_0 = s_0$ and $\gamma = 1$. In this family, the median of τ is τ_0, and values of $\gamma < 1$ result in distributions more dispersed than the default, while values of $\gamma > 1$ result in lesser dispersion.

Shrinkage Factor.   We define the quantity

$$B_i(\tau) = \frac{s_i^2}{s_i^2 + \tau^2}.$$ (17)

$B_i(\tau)$ is called the *shrinkage factor* for the ith study. It plays a key role in determining how much "borrowing strength" is appropriate, since $B_i(\tau)$ is the weight given to the population average estimate when forming the posterior estimate for study i. To denote a "typical" shrinkage proportion across studies, we define

$$B_0(\tau) = \frac{s_0^2}{s_0^2 + \tau^2}$$ (18)

based on s_0^2, the harmonic mean of the K sampling variances. Viewed as a parameter transformation of τ, $B_0(\tau)$ has certain interpretational advantages compared with τ, because it is always between 0 and 1 and does not depend on the scale of the data. The default prior distribution $\pi(\tau)$ defined in formula (15) results in a prior distribution for $1 - B_0$ with quartiles 0.1, 0.5, and 0.9.* Values of B_0 near 0 imply that virtually no shrinkage will occur and that a meta-analysis is virtually useless because there is little across-study agreement. Conversely, values of B_0 near 1 imply that τ is near 0 and that the fixed-effect model is a good approximation, yielding maximal across-study agreement, except for possible covariate effects. Use of the extended two-parameter family $\pi(\tau; \tau_0, \gamma)$ of prior distributions for τ induces a corresponding two-parameter family of prior distributions for $1 - B_0$. Rather than choose the hard-to-interpret hyperparameters τ_0 and γ subjectively, we suggest choosing instead the first and third quartiles of $1 - B_0$, denoted q_1 and q_3 respectively. Deviations from the default values of $q_1 = 0.1$ and $q_3 = 0.9$ $(0 < q_1 < q_3 < 1)$, have a simple interpretation in terms of a greater or lesser degree of belief in the amount of expected shrinkage of study estimates towards common values. For example, setting $q_1 = 0.75$ and $q_3 = 0.95$ denotes a prior probability of 50% that B_0 is between 0.75 and 0.95, which indicates a wish to bias the results toward those of a fixed-effects model, but also a desire to allow the data to overcome this bias in case the likelihood function $L(\tau)$ indicates strongly that τ is large.

---

*In the MetaGraphs software, the quantity $1 - B$ is denoted by R^2.

Bayes Factor for $\tau > 0$ versus $\tau = 0$. Because the fixed-effect model for meta-analysis is used so widely, it is appropriate to test the corresponding hypothesis that $\tau = 0$ to possibly prevent an unjustified analysis. The formula for the Bayes factor is:

$$\text{BF} = \int_{\tau > 0} L(\tau)\pi(\tau; \tau_0, \gamma)d\tau / L(0) \tag{19}$$

where $L(\tau)$ is defined in equation (11). BF is the average value of the likelihood ratio under the alternative hypothesis. We recommend not using the fixed-effect model unless BF < 1.

Computation of Integrals over the Range $0 < \tau < \infty$. The Bayes factor and many other quantities described in this chapter are defined in terms of an integral over the range $0 < \tau < \infty$. These integrals are computed by the method of Gauss–Hermite integration (16) as follows. Consider evaluation of the quantity $E[f(\tau)]$, where $\pi(\tau|Y) \propto L(\tau)(\tau; \tau_0, \gamma)$ and f is an arbitrary smooth function. For example, $f(\tau) = \mu(\tau)$, the prior mean conditional on τ. Then,

$$E[f(\tau)] = \int_{\tau > 0} f(\tau)\pi(\tau|Y)d\tau. \tag{20}$$

Make the transformation $\lambda = \log(\tau)$ and $\psi(\lambda) = \lambda + \log(\pi(e^\lambda|Y))$, so that

$$E[f(\tau)] = \int_{-\infty < \lambda < \infty} f(e^\lambda)e^{\psi(\lambda)}d\lambda. \tag{21}$$

First, a one-dimensional numerical search finds the value λ_0 that maximizes $\psi(\lambda)$ and also computes the second derivative $\psi(\lambda_0)'' = -1/\sigma^2$. This allows approximation of $e^{\psi(\lambda)}$ by the normal density $\phi[(\lambda - \lambda_0)/\sigma]$ where ϕ is the standard normal density. Next, use the Gauss–Hermite points z_h to compute $\lambda_h = \lambda_0 + \sigma z_h$ and use the corresponding weights g_h to compute the approximation

$$E[f(\tau)] \approx \frac{\sum_h g_h f(e^{\lambda_h})e^{\psi(\lambda_h)}}{\sum_h g_h e^{\psi(\lambda_h)}}. \tag{22}$$

We use a 9-point integration formula (9 values of h in the above expressions), so that the approximation would be exact if the quantity $f(e^\lambda)e^{\psi(\lambda)}/\phi[(\lambda - \lambda_0)/\sigma]$ could be represented exactly by a 17th degree polynomial in λ. (See Table 25.10 of Ref. 16, which lists values of x_h and w_h. In our notation, $z_h = \sqrt{2}x_h$ and $g_h = w_h e^{x_h^2}$.)

C. Posterior Distributions for the Hierarchical Bayes Linear Model

1. Prior Distributions

The prior distribution of τ was discussed in Sec. VIII.B. The prior distribution of the overall effect μ and of the regression coefficients β_j are

$$\mu \sim \text{normal}(b, d^2), \tag{23}$$

$$\beta_j \sim \text{normal}(b_j, d_j^2) \quad (j = 1, \ldots, J), \tag{24}$$

where b and b_j are arbitrary numbers, $d > 0, d_j > 0$, and allowing d or any of the d_j to approach ∞ describes the situation where, a priori, no value of the parameter space is preferred over any other. The prior distributions of τ, μ, and the β_j are assumed to be independent. However, the situation where the elements of $\boldsymbol{\beta}$ are correlated can be reduced to the independent case, because the specification, for arbitrary full-rank matrices X and D of sizes $K \times (J + 1)$ and $(J + 1) \times (J + 1)$ respectively,

$$\boldsymbol{\theta}|\boldsymbol{\beta}, \tau \sim \text{normal}(X\boldsymbol{\beta}, \tau^2 I), \qquad \boldsymbol{\beta} \sim \text{normal}(\mathbf{b}, D), \tag{25}$$

is equivalent to the specification

$$\boldsymbol{\theta}|\boldsymbol{\beta}, \tau \sim \text{normal}(Z\boldsymbol{\beta}, \tau^2 I), \qquad \boldsymbol{\beta} \sim \text{normal}(\mathbf{b}^*, I), \tag{26}$$

where $Z = XD^{\frac{1}{2}}$ and $\mathbf{b}^* = D^{-\frac{1}{2}}\mathbf{b}$. That is, the covariates can always be transformed so that the coefficients have uncorrelated prior distributions.

The definitions of the REML likelihood $L(\tau)$ and related quantities given in the previous section are based on the assumption that d and every d_j approach ∞. See Ref. 9 for the modifications necessary to compute $\pi(\tau|\mathbf{Y})$ when some of the ds are finite. Note that the formulae in the next subsection are all expressed in terms of D^{-1}, not D, allowing computation in situations where some of the diagonal elements of D^{-1} are 0 because the corresponding ds are infinite.

2. No Covariates

Posterior Mean and Variance of μ Conditional on τ.

$$\mu^*(\tau) \equiv E(\mu \mid \mathbf{Y}, \tau) = \frac{\sum_i Y_i(s_i^2 + \tau^2)^{-1}}{\sum_i(s_i^2 + \tau^2)^{-1}}, \tag{27}$$

$$\mu^{**}(\tau) \equiv V(\mu \mid \mathbf{Y}, \tau) = \frac{1}{\sum_i(s_i^2 + \tau^2)^{-1}}. \tag{28}$$

Posterior Mean and Variance of μ_i Conditional on τ.

$$\mu_i^*(\tau) \equiv E(\mu_i \mid \mathbf{Y}, \tau) = [1 - B_i(\tau)]Y_i + B_i(\tau)\mu^*(\tau), \tag{29}$$

$$\mu_i^{**}(\tau) \equiv V(\mu_i \mid \mathbf{Y}, \tau) = B_i(\tau)\{\tau^2 + B_i(\tau)\mu^{**}(\tau)\}. \tag{30}$$

Posterior Mean and Variance of μ.

$$\mu^* \equiv E(\mu \mid \mathbf{Y}) = \int_\tau \mu^*(\tau)\pi(\tau \mid \mathbf{Y})d\tau, \tag{31}$$

$$\mu^{**} \equiv V[\mu \mid \mathbf{Y}] = \int_\tau \{\mu^{**}(\tau) + [\mu^*(\tau) - \mu^*]^2\}\pi(\tau \mid \mathbf{Y})d\tau. \tag{32}$$

Posterior Mean and Variance of μ_i.

$$\mu_i^* \equiv E(\mu_i \mid \mathbf{Y}) = \int_\tau \mu_i^*(\tau)\pi(\tau \mid \mathbf{Y})d\tau, \tag{33}$$

$$\mu_i^{**} \equiv V(\mu_i \mid \mathbf{Y}) = \int_\tau \{\mu_i^{**}(\tau) + [\mu_i^*(\tau) - \mu_i^*]^2\}\pi(\tau \mid \mathbf{Y})d\tau. \tag{34}$$

Posterior Probability $\mu > 0$.

$$P(\mu > 0 \mid \mathbf{Y}) = \int_\tau \Phi\left(\frac{\mu^*(\tau)}{[\mu^{**}(\tau)]^{1/2}}\right)\pi(\tau \mid \mathbf{Y})d\tau, \tag{35}$$

where Φ is the cumulative standard normal distribution function.

Posterior Probability $\mu_i > 0$.

$$P(\mu_i > 0 \mid \mathbf{Y}) = \int_\tau \Phi\left(\frac{\mu_i^*(\tau)}{[\mu_i^{**}(\tau)]^{1/2}}\right)\pi(\tau \mid \mathbf{Y})d\tau. \tag{36}$$

3. Covariates

Posterior Mean and Covariance of $\boldsymbol{\beta}$ Conditional on τ.

$$\boldsymbol{\beta}^*(\tau) \equiv E(\boldsymbol{\beta} \mid \mathbf{Y}, \tau) = [X^T W(\tau)X + D^{-1}]^{-1}(X^T W(\tau)\mathbf{Y} + D^{-1}\mathbf{b}), \tag{37}$$

$$\boldsymbol{\beta}^{**}(\tau) \equiv \text{cov}(\boldsymbol{\beta} \mid \mathbf{Y}, \tau) = [X^T W(\tau)X + D^{-1}]^{-1}, \tag{38}$$

where $W(\tau) = [V + \tau^2 I_K]^{-1}$. If the components of D are such that $d_j^2 \to \infty$, for all j, then

$$\boldsymbol{\beta}^*(\tau) = [X^{\mathrm{T}} W(\tau) X]^{-1} (X^{\mathrm{T}} W(\tau) Y).$$

Posterior Mean and Variance of μ_i Conditional on τ.

$$\mu_i^*(\tau) \equiv E(\mu_i \mid \mathbf{Y}, \tau) = Y_i[1 - B_i(\tau)] + B_i(\tau)\mathbf{x}_i\boldsymbol{\beta}^*(\tau), \tag{39}$$

$$\mu_i^{**}(\tau) \equiv V(\mu_i \mid \mathbf{Y}, \tau) = B_i(\tau)\{\tau^2 + B_i(\tau)\mathbf{x}_i\boldsymbol{\beta}^{**}(\tau)\mathbf{x}_i^{\mathrm{T}}\}. \tag{40}$$

Posterior Mean and Variance of β_j.

$$\beta_j^* \equiv E(\beta_j \mid \mathbf{Y}) = \int_\tau \beta_j^*(\tau)\pi(\tau \mid \mathbf{Y})d\tau, \tag{41}$$

$$\boldsymbol{\beta}^{**} \equiv \mathrm{cov}(\boldsymbol{\beta} \mid \mathbf{Y}) = \int_\tau \{\boldsymbol{\beta}^{**}(\tau) + [\boldsymbol{\beta}^*(\tau) - \boldsymbol{\beta}^*][\boldsymbol{\beta}^*(\tau) - \boldsymbol{\beta}^*]^{\mathrm{T}}\}\pi(\tau \mid \mathbf{Y})d\tau.$$
$$\tag{42}$$

Posterior Mean and Variance of Linear Predictor $\mathbf{x}\boldsymbol{\beta}$.

$$E(\mathbf{x}\boldsymbol{\beta} \mid \mathbf{Y}) = \mathbf{x}\boldsymbol{\beta}^*, \tag{43}$$

$$V(\mathbf{x}\boldsymbol{\beta} \mid \mathbf{Y}) = \mathbf{x}\boldsymbol{\beta}^{**}\mathbf{x}^{\mathrm{T}}. \tag{44}$$

Posterior Mean and Variance of μ_i.

$$\mu_i^* \equiv E(\mu_i \mid \mathbf{Y}) = \int_\tau \mu_i^*(\tau)\pi(\tau \mid \mathbf{Y})d\tau, \tag{45}$$

$$\mu_i^{**} \equiv V(\mu_i \mid \mathbf{Y}) = \int_\tau \{\mu_i^{**}(\tau) + [\mu_i^*(\tau) - \mu_i^*]^2\}\pi(\tau \mid \mathbf{Y})d\tau. \tag{46}$$

Posterior Probability $\beta_j > 0$.

$$P(\beta_j > 0 \mid \mathbf{Y}) = \int_\tau \Phi\left(\frac{\beta_j^*(\tau)}{\sqrt{\beta_j^{**}(\tau)}}\right)\pi(\tau \mid \mathbf{Y})d\tau. \tag{47}$$

Posterior Probability $\mu_i > 0$.

$$P(\mu_i > 0 \mid \mathbf{Y}) = \int_\tau \Phi\left(\frac{\mu_i^*(\tau)}{\sqrt{\mu_i^{**}(\tau)}}\right)\pi(\tau \mid \mathbf{Y})d\tau. \tag{48}$$

D. Cross-validation

1. Cross-validated Residual

$$u_k = P(Y_k < y_k \mid y_i, i \neq k) = \int_\tau P(Y_k < y_k \mid \tau, y_i, i \neq k)$$

$$\pi(\tau \mid y_i, i \neq k)d\tau, \tag{49}$$

$$\text{with } P(Y_k < y_k \mid \tau, y_i, i \neq k) = \Phi\left[\frac{(y_k - \mathbf{x}_k \boldsymbol{\beta}^*_{-k}(\tau))}{(\mathbf{x}_k \boldsymbol{\beta}^{**}_{-k}(\tau)\mathbf{x}_k^\mathrm{T} + \tau^2 + s_k^2)^{1/2}}\right], \tag{50}$$

where $\mathbf{x}_k$ is the kth row of X, $\boldsymbol{\beta}^{**}_{-k}(\tau) = \mathrm{cov}[\boldsymbol{\beta} \mid \tau, y_i, i \neq k] = [X^\mathrm{T}_{-k}W_{-k}(\tau)X_{-k} + D^-1]^{-1}$, and $\boldsymbol{\beta}^*_{-k}(\tau) = \boldsymbol{\beta}^{**}_{-k}(\tau)(X^\mathrm{T}_{-k}W_{-k}(\tau)X_{-k}\mathbf{Y}_{-k}+ D^{-1}\mathbf{b})$ is the estimate of $\boldsymbol{\beta}$ when dropping the kth study.

2. The Bonferroni Statistic

$$P = K \min_k(1 - \mid 2u_k - 1 \mid). \tag{51}$$

3. Computation of Cross-validation Statistics

It is not necessary to completely redo all computations K times when computing the leave-one-out estimates. For fixed values of τ, there are well-known regression updating formulae (see, e.g., Ref. 17) for computing all the leave-one-out regressions based on the results of the initial regression (it is necessary to use versions of these formulas valid for weighted regression). Now recall that, in our Sec. VIII.B discussion of the calculations, we stated that only 9 separate values of τ (based on 9 values of λ) are used in the Gauss–Hermite approximations of all integrals with respect to τ that occur in Sec. VIII.C. Rather than maximize the posterior distributions $\pi(\tau \mid \mathbf{Y}_{-k})$ separately for each k, the values of $\pi(\tau \mid \mathbf{Y}_{-k})$ are merely recomputed for each k at each of the 9 values of τ previously selected during the complete-data analysis. This allows us to make use of the regressions updating formulae that require unchanged values of τ across all the leave-one-out regressions. The result is a reweighting of the 9 values of τ in the summation that approximate the many integrals in Sec. VIII.C, and a great saving in computation compared to making K complete recalculations. This shortcut loses more accuracy if K is small, because dropping a single Y_k is more likely to significantly modify $\pi(\tau \mid \mathbf{Y})$; but when K is small, it is not such a burden to perform K sets of computations.

References

1. Cochran Database of Systematic Reviews. 1997, Issue 3.
2. HM Cooper, LV Hedges, eds. The Handbook of Research Synthesis. New York: Russell Sage Foundation, 1994, Chapter 18.
3. JD Emerson, JC Hoaglin, F Mosteller. A modified random-effect procedure for combining risk difference in sets of 2×2 tables from clinical trials. Journal of the Italian Statistical Society 2:269–290, 1992.
4. KA L'Abbe, AS Detsky, K O'Rourke. Meta-analysis in clinical research. Annals of Internal Medicine 107:224–233, 1987.
5. GH Givens, DD Smith, RL Tweedie. Publication bias in meta-analysis: A Bayesian data-augmentation approach to account for issues exemplified in the passive smoking debate. Statistical Science. 12(4):221–250, 1997.
6. J Lau, CH Schmid, TC Chalmers. Cumulative meta-analysis of clinical trials builds evidence for exemplary medical care. Journal of Clinical Epidemiology 48(1):45–57, 1995.
7. R DerSimonian, N Laird. Meta-analysis in clinical trials. Controlled Clinical Trials 7:177, 1986.
8. HD Patterson, R Thompson. Recovery of interblock information when block sizes are unequal. Biometrika 58:545–554, 1971.
9. WH DuMouchel, JE Harris. Bayes methods for combining the results of cancer studies in humans and in other species. Journal of the American Statistical Association 78:293–315, 1983.
10. CN Morris and SL Normand. Hierarchical models for combining information and for meta-analysis. In: JM Bernardo, JO Berger, AP David, AFM Smith, eds. Bayesian Statistics 4. Oxford: Clarendon Press, 1992, pp 321–344 (with discussion).
11. WH DuMouchel. Bayesian Meta-analysis. In: D Berry, ed. Statistical Methodology in the Pharmaceutical Sciences. New York: Marcel Dekker, 1990, pp 509–529.
12. WH DuMouchel, PG Groer. A Bayesian methodology for rescaling radiation studies from animals to man. Health Physics 57(suppl 1):411–418, 1989.
13. RE Kass, AE Raftery. Bayes factors. Journal of the American Statistical Association 90:773–795, 1995.
14. WH DuMouchel. Predictive cross-validation of Bayesian meta-analysis. In: JM Bernardo, JO Berger, AP David, AFM Smith, eds. Bayesian Statistics 5. Oxford: Clarendon Press, 1996, pp 107–127 (with discussion).

15. TC Chalmers, H Smith Jr., B Blackburn, B Silverman, B Schroeder, D Reitman, A Ambroz. A method for assessing the quality of a randomized control trial. Controlled Clinical Trials 2:31–49, 1981.

16. M Abramowitz, I Stegun. Handbook of Mathematical Functions, Applied Mathematics Series, 55. Washington DC: National Bureau of Standards, 1970.

17. RD Cook, S Weisberg. Residuals and Influence in Regression. New York: Chapman and Hall, 1982.

7

Meta-analysis for 2 × 2 Tables with Multiple Treatment Groups

Leon J. Gleser
University of Pittsburgh, Pittsburgh, Pennsylvania

Ingram Olkin
Stanford University, Stanford, California

Abstract

Meta-analytic methods have been developed for determining the effectiveness of a single treatment versus a control. However, in large studies, more than one competing treatment may be tested for its effectiveness. Further, different sites may use different subsets of treatments. This leads to a model with missing data and with correlated effect sizes. A regression procedure is developed that yields an estimate of the overall effect.

I. PRELIMINARIES

Meta-analytic methods have been developed for determining the effectiveness of a single treatment versus a control or standard by combining such comparisons over a number of studies. When the endpoint is dichotomous (e.g., effective or not effective, side effects present or absent, life or death), effectiveness is typically measured in terms of differences in risks (proportions) or odds ratios. In large studies, more than one treatment may be involved, with each treatment being compared to the common control. This is particularly true of pharmaceutical studies, in which the effects of several drugs or drug doses are compared in order to identify the most promising choices. Because investigators may have different goals, or are prevented by financial or other constraints from testing all possible treatments, different studies may involve different treatments. When later a meta-analytic review is attempted of all studies that involve the treatments of interest to the researcher, the facts that some studies may be missing one or more of the treatments and that (because of the common control) the effect sizes within studies are correlated need to be accounted for in the statistical analysis.

The present chapter illustrates, in the context of an example, how information from studies can be combined to estimate increments in proportions (or in log odds ratios) due to various treatments. Also shown is how to construct appropriate simultaneous confidence intervals for such increments (and for contrasts of increments). The approach illustrated is approximate, with the approximation being best when studies have large sample sizes. It can be regarded as the specialization to sample proportions of the general approach in Gleser and Olkin (1994).

II. EXAMPLE

Suppose that one is interested in combining the results of several studies of the effectiveness of one or more of three anti-hypertension therapies in preventing heart disease in males or females considered "at risk" because of excessively high blood-pressure readings. For specificity, suppose that the therapies are T_1 = use of a new beta-blocker, T_2 = use of a drug that reduces low-density cholesterol, and T_3 = use of a calcium channel blocker, with the endpoint being the occurrence of coronary heart disease. In every study, the effects of these therapies are compared to that of using only a diuretic (the control). Although the data are hypothetical, they represent an actual therapy.

A statistical problem arises from the fact that a comparison of therapy T_i with the control and a comparison of therapy T_j with the control, in the same study, have the control in common. If the control rate of occurrence of heart disease is high, then it is likely that both therapies will show an effect: if the control rate is low, it will be difficult for the therapies to do better (be effective). Consequently, the comparisons of the therapies with the control are positively correlated, and information about the effect size for therapy T_i can be used to predict the effect size for any therapy T_j in the same study, and vice versa.

To illustrate the model, suppose that a thorough search of published and unpublished studies (using various medical registers) yields five studies. Although all of these studies use a similar control, each study need not include all of the treatments, as exemplified by the hypothetical data given in Table 1.

For our illustration, effect sizes will be measured by differences in proportions:

effect size = proportion control − proportion therapy $\equiv p_c - p_t$.

Modification of the methods for other measures of effect sizes are discussed in Secs 4 and 5. The effect size estimates for risk differences, arranged by study, are:

$$d_1 = (-0.0050, -0.0175, ---) \qquad d_2 = (0.0125, ---, ---)$$
$$d_3 = (---, 0.0139, - - -), \qquad d_4 = (---, -0.0050, 0.0250), \qquad (1)$$
$$d_5 = (---, ---, 0.0250),$$

where the dashes indicate effect sizes that are unavailable from the particular study.

Table 1 Number and Proportions of Subjects Exhibiting Heart Disease for a Control and Three Therapies

Study	Control	T_1	T_2	T_3
1	20/1000 (0.0200)	100/4000 (0.0250)	150/4000 (0.0375)	—
2	10/200 (0.0500)	15/400 (0.0375)	—	—
3	40/450 (0.0889)	—	30/400 (0.0750)	—
4	150/2000 (0.0750)	—	80/1000 (0.0800)	50/1000 (0.0500)
5	60/400 (0.1500)	—	—	50/400 (0.1250)

The (estimated) variance of a difference of proportions $p_0 - p_i$ is

$$\widehat{var}(p_0 - p_i) = \frac{p_0(1 - p_0)}{n_0} + \frac{p_i(1 - p_i)}{n_i}, \tag{2}$$

where n_0 is the number of subjects given the control treatment and n_i is the number of subjects given therapy T_i. The covariance between two such differences $p_0 - p_i$ and $p_0 - p_j$ is $p_0(1 - p_0)/n_0$. The following are estimated covariance matrices for each study, where only entries corresponding to effect sizes that can be estimated in the study are given:

$$\hat{\Psi}_1 = 10^{-9}\begin{bmatrix} 25694 & 19600 \\ 19600 & 28623 \end{bmatrix}, \quad \hat{\Psi}_4 = 10^{-9}\begin{bmatrix} 108288 & 34688 \\ 34688 & 82188 \end{bmatrix},$$

$$\hat{\Psi}_2 = 10^{-9}(327734), \quad \hat{\Psi}_3 = 10^{-9}(353410), \quad \hat{\Psi}_5 = 10^{-9}(592188). \tag{3}$$

III. A REGRESSION PROCEDURE FOR RISK DIFFERENCES

Following the approach in Gleser and Olkin (1994), we combine the vectors d_i of effect sizes into a single vector, omitting all missing comparisons. We thus obtain from (1) the vector

$$d = (-0.0050, -0.0175, 0.0125, 0.0139, -0.0050, 0.0250, 0.0250).' \tag{4}$$

The estimated covariance matrix of d is obtained from (3), where the blocks represent the studies.

$$\hat{\Psi} = 10^{-9}\begin{bmatrix} 25694 & 19600 & 00000 & 00000 & 00000 & 00000 & 00000 \\ 19600 & 28623 & 00000 & 00000 & 00000 & 00000 & 00000 \\ 00000 & 00000 & 327734 & 00000 & 00000 & 00000 & 00000 \\ 00000 & 00000 & 00000 & 353410 & 00000 & 00000 & 00000 \\ 00000 & 00000 & 00000 & 00000 & 108288 & 34688 & 00000 \\ 00000 & 00000 & 00000 & 00000 & 34688 & 82188 & 00000 \\ 00000 & 00000 & 00000 & 00000 & 00000 & 00000 & 592188 \end{bmatrix}. \tag{5}$$

Let β_i be the effect size (assumed common to all studies) for therapy T_i ($i = 1, 2, 3$), and let $\beta = (\beta_1, \beta_2, \beta_3).'$ We can now write a regression model for d, namely

$$d = X\beta + \text{error},$$

with design matrix

$$X = \begin{bmatrix} 1 & 0 & 0 \\ 0 & 1 & 0 \\ \cdots & \cdots & \cdots \\ 1 & 0 & 0 \\ 0 & 1 & 0 \\ \cdots & \cdots & \cdots \\ 0 & 1 & 0 \\ 0 & 0 & 1 \\ \cdots & \cdots & \cdots \\ 0 & 0 & 1 \end{bmatrix}, \tag{6}$$

in which the columns represent the three therapies and the rows correspond to groups of individuals who receive one of the therapies, two groups in studies 1 and 4 and one group in studies 2, 3, and 5.

The estimates of effect sizes (here, rate differences) are obtained from a weighted least squares fit of this regression model:

$$\hat{\beta} \equiv (\hat{\beta}_1, \hat{\beta}_2, \hat{\beta}_3)' = (X'\hat{\Psi}^{-1}X)^{-1}X'\hat{\Psi}^{-1}d, \tag{7}$$

where $\hat{\Psi}$ is the sample covariance matrix given in (5). The needed vector–matrix operations can easily be carried out using most standard statistical software packages. For our example,

$$\hat{\beta}_1 = -0.0010808, \quad \hat{\beta}_2 = -0.0125192, \quad \hat{\beta}_3 = 0.0228495 \quad .$$

The estimated variance–covariance matrix of the effect size estimates is

$$C \equiv \text{cov}(\hat{\beta}) = (X'\hat{\Psi}^{-1}X)^{-1} = 10^{-9} \begin{bmatrix} 20806 & 13599 & 3889 \\ 13599 & 20604 & 5893 \\ 3889 & 5893 & 65145 \end{bmatrix}.$$

Let c_{ij} be the (i, j)th element of $C = \text{cov}(\hat{\beta})$. We will need C to construct approximate simultaneous confidence intervals for the effect sizes or linear combinations of the effect sizes.

Simultaneous 95% confidence intervals for the individual therapy effects are of the form:

$$\hat{\beta}_i \pm [\chi_q^2(0.95)c_{ii}]^{1/2} \quad (i = 1, 2, 3),$$

where $\chi_q^2(a)$ is the $100a$th percentile of the chi-squared distribution with q degrees of freedom, and q is the dimension of the vector β; in our example, $q = 3$. Table 2 gives these intervals for our example. (Alternatively, the Bonferroni or maximum-modulus methods could be used; see Seber (1977). For multiple-comparison methods, see Hochberg and Tamhane (1987).)

Simultaneous 95% confidence intervals for the three possible comparisons $\beta_1 - \beta_2$, $\beta_1 - \beta_3$ and $\beta_2 - \beta_3$ of therapy effect sizes are given by

$$\hat{\beta}_i - \hat{\beta}_j \pm [\chi_3^2(0.95)(c_{ii} + c_{jj} - 2c_{ij})]^{1/2} \quad (1 \le i < j \le 3).$$

Table 3 gives these intervals for our example.

These confidence intervals can also be used for simultaneous two-sided tests of hypotheses about the values of the effect sizes β_i, or of comparisons between the effect sizes. Thus, for example, on the basis of Table 2, we reject the null hypothesis that the effect size for therapy 3 is 0, but do not reject null hypotheses that the effect sizes for therapies 1 and 2 are 0, at the simultaneous 5% level of significance. Similarly, from Table 3, we conclude that only the pair of therapies 2 and 3 have significantly different effect sizes at the simultaneous 5% level of significance.

Table 2 Simultaneous 95% Confidence Intervals for Overall Effect Sizes $\beta_i = p_0 - p_i$ $(i = 1, 2, 3)$

Effect size	Confidence interval
β_1	-0.00108 ± 0.01275 or $[-0.0138, 0.0117]$
β_2	-0.01252 ± 0.01269 or $[-0.0252, 0.0002]$
β_3	0.02285 ± 0.02256 or $[0.0003, 0.0454]$

Table 3 Simultaneous 95% Confidence Intervals for $\beta_i - \beta_j$ $(1 \le i < j \le 3)$

Comparison	Confidence interval
$\beta_1 - \beta_2$	0.01144 ± 0.01054 or $[-0.0096, 0.0220]$
$\beta_1 - \beta_3$	-0.02393 ± 0.02472 or $[-0.0486, 0.0008]$
$\beta_2 - \beta_3$	-0.03537 ± 0.02404 or $[-0.0594, -0.0113]$

IV. REGRESSION FOR LOG ODDS RATIOS

If the effect size used for the therapies is the log odds ratio

$$d_i^* = \log(p_0/(1 - p_0)) - \log(p_i/(1 - p_i))$$

instead of the risk differences $p_i - p_0$, then the large-sample variance of d_i^* is

$$\psi_{ii} = \frac{1}{n_0 p_0 (1 - p_0)} + \frac{1}{n_i p_i (1 - p_i)},$$

and the large-sample covariance between d_i and d_j is

$$\psi_{ij} = \frac{1}{n_0 p_0 (1 - p_0)}.$$

A regression model of the form given in Sec. III for the vector d of estimated effect sizes can now be used to estimate the common effect sizes for the therapies. The design matrix X given by (6) remains the same, and the computation of $\hat{\beta}$ in (7) uses the new values of ψ_{ij}.

Here

$$d = (-0.2283, -0.6466, 0.3008, 0.1850, -0.0700, 0.4321, 0.2113)',$$

$$\hat{\Psi}_1 = \begin{bmatrix} 0.06128 & 0.05102 \\ 0.05102 & 0.05795 \end{bmatrix}, \qquad \hat{\Psi}_2 = 0.17403, \qquad \hat{\Psi}_3 = 0.06348,$$

$$\hat{\Psi}_4 = \begin{bmatrix} 0.02051 & 0.00693 \\ 0.00693 & 0.02798 \end{bmatrix}, \qquad \hat{\Psi}_5 = 0.04246,$$

which from (7) yields

$$(\hat{\beta}_1, \hat{\beta}_2, \hat{\beta}_3) = (0.2169, -0.1500, 0.3315)$$

with approximate covariance matrix

$$C = (X'\hat{\Psi}X)^{-1} = 10^{-7} \begin{bmatrix} 2240 & 9258 & 1997 \\ 9258 & 11503 & 2482 \\ 1997 & 2482 & 1657 \end{bmatrix}.$$

Consequently, simultaneous 95% confidence intervals for the βs are:

$$\beta_1: [-0.2015, 0.6353],$$
$$\beta_2: [-0.4498, 0.1498],$$
$$\beta_3: [-0.0284, 0.6913],$$

and simultaneous 95% confidence intervals for paired differences are:

$$\beta_1 - \beta_2 : [0.020, 0.714],$$
$$\beta_1 - \beta_3 : [-0.637, 0.408],$$
$$\beta_2 - \beta_3 : [-0.906, -0.057].$$

Confidence intervals on the βs (log odds ratios) can be converted to confidence intervals for the odd ratios, $\omega = e^{\beta}$:

$$\omega_1 : [0.818, 1.888],$$
$$\omega_2 : [0.638, 1.162],$$
$$\omega_3 : [0.972, 1.996].$$

V. REGRESSION USING A VARIANCE-STABILIZING TRANSFORMATION

It is well known that the transformation $2 \arcsin \sqrt{p}$ stabilizes the variance. For large samples, the variance of $2 \arcsin \sqrt{p}$ (measured in radians) is $1/n$. With this transformation, the effect sizes are

$$d_i = 2 \arcsin \sqrt{p_0} - 2 \arcsin \sqrt{p_i},$$

in which case the large-sample variance of d_i is $\psi_{ii} = (1/n_0) + (1/n_i)$ and the large-sample covariance between d_i and d_j is $\psi_{ij} = 1/n_0$. The regression analysis now proceeds as in Sec. III with these values of ψ_{ij}.

For this model,

$$d = (-0.0338, -0.1060, 0.0613, 0.0507, -0.0187, 0.1038, 0.0727)',$$

$$\hat{\Psi}_1 = \begin{bmatrix} 0.00125 & 0.00100 \\ 0.00100 & 0.00125 \end{bmatrix}, \qquad \hat{\Psi}_2 = 0.00750, \qquad \hat{\Psi}_3 = 0.04722,$$

$$\hat{\Psi}_4 = \begin{bmatrix} 0.00150 & 0.00050 \\ 0.00050 & 0.00150 \end{bmatrix}, \qquad \hat{\Psi}_5 = 0.00500,$$

which from (7) yields

$$(\hat{\beta}_1, \hat{\beta}_2, \hat{\beta}_3) = (0.0058, -0.0612, 0.0860)$$

with approximate covariance matrix

$$C = (X'\hat{\Psi}X)^{-1} = 10^{-8} \begin{bmatrix} 78369 & 47589 & 12524 \\ 47589 & 63056 & 16594 \\ 12524 & 16594 & 109630 \end{bmatrix}.$$

Consequently, simultaneous 95% confidence intervals for the βs are:

β_1: $[-0.0725, 0.0840]$,

β_2: $[-0.1314, 0.0090]$,

β_3: $[-0.0065, 0.1786]$,

and simultaneous 95% confidence intervals for paired differences are:

$\beta_1 - \beta_2$: $[0.0065, 0.1267]$,

$\beta_1 - \beta_3$: $[-0.1935, 0.0322]$,

$\beta_2 - \beta_3$: $[-0.2517, -0.0429]$.

VI. DISCUSSION

There were five studies, each involving a subset of the three therapy treatments. We now summarize the confidence intervals for the effect sizes (versus a control) of the three therapies under the following definitions of effect size: difference of proportions (Sec. 3), arc sin transformation of proportions (Sec. 5), and the log odds ratio (Sec. 4); see Table 4.

The three effect size methods agree that therapies 1 and 2 are not efficacious with respect to the control. The arcsin and log odds ratio (or odds ratio) methods agree that therapy 3 is not efficacious, whereas the difference method shows a slight positive effect.

With respect to a comparison of the effectiveness of the treatments, the majority of the analyses show that therapies 2 and 3 are not significantly different from therapy 1, but that these therapies are different from each other.

The three effect size methods are based on large-sample theory. The difference method and the log odds ratio also require using sample estimates in the variance, whereas the arcsin transformation does not. Thus, in terms of a testing procedure, in contrast to an estimation procedure, the arcsin method is perhaps the most reliable.

Table 4 Confidence Intervals for Effect Sizes of the Three Therapies

Therapy	Differences	Arcsin	Log odds ratio	Odds ratio
1	(−0.0138, 0.0117)	(−0.0725, 0.0840)	(−0.2015, 0.6353)	(0.818, 1.888)
2	(−0.0252, 0.0002)	(−0.1314, 0.0090)	(−0.4498, 0.1498)	(0.638, 1.162)
3	(0.0003, 0.0454)	(−0.0065, 0.1786)	(−0.0284, 0.6913)	(0.972, 1.996)

Acknowledgments

The research of Leon Gleser was partially supported under Grant DMS-9504924 from the National Science Foundation. The research of Ingram Olkin was supported in part by the Centers for Disease Control and Prevention and the National Science Foundation, Grant DMS-9301366.

References

Gleser LJ, Olkin I. Stochastically dependent effect sizes. In: H Cooper, LV Hedges, eds. The Handbook of Research Synthesis. New York: Russell Sage Foundation, pp 339–355.

Hochberg Y, Tamhane AC. Multiple Comparison Procedures. John Wiley and Sons.

Seber GAF. Linear Regression Analysis. New York: John Wiley, pp 125–133.

8

A Bayesian Meta-analysis of the Relationship between Duration of Estrogen Exposure and Occurrence of Endometrial Cancer

Daniel T. Larose
Central Connecticut State University, New Britain, Connecticut

Abstract

A Bayesian random-effect model is presented for combining exposure–response information from several studies examining possible association between estrogen exposure and endometrial cancer. A Wishart distribution is used to model the within-study dependence of the vector of log relative risks. Data from 17 published studies which provided exposure–response information were combined. Samples from the joint posterior were generated by the Gibbs sampler (1). Results indicate evidence of a significant exposure–response relationship between estrogen and endometrial cancer.

I. INTRODUCTION

Cancers of the breast, corpus uteri, and ovary account for more than a quarter of all cancer deaths among women (2). In the United States from the mid 1960s until 1975, rates of endometrial cancer showed an unprecedented climb; replacement estrogen use during this time also rose

191

sharply (3). The suspicion that estrogen stimulation may contribute to the development of endometrial cancer goes back at least to 1946 (4), while the role of estrogens in the genesis of breast cancer was first suggested as long ago as 1836 (5). In 1975, two articles were published which showed large relative risks of five-fold to seven-fold for endometrial cancer among ever-users of estrogen compared to never-users (6, 7). The effect of these publications was to reduce dispensed prescriptions of noncontraceptive estrogens by 50% between 1975 and 1980; concurrently, the incidence of endometrial cancer fell as well. (3).

More recently, the prophylactic effect of estrogen on cardiovascular disease and osteoporosis has again led to an increase in replacement estrogen use, especially when opposed by a progestational agent. By 1986, 20.3 million prescriptions were dispensed for noncontraceptive estrogen (8). Today, about 32% of women aged 50–65 years use menopauasal estrogens, making it one of the most widely used drugs among American women (9). Because of this, and because women are postmenopausal on average for a third of their lives, estrogen replacement is, as one investigator puts it ". . . . potentially one of the quantitatively most important means of preventing major causes of morbidity and mortality in older women . . . ," and that, therefore, "full assessment . . . of the established and suggested risks is urgently required" (10).

The causal role of estrogens in cancer of the endometrium has now been generally accepted (10). However, quantitative conclusions concerning the exposure–response effect of estrogen use on endometrial cancer risk remain less clear. Investigators have turned to meta-analysis for help in clarifying this situation. This paper presents a Bayesian random-effects model for combining exposure–response information from several studies examining possible association between estrogen and endometrial cancer.

Searches were made of Medline, International Pharmaceutical Abstracts, and other medical databases for articles regarding possible links between estrogen and endometrial cancer especially review articles and existing meta-analyses. Then, from the bibliographies of these articles, further citations were obtained. The treatment under investigation consists of conjugated estrogens (such as Premarin), which do not include oral contraceptives.

In the end, 17 studies (4, 6, 9, 11–24) were found which provided sufficient exposure–response information to be included in the present analysis. Sixteen of these studies were retrospective, while one (20) was prospective. Descriptive information on the designs of these studies, patient characteristics, and control group characteristics are listed in

Table 1. Every study but one (13) required controls not to have undergone a hysterectomy. Of course, no two study designs are exactly alike, a situation which introduces hetreogeneity into any subsequent synthesis. A meta-analysis which posits a fixed-effect model will likely miss this heterogeneity, thus underestimating the variability of the effects, and possibly overstating their significance. A random-effect model is thus the more conservative approach. If in fact the parameter modeling the heterogeneity (σ^2 in Sec. II below) turns out not to be significant, then the parameter may be omitted and a fixed-effect model undertaken. Of course, there are situations where groups of studies are too dissimilar to be combined. See Ref. 25 for a discussion of modeling this situation.

Relevant exposure–response information was extracted from the 17 studies, and is shown in Table 2, including the first author of the study, the vector of midpoints of the intervals for the duration of exposure to conjugated estrogen, the log relative risks for each duration level, and the corresponding standard error. For example, the Antunes (11) study reported results for durations of 0 to 1 year, 1 to 5 years, and more than 5 years. Our first two midpoints are thus 0.5 and 3.0, while the third midpoint of 6.0 follows Berlin, Longnecker, and Greenland (26) in reporting midpoints of open-ended interals to be 20% greater than the lower limit. The standard errors for the prospective study were estimated as the width of the 95% confidence interval divided by 3.92.

Kelsey et al. (17) remark that "the longer estrogen replacement therapy was used, the higher the risk . . . ," noting that their results were similar to those obtained by other studies. The motivation therefore exists for an investigation into the relationship between the length of exposure to conjugated estrogens and the risk of developing cancer of the endometrium. Support for the linear nature of such a relationship is found in Fig. 1, which plots the log relative risks of developing endometrial cancer against the various midpoints of exposure to conjugated estrogens for all seventeen studies. Clearly, *as estrogen exposure increases, there is a strong tendency for the log relative risk to increase*. The present paper seeks to quantify this relationship via meta-analysis of the seventeen studies.

In any meta-analysis based on published monographs, the analyst should address the possible presence of publication bias: the tendency of investigators or editors to base decisions regarding submission or acceptance of manuscripts for publication on the strength of the investigator's study findings (27); see also Smith et al. (Chap. 12). The funnel plot of Light and Pillemer (28) can serve as a useful diagnostic for assessing the

Table 1 Descriptive Characteristics of the 17 Studies

Study	Location	Case dates	Patient characteristics	Controls
Antunes	Md	1973–77	Mean age 60.6	Non-gynecological hospital controls
Brinton	Varied	1987–90	Ages 20–74	Matched population controls
Buring	Ma	1970–75	Ages 40–80	Same-hospital non-gynecological
Gray	Ky	1947–76	Mean age 56.25	Matched, with for hysterectomies
Hoogerland	Wi	1960–74	Matched pairs	Gynecological controls
Hulka	NC	1970–76	Mean age 59.7	Two control groups: gynecological and community
Jelvosek	NC	1940–75	Mean age 58.9	One per patient matched from registration files
Kelsey	Ct	1977–79	Ages 45–74	Non-gynecological hospital controls
Levi	Switz.	1988–92	Ages 30–74	Non-gynecological hospital controls
Mack	L.A.	1971–75	Retirement community	Age-matched from community roster
McDonald	Mn	1945–74	Controls had more children	Four age-matched to each patient
Paganini-Hill	Ca	1981–85	Retirement community of 5160 women, 44–100	Prospective study
Shapiro	Varied	1976–79	Ages 50–69	Non-gynecological hospital controls
Spengler	Ont.	1977–78	Ages 40–74	Neighborhood matched
Stavracky	Ont.	1976–78	Ages 40–80	Two control groups, gynecological and non-gynecological
Weiss	Wash.	1975–76	Ages 51–74	Selected from area surveys
Ziel	L.A.	1970–74	Health plan members	Two per patient

Table 2 Duration Midpoints, Log Relative Risks, and Standard Errors from the Seventeen Studies

Study	Midpoints	Log R.R.	S.E.	Study	Midpoints	Log R.R.	S.E.
Antunes	0.5	0.79	0.24	Mack	0.5	1.03	0.22
	3.0	1.06	0.19		3.0	1.50	0.10
	6.0	2.71	0.37		6.5	2.23	0.26
Brinton	2.5	0.34	0.25		9.6	2.17	0.11
	6.0	1.79	0.15	McDonald	0.25	−0.45	0.24
Buring	0.5	0.15	0.11		0.75	1.15	0.46
	2.5	0.69	0.12		2.0	0.86	0.55
	7.0	1.86	0.15		3.6	1.78	0.29
	12.0	2.03	0.34	Paganini-Hill	1.0	1.65	0.52
Gray	2.0	0.18	0.26		5.0	1.95	0.52
	7.0	1.41	0.66		11.0	1.39	0.52
	12.0	2.45	1.11		18.0	3.00	0.52
Hoogerland	0.25	0.18	0.22	Shapiro	0.5	−0.11	0.13
	0.75	0.59	0.21		2.5	1.06	0.07
	2.0	1.16	0.15		7.0	1.72	0.08
	4.0	1.36	0.22		12.0	2.30	0.09
	7.5	1.22	0.19				
	12.0	1.90	0.59	Spengler	0.3	0.34	0.33
Hulka	1.75	−0.22	0.12		1.25	0.96	0.22
	4.2	1.41	0.24		3.5	0.79	0.32
					6.0	2.15	0.25
Jelovsek	1.75	0.34	0.20	Stavracky	1.0	−0.36	0.18
	4.0	0.34	0.59		3.0	0.00	0.12
	7.5	1.57	0.31		7.0	0.53	0.13
	12.0	0.96	0.19		12.0	1.86	0.25
Kelsey	0.5	0.10	0.11	Weiss	1.5	0.18	0.81
	1.75	0.00	0.30		3.5	1.69	0.45
	3.75	1.06	0.14		6.0	1.55	0.24
	6.25	1.46	0.18		9.0	2.46	0.28
	8.75	2.10	0.19		12.5	3.19	0.44
	12.0	0.99	0.15		17.0	2.32	0.39
Levi	2.5	0.53	0.06		24.0	2.12	1.31
	6.0	1.44	0.09	Ziel	2.0	1.53	0.18
					14.0	2.22	0.13

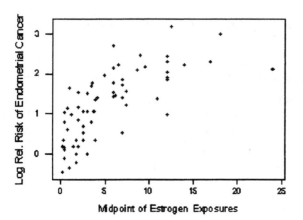

Figure 1   Plot of log relative risks vs. estrogen exposures.

presence of publication bias by plotting study sample size against effect size. Since effect size is more highly variable with small samples, the plot should resemble a funnel with the spout pointing up. Since mean effect size should be independent of sample size, the points should be evenly distributed on both sides of the mean for every sample size. That is, a skewed funnel is indicative of bias.

However, the response for each study in the present meta-analysis is a vector of log relative risks at various exposures. As a surrogate, the effect size used for the funnel plot in Fig. 2 is the log relative risk for ever-users. Figure 2 contains the funnel plot for the retrospective studies in the present meta-analysis. The plot shows the greater variability at smaller sample sizes, and does appear to be slightly skewed. However, this skewness is due in large part to the study in the upper left (sample size of 881, log relative risk of 0.47). In that study (26) the authors write: "The magnitude of the odds ratio in the current study is less than that in most other studies; the reason for this difference is not apparent." Thus, at least with respect to the surrogate effect size, this diagnostic plot presents insufficient evidence of bias, though judgements may vary. Finally, note that the funnel plot seems to be sensitive to scale, since a plot (not shown) of the sample sizes against the relative risks (not log relative risks) did seem to be skewed. DuMouchel and Normand (Chap. 6) address the importance of checking alternative parametrizations for this and other reasons.

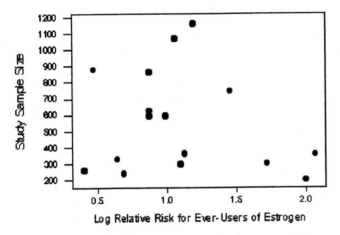

Figure 2   Funnel plot to assess publication bias.

In Sec. II, the Bayesian random-effects hierarchical model will be motivated and formulated to quantify the exposure–response relationship between conjugated estrogens and endometrial cancer. Computational aspects, necessary in many of today's sampling-based methodologies are discussed in Sec. III. Finally, Sec. IV presents the results and offers some concluding remarks.

II. THE MODEL

Why do we adopt a Bayesian perspective in fitting meta-analysis models? We are primarily concerned with the information and modeling incorporated in the likelihood. As a result, we take an automatic Bayesian stance employing appropriate diffuse prior specifications. The use of Bayesian hierarchical models means that, for the small sample sizes typical in meta-analysis, estimates of variability should be more appropriate than asymptotic ones arising from maximum likelihood. Also, Bayesian models can explicitly incorporate available background information such as expert medical opinion.

The response y_i ($i = 1, \ldots, 17$) is, for each study, the vector of log relative risks of contracting cancer for users of estrogen at the various duration levels compared to never-users. Relative risk is a measure of the strength of association between estrogen and cancer. It is often approximated by the relative odds: ad/bc, where $a =$ the number of exposed cases, $b =$ the number of exposed controls, $c =$ the number of unexposed cases, and $d =$ the number of unexposed controls. The standard error of the log relative risk is usually estimated as $a^{-1} + b^{-1} + c^{-1} + d^{-1}$ (29).

The mean of each $\mathbf{y}_i$ is assumed to be $\mathbf{D}_i\beta_i$, where $\mathbf{D}_i$ represents the known duration vector for the ith study. These durations are the midpoints of the intervals corresponding to each duration level published in a particular study (26). For each study, the scalar β_i represents the unknown exposure–response slope which is to be estimated.

Since the relative-risk estimates from different durations use the same reference group, they are clearly not independent. This within-study dependence is modeled as follows. The vector $\mathbf{y}_i$ of log relative risks is assumed to vary according to the random covariance matrix Σ_i, calibrated by the diagonal matrix of known standard errors $\mathbf{S}_i$. In turn, Σ_i^{-1} is assumed to be Wishart, with both degrees of freedom v_i and precision matrix $\mathbf{R}_0$ known. The values of v_i and $\mathbf{R}_0$ are chosen so that the prior $\pi(\Sigma_i^{-1})$ is diffuse. This is accomplished by setting $\frac{1}{2}(v_i - d - 1)$ (where d is the dimension of Σ_i^{-1}) equal to 1, and solving for v_i, and by taking the precision matrix to be some large scalar times the identity matrix. The covariance within studies is accounted for by the random covariance matrix Σ_i, which is calibrated by the diagonal matrix of known standard errors $\mathbf{S}_i$.

The hyperparameter μ represents the underlying mean slope, and σ^2 represents the between-study variation. The proper hyperpriors are chosen to approximate diffuse forms, while maintaining the propriety of the posterior.

The model follows:

$$(\mathbf{y}_i | \mathbf{D}_i, \beta_i, \mathbf{S}_i, \Sigma_i) \overset{\text{ind.}}{\sim} \text{normal}(\mathbf{D}_i\beta_i, \mathbf{S}_i\Sigma_i\mathbf{S}_i'),$$

$$(\beta_i | \mu, \sigma^2) \sim \text{normal}(\mu, \sigma^2), \qquad (\Sigma_i^{-1} | v_i, \mathbf{R}_0) \sim \text{Wishart}(v_i, \mathbf{R}_0),$$

$$(\mu) \sim \text{normal}(0, V(\mu)), \qquad (\sigma^2) \sim \text{inverse-gamma}(0.001, 1000),$$

where, for the ith study, $\mathbf{y}_i$ is the vector of log relative risks, $\mathbf{D}_i$ is the vector of midpoints of the durations, $\mathbf{S}_i$ is the diagonal matrix of the standard errors, v_i and $\mathbf{R}_0$ are the assumed known degrees of freedom and precision matrix of the Wishart distribution chosen so that the prior is diffuse, and $V(\mu)$ is a constant chosen so that the prior for μ is also diffuse.

III. COMPUTATIONAL ASPECTS

The development of sampling-based methods has made possible the analysis of posterior densities hitherto intractable. The Gibbs sampler (1, 30) is a Markov chain Monte Carlo updating scheme that produces samples from the joint posterior, using iterated sampling from the full or complete conditional distributions. The complete conditional distributions are easily found from the complete Bayesian model (likelihood times prior structure).

The complete conditional for Σ_i^{-1} takes the form of a Wishart distribution of dimension equal to the dimension of Σ_i^{-1}, degrees of freedom $\nu_i + 1$, and precision matrix $\mathbf{C}$, where $\mathbf{C} = [(\mathbf{y}_i - \mathbf{D}_i\beta_i)(\mathbf{y}_i - \mathbf{D}_i\beta_i)^{\mathrm{T}} + \mathbf{R}_0^{-1}]^{-1}$. Sample matrices from the requisite Wishart distribution are generated using the algorithm of Odell and Feiveson (31).

After an initial burn-in period of 100 cycles, the Gibbs sampler generated a further 1000 iterations, each iteration generating variates from the marginal posterior distribution of each of the parameters. A single chain was used, retaining only the variates of every tenth cycle to insure independence. Convergence of each distribution was checked using the convergence diagnostics of Gelman and Rubin (32). Their software reports 50% and 97.5% quantiles of an upper bound on the shrinkage of the interval between the 2.5% and 97.5% quantiles of the target distribution will shrink if the iterative simulation is allowed to continue indefinitely. The former quantiles should be near one. Since the algorithm of Gelman and Rubin is tailored for scalar quantities, for the covariance matrix Σ_i, we performed the diagnostics on the log likelihood evaluated at the posterior sample values of Σ_i.

IV. RESULTS

Table 3 contains the posterior mean, median, standard deviation, and 2.5th and 97.5th percentiles for the parameters of interest. Figure 3 presents these results graphically. The 2.5th percentile for the posterior of μ is greater than zero. This is evidence that the underlying mean slope across all studies (the change in risk per year of estrogen use) is positive. Thus, the meta-analysis has uncovered a significant exposure–response

Table 3 Posterior Statistics for Overall Mean Slope μ, Between-study Variance σ^2, and Individual Study Slopes β_i ($i = 1,\ldots,17$)

Parameter	2.5th Percentile	Median	97.5th Percentile	Mean	SD
Overall mean slope μ	0.013	0.220	0.442	0.222	0.106
Between-study var σ^2	0.028	0.104	0.759	0.174	0.257
1. Antunes	0.124	0.398	0.664	0.387	0.142
2. Brinton	-0.004	0.263	0.535	0.258	0.140
3. Buring	0.069	0.195	0.330	0.196	0.069
4. Gray	0.088	0.195	0.334	0.197	0.061
5. Hoogerland	0.069	0.171	0.310	0.180	0.062
6. Hulka	-0.113	0.257	0.602	0.252	0.177
7. Jelovsek	-0.007	0.106	0.281	0.109	0.059
8. Kelsey	0.059	0.166	0.266	0.160	0.045
9. Levi	-0.039	0.243	0.537	0.244	0.141
10. Mack	0.087	0.290	0.460	0.281	0.098
11. McDonald	0.063	0.366	1.137	0.390	0.230
12. Paganini-Hill	0.084	0.174	0.260	0.173	0.045
13. Shapiro	0.048	0.211	0.365	0.202	0.068
14. Spengler	0.026	0.325	0.571	0.300	0.122
15. Stavracky	-0.003	0.134	0.259	0.132	0.068
16. Weiss	0.078	0.154	0.188	0.148	0.029
17. Ziel	0.055	0.176	0.316	0.176	0.067

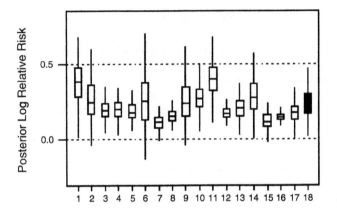

Figure 3   Posterior distributions of study slope and overall mean slope.

relationship: the longer a woman's exposure to estrogen, the greater her risk of endometrial carcinogenesis.

How is the point estimate (median = 0.22) for μ to be interpreted? To obtain specific relative risk estimates, multiply the point estimate of the slope (0.22) by the number of years duration (say, 5), and then take e raised to the power of this product: $e^{0.22 \times 5} = 3.004$. After five years of exposure to estrogen, the estimated increase in cancer risk is threefold.

The between-study variation is non-negligible, as seen in the posterior statistics for σ^2, recalling that σ^2 models the variability on the log scale. It seems likely that a fixed-effect formulation for this data set would have missed some between-study variation.

Finally, though all the individual study slope medians associated are positive, five of the 2.5th percentiles are negative, indicating that not all are significantly greater than zero. Despite this, the underlying overall expected slope μ was found to be significantly greater than zero.

Endometrial cancer poses a significant health risk. The present meta-analysis uncovered a significant exposure–response association between estrogen exposure and endometrial cancer while accounting for within-study dependence. Bayesian hierarchical models present a natural framework for studying questions such as these.

References

1. S Geman, D Geman. Stochastic relaxation, Gibbs distributions, and the Bayesian restoration of images. IEEE Transactions on Pattern Analysis and Machine Intelligence 6:721–741, 1984.

2. J Kelsey, A Whittemore. Epidemiology and primary prevention of cancers of the breast, endometrium, and ovary. Annals of Epidemiology 4:89–95, 1994.

3. S Jick, A Walker, H Jick. Estrogens, Progesterone and Endometrial Cancer. Epidemiology 4:20-24, 1993.

4. T McDonald et al. Exogenous estrogen and endomentrial carcinoma: Case control and incidence study. American Journal of Obstetrics and Gynecology 127:572–579, 1977.

5. K McPherson, H Dahl. Oestrogens and breast cancer: Exogenous hormones. British Medical Bulletin 47:484–492, 1991.

6. H Ziel, W Finkle. Association of estrone with the development of endometrial carcinoma. American Journal of Obstetrics and Gynecology 124:735–740, 1975.

7. D Smith et al. Association of exogenous estrogen and endometrial carcinoma. New England Journal of Medicine 293:1164, 1975.

8. R Gorsky. Relative risks and benefits of long-term estrogen replacement therapy: a decision analysis. Obstetrics and Gynecology 83:161–166, 1994.

9. L Brinton, C Schairer. Estrogen replacement therapy and breast cancer risk. Epidemiologic Reviews 15:66–79, 1993.

10. H Adami. Long-term consequences of estrogen and estrogen–progestin replacement. Cancer Causes and Control 3:83–90, 1992.

11. C Antunes et al. Endometrial cancer and estrogen use. New England Journal of Medicine 300:9–13, 1979.

12. J Buring et al. Conjugated estrogen use and risk of endometrial cancer. American Journal of Epidemiology 124:434–441, 1986.

13. L Gray et al. Estrogens and endometrial carcinoma. Obstetrics and Gynecology 49:385–389, 1977.

14. D Hoogerland. Estrogen use—risk of endometrial carcinoma. Gynecologic Oncology 6:451–458, 1978.

15. B Hulka. Estrogen and endometrial cancer: Cases and two control groups from North Carolina. American Journal of Obstetrics and Gynecology 137:92–101, 1980.

16. F Jelovsek et al. Risk of exogenous estrogen therapy and endometrial cancer. American Journal of Obstetrics and Gynecology 137:85–91, 1980.

17. J Kelsey et al. A case-control study of cancer of the endometrium. American Journal of Epidemiology 116:333–342, 1982.

18. F Levi et al. Oestrogen replacement treatment and the risk of endometrial cancer: an assessment of the role of covariates. European Journal of Cancer 29A:1445–1449, 1993.

19. T Mack. Estrogens and Endometrial Cancer in a Retirement Community. New England Journal of Medicine 294:1262–1267, 1976.

20. A Paganini-Hill et al. Endometrial cancer and patterns of use of oestrogen replacement therapy: a cohort study. British Journal of Cancer 59:445–447, 1989.

21. S Shapiro. Risk of localized and wide-spread endometrial cancer in relation to recent and discontinued use of congugated estrogens. New England Journal of Medicine 313:969–972, 1985.

22. R Spengler et al. Exogenous estrogens and endometrial cancer: a case-control study and assessment of potential biases. American Journal of Epidemiology 114:497–506, 1981.

23. K Stavraky et al. A comparison of estrogen use by women with endometrial cancer, gynecologic disorders, and other illnesses. American Journal of Obstetrics and Gynecology 141:547–555, 1981.

24. N Weiss et al. Endometrial cancer in relation to patterns of menopausal estrogen use. Journal of the American Medical Association 242:261–264, 1979.

25. D Larose, D Dey. Grouped random effects models for Bayesian meta-analysis. Statistics in Medicine 16:1817–1829, 1997.

26. J Berlin, M Longnecker, S Greenland. Meta-analysis of epidemiologic dose-response data. Epidemiology 4:218–228, 1993.

27. D Larose, D Dey. Modeling publication bias using weighted distributions in a Bayesian framework. Computational Statistics and Data Analysis 26:279–302, 1998.

28. R Light, D Pillemer. Summing up: the science of reviewing research. Cambridge, Ma: Harvard University Press, 1984.

29. A Agresti. Categorical Data Analysis. New York: Wiley Series in Probability and Mathematical Statistics, 1990.

30. A Gelfand, A Smith. Sampling-based approaches to calculating marginal densities. Journal of the American Statistical Association 85:398–409, 1990.

31. P Odell, A Feiveson. Numerical procedure to generate a sample covariance matrix. Journal of the American Statistical Association 199–205, 1966.

32. A Gelman, D Rubin. Inference from iterative simulation using multiple sequences (with discussion). Statistical Science 7:457–511, 1992.

9
Modeling and Implementation Issues in Bayesian Meta-analysis

Donna K. Pauler
Biostatistics Center, Massachusetts General Hospital, Boston, Massachusetts

Jon Wakefield
University of Washington, Seattle, Washington

Summary

In this chapter we outline a general hierarchical model for meta-analysis, introduce a new sampling algorithm for estimation of its parameters, and provide diagnostics for examination of distributional assumptions. We derive a simple rejection algorithm which generates independent samples from the posterior distribution under arbitrary priors, and can be easily programmed in popular statistical packages. Although flat or diffuse priors are standard for fixed-effect parameters, which describe the overall mean or effect of covariates, the choice of prior distribution for the hyperparameters of a random-effect distribution for unobserved factors requires attention; see DuMouchel and Normand (Chap. 6). We introduce a proper prior for the variance component of the random-effect distribution based on the unit-information concept of Kass and Wasserman (1), which we recommend be used in a sensitivity analysis when subjective priors are available or as a default choice. We show how a test for heterogeneity can be performed based on the Bayes factor, directly calculable from our rejection algorithm. To demonstrate the fully Bayesian approach, we implement our procedures in three previously analyzed examples, compare results to those obtained originally, and present novel diagnostics for examining such modeling assumptions as the effect of the prior and the appropriateness of the normality assumption at the second stage of the model.

I. INTRODUCTION

Consider the situation in which k studies have been conducted to investigate the differences between two treatments, and denote by Y_i the summary statistic of study i $(i = 1, \dots, k)$. For example, Y_i may denote the treatment difference for normal data, the log odds ratio for binary data, the log hazard ratio in a proportional-hazards regression model for survival data, or the odds ratio in a proportional odds model for ordinal categorical data; see for example, Refs 2 and 3. If the precision associated with each individual study estimate is high, or the effect of interest is normally distributed, we may assume:

$$Y_i | \theta_i \sim N(\theta_i, w_i^{-1}), \tag{1}$$

where w_i denotes the (known) precision of the observed estimate, θ_i the treatment effect in study i, and $N(a, b)$ the normal density with mean a and variance b. If we further assume the individual study effects are exchangeable and that there is no systematic bias in the selection of studies, as might arise due to publication bias (Smith et al., Chap. 13), we may model the distribution of θ_is as a normal density with mean μ, representing the population treatment effect, and variance τ^2, representing the extent of between-study heterogeneity:

$$\theta_i | \mu, \tau^2 \sim N(\mu, \tau^2). \tag{2}$$

Model (1) combined with (2) implies that marginally the Y_i follow independent $N(\mu, \tau^2 + w_i^{-1})$ distributions.

The inclusion of study-level covariates that contribute to heterogeneity can make the exchangeability assumption more appropriate. Covariates are formally introduced into the model by replacing (2) with

$$\theta_i | \mu, \beta, \tau^2 \sim N(\mu + x_i'\beta, \tau^2), \tag{3}$$

where $x_i' = (x_{i1}, \dots, x_{ip})$ is a p-dimensional vector of covariates (with $p < k - 1$) for study i $(i = 1, \dots, k)$, and β is a p-dimensional vector of population regression parameters. We assume that the x_{ij} have zero mean so that μ still measures the average treatment effect. Often the effects of covariates are of interest in their own right. For example, in their meta-analysis of nicotine-replacement therapy, DuMouchel and Normand (Chap. 6) analyze the effect of two covariates on smoking cessation: type of therapy and intensity of support.

Although primary interest centers on the average treatment effect μ, accurate assessment of between-study variability τ^2 requires careful attention, since its estimated magnitude affects widths of both posterior and predictive intervals for μ. To see this, note that the predictive distribution for the treatment effect in a new study exchangeable with those which supplied the observed data has variance $E[\tau^2 \mid y] + \text{var}[\mu \mid y]$, which depends on τ^2. Despite this dependency, many meta-analyses report tests of between-study heterogeneity, and then set $\tau^2 = 0$ and proceed with a fixed-effects analysis if the null hypothesis of homogeneity is not rejected (2). While we believe conclusions from such tests are useful components of the final report of a meta-analysis, for prediction purposes, even if the null hypothesis is not rejected, results from the fixed analysis should be contrasted with those from the full model. The posterior distribution of τ^2 provides added information not captured by the test, such as on the *clinical* significance of the size of between-study variability.

A statistic that has frequently been used to test for heterogeneity is given by

$$Q = \sum_{i=1}^{k} w_i (Y_i - \bar{Y})^2,$$

where $\bar{Y} = \sum w_i Y_i / \sum w_i$. When no heterogeneity is present, Q follows an asymptotic chi-squared distribution with $k - 1$ degrees of freedom. One of the drawbacks of this test is that it has low power (2, 6, 7). The Bayesian analogue to frequentist tests is based on the Bayes factor, which is given by the ratio of posterior to prior odds. The Bayes factor provides direct evidence in favor of a model specifying between-study heterogeneity. In Ref. 8 Bayes factors and asymptotic approximations are derived for variance component models.

Because the marginal likelihood arising from (1) and (2) cannot be maximized in closed form, a number of alternatives have been suggested, including a method-of-moments estimator (9, 2) and a profile likelihood approach (10). From a Bayesian perspective, the integrals required to derive the joint posterior distribution for μ and τ^2 are similarly intractable. Reference 11 considers analytic approximations, while a number of authors (12, 13, 14) have resorted to Markov chain Monte Carlo (MCMC) methods in which *dependent* samples are generated from the posterior distribution, often using the BUGS software (15). Because standard errors of dependent samples are more difficult to calculate, experience is required to assess convergence of the Markov chain and to determine the number of samples required. In variance-component models, numerical problems also arise in MCMC procedures for certain types of prior when the variance component is close to zero (16: p 96). Methods for calculating the Bayes factor based on MCMC (17) fail under the same circumstances. Recently (8), a rejection algorithm has been defined which generates independent samples from the posterior distribution and hence avoids many of these problems.

The advantage of sampling-based methods is that they allow the examination of both the posterior distribution of parameters of interest, and of diagnostics that allow the appropriateness of modeling assumptions to be investigated. For hierarchical models in general, definition and interpretation of appropriate diagnostics are the subject of ongoing research (18).

In this chapter we outline a simple Bayesian approach for performing a complete meta-analysis, which includes a sampling algorithm for obtaining posterior and predictive summaries of interest under arbitrary priors, a recommendation for a default choice of the prior distribution for τ^2, methods for computing the Bayes factor in favor of heterogeneity, and diagnostics for assessing the normality assumption (2). In Sec. II, we describe a rejection algorithm that may be used to generate independent samples from the posterior distribution. In Sec. III, we discuss the use of Bayes factors to assess the level of between-study heterogeneity, and demonstrate how they may be calculated directly from the output of the rejection algorithm. In Sec. IV, we highlight the sensitivity of conclusions and predictions to the prior for τ^2, and offer a default choice based on the unit-information concept of Kass and Wasserman (1). In Sec. V, we analyze three previous meta-analyses to illustrate our methods and compare results to the usual frequentist procedures. Section VI contains a concluding discussion.

II. IMPLEMENTATION

We first describe how posterior samples for all fixed and random parameters in model (1)–(2) may be obtained via a simple rejection algorithm. We extend the algorithm to cover study-level covariates in Sec. II.C.

Given a prior $\pi(\mu, \tau^2)$ for the overall treatment effect, μ, and between-study variability, τ^2, the Bayesian approach is based on inference from the posterior distribution $\pi(\mu, \tau^2 \mid y) \propto L(\mu, \tau^2) \times \pi(\mu, \tau^2)$ where $L(\mu, \tau^2)$ denotes the likelihood for (μ, τ^2). It is common to assume the improper prior $\pi(\mu) \propto c$ for μ so that the prior specification becomes $\pi(\mu, \tau^2) \propto \pi(\tau^2)$. The posterior distribution for μ is fairly insensitive to the choice of prior for μ, as long as it is smooth and regular, so the particular specification is not of primary concern. To obtain the marginal posterior distribution for τ^2 under this prior, we analytically integrate out μ to give

$$\pi(\tau^2 \mid y) \propto L_R(\tau^2) \times \pi(\tau^2), \tag{4}$$

where $L_R(\tau^2)$ denotes the restricted likelihood, given by

$$L_R(\tau^2) = (2\pi)^{-(k-1)/2} \left(\sum_{i=1}^{k} \frac{1}{\tau^2 + w_i^{-1}} \right)^{-1/2} \prod_{i=1}^{k} (\tau^2 + w_i^{-1})^{-1/2}$$
$$\times \exp\left\{ -\frac{1}{2} \sum_{i=1}^{k} \frac{1}{\tau^2 + w_i^{-1}} \left(y_i - \mu^*(\tau^2) \right)^2 \right\}$$

and

$$\mu^*(\tau^2) = \sum_{i=1}^{k} \frac{y_i}{\tau^2 + w_i^{-1}} \left(\sum_{i=1}^{k} \frac{1}{\tau^2 + w_i^{-1}} \right)^{-1}. \tag{5}$$

Note that $\mu^*(\tau^2)$ is a weighted combination of the study effects with weights proportional to the precision of each data point.

The procedure for obtaining posterior samples for μ, τ^2, and θ_i, for $i = 1, \ldots, k$, is accomplished in two steps. First, a rejection algorithm is used to draw samples of τ^2 from the posterior distribution (4). Conditional on τ^2, the average treatment effect μ and the random effects θ_i follow a normal distribution.

A. Rejection Algorithm for τ^2

To obtain a sample from (4), the rejection algorithm (19) requires finding a suitable proposal density $g(\tau^2)$ from which it is easy to sample and for which $L_R(\tau^2)\pi(\tau^2)/g(\tau^2)$ is bounded. Define $M = \sup_{\tau^2} L_R(\tau^2)\pi(\tau^2)/g(\tau^2)$; then the algorithm proceeds as follows:

1. generate τ^2 from $g(\tau^2)$ and U independently from $U(0, 1)$, the uniform density on $[0,1]$,
2. accept τ^2 if $U \leq L_R(\tau^2)\pi(\tau^2)/(Mg(\tau^2))$.

Let $w^* = \min_i\{w_i\}$ denote the minimum weight and $SS_y = \sum_{i=1}^{k}(y_i - \bar{y})^2$, where $\bar{y} = 1/k \sum_{i=1}^{k} y_i$ denotes the sample mean. Then, if the number of studies is greater than 2 and the prior $\pi(\tau^2)$ is bounded, a proposed density $g(\tau^2)$ which satisfies the boundedness condition is

$$g(\tau^2) = IG(\tau^2 + w^{*-1} \mid a, b), \tag{6}$$

where $a = (k - 3)/2$, $b = SS_y/2$, and $IG(x \mid a, b)$ denotes the inverse gamma distribution for x with mean $b/(a - 1)$ and variance $b^2/((a - 1)^2(a - 2))$. The density given by (6) has the same form as the restricted likelihood that would result from k studies with observations y_i $(i = 1, \ldots, k)$, each with precision w^*. The intuition behind this choice is the desire for a proposed density with flatter tails than the likelihood $L_R(\tau^2)$. To achieve this we use the same data but replace each of the precisions with the minimum precision. It is straightforward to show that a finite bound M exists for this choice of g. An analytical conservative bound is given by:

$$M = \frac{(2\pi)^{-(k-1)/2}\Gamma(\frac{k-3}{2})k^{-1/2}}{(SS_y/2)^{(k-3)/2}} \prod_{i=1}^{k}\left(\frac{w_i}{w^*}\right)^{1/2} \sup_{\tau^2} \pi(\tau^2), \tag{7}$$

but a much tighter bound can be achieved by numerical maximization of the one-dimensional function $L_R \times \pi/g$. It is often sufficient simply to plot this function for a grid of τ^2 values and visually determine an approximate maximum.

Alternative choices of g may lead to tighter bounds and hence more efficient algorithms. For example, if $\pi(\tau^2)$ is taken to be the density $IG(a', b')$, then an efficient proposal is given by (6) with $a = (k - 3)/2 + a'$ and $b = SS_y/2 + b'$. If $\pi(\tau^2)$ is unbounded, the proposal density (6) will not provide a bounding envelope and so will have to be modified in some way, such as by increasing a.

B. Samples for Population Mean and Individual Effects

Given samples for τ^2, posterior samples for μ and $\theta = (\theta_1, \ldots, \theta_k)$ can be obtained via $p(\mu, \theta \mid \tau^2, Y) \propto p(\theta \mid \mu, \tau^2, Y) \times p(\mu \mid \tau^2, Y)$, where

$$p(\mu \mid \tau^2, Y) = N\left(\mu^*(\tau^2), \left[\sum_{i=1}^{k} \frac{1}{\tau^2 + w_i^{-1}}\right]^{-1}\right) \tag{8}$$

with $\mu^*(\tau^2)$ given by (5), and

$$p(\theta_i \mid \mu, \tau^2, Y) = N\left(\frac{y_i w_i + \mu/\tau^2}{w_i + 1/\tau^2}, \frac{1}{w_i + 1/\tau^2}\right) \tag{9}$$

for $i = 1, \ldots, k$. Once the more difficult task of generating samples from the posterior distribution of τ^2 has been completed, sampling the fixed and individual random effects only requires sampling from normal densities.

C. Extension to Covariates

The above sampling algorithm can be easily extended to accommodate covariates in the model. Let $\gamma' = (\mu, \beta)$ and assume the improper prior specification $\pi(\gamma) \propto c$ for γ. In this case, (4) holds with L_R replaced by

$$L_{R,\gamma}(\tau^2) = (2\pi)^{-(k-p)/2} \left| \sum_{i=1}^{k} (\tau^2 + w_i^{-1})^{-1} x_i x_i' \right|^{-1/2} \prod_{i=1}^{k} (\tau^2 + w_i^{-1})^{-1/2}$$

$$\times \exp\left\{ -\frac{1}{2} \sum_{i=1}^{k} \frac{1}{\tau^2 + w_i^{-1}} (y_i - x_i' \gamma^*(\tau^2))^2 \right\}$$

and

$$\gamma^*(\tau^2) = \left[\sum_{i=1}^{k} (w_i^{-1} + \tau^2)^{-1} x_i x_i' \right]^{-1} \sum_{i=1}^{k} (\tau^2 + w_i^{-1})^{-1} y_i x_i. \tag{10}$$

The proposal density g in (5) may still be used, replacing b by $\sum_{i=1}^{k} (y_i - x_i' \hat{\gamma})^2$, where $\hat{\gamma}$ is the usual least-squares estimator: $\hat{\gamma} = (\sum_{i=1}^{k} x_i x_i')^{-1} \sum_{i=1}^{k} y_i x_i$. Again, we recommend evaluating $L_{R,\gamma} \times \pi/g$ on a grid of τ^2 values to find M. Given τ^2, the random effects θ_i and the regression coefficients γ can be simulated as in (8) and (9) via

$$p(\gamma \mid \tau^2, Y) = N\left(\gamma^*(\tau^2), \left[\sum_{i=1}^{k} (\tau^2 + w_i^{-1})^{-1} x_i x_i'\right]^{-1}\right)$$

with $\gamma^*(\tau^2)$ given by (10), and

$$p(\theta_i \mid \gamma, \tau^2, Y) = N\left(\frac{y_i w_i + x_i' \gamma/\tau^2}{w_i + 1/\tau^2}, \frac{1}{w_i + 1/\tau^2}\right),$$

for $i = 1, \ldots, k$.

III. TESTS FOR HETEROGENEITY

As discussed earlier, the between-study variability parameter τ^2 plays a fundamental role in prediction. Examination of the posterior distribution for τ^2 can give some indication of whether or not τ^2 is close to zero, but such an inspection cannot be easily calibrated. It is useful therefore to report a single-number summary of the evidence of between-study heterogeneity. A formal Bayesian approach compares the models $M_0 : \tau^2 = 0$ and $M_A : \tau^2 > 0$ via the Bayes factor.

The Bayes factor is defined as the ratio of posterior to prior odds of M_0 to M_A:

$$\text{BF} = \frac{P(M_0 \mid Y)/P(M_A \mid Y)}{P(M_0)/P(M_A)} = \frac{P(Y \mid M_0)}{P(Y \mid M_A)} = \frac{L_R(0)}{\int L_R(\tau^2)\pi(\tau^2)d\tau^2}, \quad (11)$$

where the first equality follows by Bayes' theorem and the second by integrating the unknown parameters under each model. An advantage of Bayes factors over frequentist measures such as p-values is that they have a natural interpretation, on the odds scale. Jeffreys(1961) interpreted Bayes factors on the log-base-10 scale, with values less than 0.5 indicating evidence in favor of H_0 not worth more than a bare mention, values between 0.5 and 1, substantial evidence, between 1 and 2 strong evidence, and greater than 2, decisive evidence. Bayes factors have been used in the context of meta-analysis by a number of authors, including S Berry (21) and Abrams et al. (Chap. 2).

The Bayes factor can be calculated directly from draws $\tau^{2(t)}$ $(t = 1, \ldots, T)$, from the posterior distribution. The integral in the denominator of (11), which we shall denote by I, is approximated from these draws via the importance sampling estimator

$$\hat{I} = \frac{1}{T}\sum_{t=1}^{T}\frac{L_R(\tau^{2(t)})\pi(\tau^{2(t)})}{g(\tau^{2(t)})} \quad (12)$$

and has variance $V(\hat{I}) = (1/T)\text{var}\left(L_R(\tau^2)\pi(\tau^2)/g(\tau^2)\right)$, which can be estimated by the sample variance of the terms in (12). The choice of proposal density g is crucial, since there is no guarantee that the variance will be finite. Reference 8 shows that the variance is finite when the samples arise as the result of a rejection algorithm.

IV. PRIOR DISTRIBUTIONS

The choice of prior distribution for τ^2 is important, since it affects the width of interval estimates for μ, the amount of shrinkage that is imposed on study-level estimates θ_i, and the size of the Bayes factor for assessing heterogeneity. In this section, we define a specific proper prior for τ^2 that we recommend as a default choice for use in estimation and for Bayes factors.

There are several problems with obvious or commonly used priors for variance components. The usual prior for scale parameters, $\pi(\tau^2) \propto 1/\tau^2$, leads to an improper posterior distribution (22). As discussed by DuMouchel and Normand (Chap. 6), a flat prior for τ yields a proper posterior but may impose between-study variability. For convenience, inverse gamma densities which are almost flat (*e.g.* $\tau^2 \sim$ IG (a, a) with a small) are often used, but these probability densities have the undesirable property of placing the majority of their probability on large values of τ^2, which may induce heterogeneity in the posterior distribution. In Ref. 23 this prior is discussed in the context of disease mapping, and it is pointed out that, for the IG(0.001, 0.001) distribution, the 1% quantile for the standard deviation τ is 6.4, which would clearly be too large for certain problems.

Choosing a prior for the parameter of interest under the alternative for the Bayes factor, such as for τ^2 in the test for heterogeneity, is nontrivial for several reasons. First, unlike in estimation, Bayes factors are sensitive, even asymptotically, to this choice of prior (24). Second, priors with large variances for the parameter under the alternative lead to Bayes factors which overestimate evidence in favor of the null model. This occurs because the denominator integral I in (11) averages over ever smaller values of the likelihood. Therefore, although careful elicitation of subjective priors based on other data (11) or expert opinion (25) is recommended, it is useful to assess sensitivity across a range of sensible prior distributions. Unfortunately, in practical problems, well-defined prior information is often not available, and experts are reluctant to formulate opinions without reference to a standard method.

Regulatory authorities are also likely to require a default 'reference' choice if a meta-analysis is contained in a submission document for a new compound. Hence there is a need to define a default conventional prior for τ^2.

Our recommended prior is a normal distribution centered at zero and restricted to positive values. If homogeneity of treatment effect is a possibility, then it is sensible for a prior for τ^2 to place substantial mass near zero, a property possessed by our choice (the log-logistic prior of DuMouchel and Normand (Chap. 6) also satisfies this criterion). Under this prior, the Bayes factor is sensitive to the choice of scale, or the variance of the normal density. As this variance increases, the Bayes factor in favor of H_0 will generally increase. Therefore, it is important that the scale be calibrated with the likelihood. A simple way to achieve this is to use as the prior information the information in the likelihood that would have arisen from a single observation. In other words, treat the prior information as arising from a single observation exchangeable with those in the current sample. This is the principle behind the unit-information prior of Kass and Wasserman (1).

In hierarchical models, the concept of a unit of information is ambiguous (26). We define a unit of information as a single observation in one of the k studies; that is, we regard a sample of k studies with precisions w_i, for $i = 1, \ldots, k$, as composed of $\sum_{i=1}^{k} w_i$ observations. Justification for this choice can be found in Ref. 8. Let $I^*(\hat{\tau}^2)$ denote the observed information for τ^2, i.e. the negative second derivative of $\log L_R(\tau^2)$, evaluated at the restricted maximum likelihood estimator, $\hat{\tau}^2$. We specify the normal variance as

$$V = \left(\frac{I^*(\hat{\tau}^2)}{\Sigma_{i=1}^{k} w_i} \right)^{-1} \tag{13}$$

and the prior for τ^2 as

$$\pi(\tau^2) = 2(2\pi V)^{-1/2} \exp\left(-\frac{1}{2V} \tau^4 \right) I(\tau^2 > 0) \tag{14}$$

with V given by (12). We denote this "half-normal" prior by HN(0, V).

We note that the expected information for τ^2 from the full likelihood $L(\mu, \tau^2)$ will often be similar to that based on L_R and so may be substituted in (13). Information from the full likelihood can be calculated analytically. Otherwise, statistical packages such as SAS or Splus return the observed or expected information as part of their standard output. If no heterogeneity is present, the unrestricted maximum-likelihood estimate may be negative. Evaluation of information at this negative value will often be similar to that obtained from evaluating the likelihood at zero and so may also be used in (14). Under prior (14), the final term in the bound for the rejection algorithm (7) is $(2\pi V)^{-1/2}$.

V. EXAMPLES

In order to illustrate our fully Bayesian procedure and to make comparisons with standard approaches, we re-analyze three previous meta-analyses. We emphasize implementation of the algorithm outlined in Sec. II, calculation and interpretation of the Bayes factor in Sec. III, and the use of familiar diagnostics for checking the normality assumption of the study random effects. Table 1 contains the study-level estimates and

Table 1 Treatment Effects and Precisions for Three Examples*

	Example 1		Example 2		Example 3	
Study i	Y_i	w_i	Y_i	w_i	Y_i	w_i
1	−0.86	3.07	−0.40	36.1	0.04	6.27
2	−0.33	3.25	−0.53	8.7	−0.92	8.49
3	−0.47	8.09	−0.59	41.8	−1.11	5.62
4	−0.50	15.88	−1.24	5.9	−1.47	3.35
5	0.28	3.43	−1.44	1.7	−1.39	8.75
6	−0.04	13.21	−0.42	14.8	−0.30	68.34
7	−0.80	1.63	−0.32	18.8	−0.26	8.29
8	−0.19	55.97	1.10	1.0	1.09	1.46
9	−0.49	12.92	0.61	0.7	0.13	14.73
10	—	—	−1.06	5.3	—	—
11	—	—	−0.66	12.4	—	—
12	—	—	−0.42	18.1	—	—
13	—	—	−0.58	13.7	—	—

*Example 1 compares the dentifrices NaF and SMFP with outcome measure $Y_i = \text{NaF}_i - \text{SMFP}_i$, the difference in changes from baseline in the decayed/missing (due to caries)/filled-surface dental index between NaF and SMFP. Examples 2 and 3 compare log odds ratios for treatment versus control (Y_i). In example 2 the outcome is stroke in studies of anti-hypertensive drugs and in example 3, pre-eclampsia in studies considering diuretic use during pregnancy.

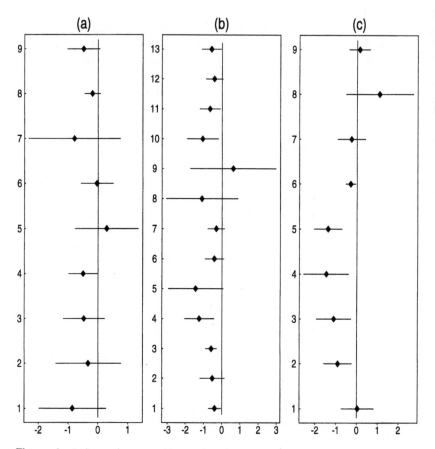

Figure 1 Point estimates and associated 95% interval estimates for (a) denti-frice, (b) anti-hypertensive, and (c) pre-eclampsia examples.

precisions for the three examples to be analyzed below, and Fig. 1 shows these estimates along with associated 95% confidence intervals in a *ladder plot*.

A. Example 1: Dentifrice

Our first example concerns a series of randomized controlled clinical trials comparing sodium monofluorophosphate (SMFP) to sodium fluor-ide (NaF) dentifrice in the prevention of caries development. The data consist of treatment differences, $Y_i = \mathrm{NaF}_i\text{–}\mathrm{SMFP}_i$, where NaF_i is the change from baseline in the decayed/missing (due to caries)/filled-surface dental index at three years follow up for regular use of NaF, and SMFP_i is defined similarly for $i = 1, \ldots, k = 9$. Also presented are the associated weights, w_i, assumed known. These data were first presented in Ref. 27 before being re-analysed in Refs 28 and 11. The ladder plot in Fig. 1(a) shows that all but one of the estimated treatment differences are negative but that all confidence intervals include the value zero.

In Ref. 11, two priors were considered for τ^2. The first is IG(0.5, 0.11) and is constructed from a meta-analysis of three similar trials. The second is a so-called reference prior, IG(0, 0.5). The unrestricted maximum-likelihood estimate is negative and is given by $\hat{\tau}^2 = -0.0078$. We used the observed information based on L_R evaluated at this value for our default prior of Sec. IV. This gave a half-normal prior with variance $V = 0.058$.

We implemented the rejection algorithm described in Sec. III in Splus to obtain posterior samples of τ^2 for the two inverse gamma priors and the default half-normal prior. Based on 100,000 candidate points, each analysis took roughly 2 minutes on a Sun Sparc station, and had an acceptance rate near 0.02, yielding approximately 2,000 samples from the posterior for each prior. Kernel density estimates of the marginal posterior density for τ^2 based on each prior, along with the corresponding prior, are displayed in Fig. 2.

Figure 2 shows the sensitivity of the posterior distribution to the choice of prior. The half-normal prior places more mass near zero, and hence yields a marginal posterior density for τ^2 that is concentrated close to zero. Conversely, the reference inverse gamma prior places little prior mass near zero, which translates the posterior distribution towards larger values of τ^2.

Posterior and predictive summaries, and Bayes factors in favor of homogeneity for the three priors, which are displayed in Table 2, are also sensitive to the choice of prior distribution. The Bayes factor based on the half-normal prior indicates substantial evidence in favor of the null hypothesis, and those based on the inverse-gamma priors suggest even stronger evidence. With only nine studies in our sample, it appears that the Bayes factor based on the IG(0, 0.5) reference prior overestimates evidence in favor of homogeneity. Posterior medians for μ are constant across the range of analyses, but interval estimates are very sensitive to the choice of prior distribution. The width of the predictive intervals for a new study treatment effect display a five-fold increase in variability

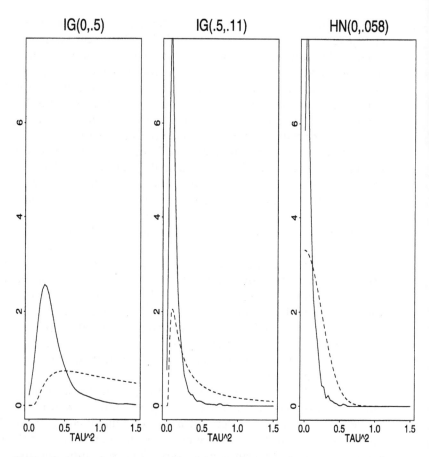

Figure 2   Kernel density estimates of marginal posterior densities for τ^2 (solid line) and associated prior densities (broken line) under three priors.

between the analysis based on the $IG(0, 0.5)$ prior and the fixed-effect analyses. Among the heterogeneity analyses, there is a two-fold range of variability, highlighting why it is sensible to report predictive intervals for a range of plausible prior distributions for τ^2.

Diagnostics determining the adequacy of modeling assumptions are an essential part of any analysis. Here we assess the second-stage normality assumption. Since the θ_is are unobserved, it is not straightforward to interpret plots of normal scores (29–30). Figure 3 shows the ordered

Table 2 Posterior Medians and 95% Equal-tail Credible Intervals for μ and τ^2, Bayes factors on the Log-base-10 Scale in Favor of Homogeneity with Standard Errors, and 95% Predictive Intervals for a New Study for Analyses Based on a Range of Priors for τ^2 and a Fixed Analysis ($\tau^2 = 0$) in Example 1.

Analysis	Posterior median (95% interval)		$\log_{10} BF$ (s.e.)	95% Predictive interval
	μ	τ^2		
IG(0, 0.5) prior	−0.35 (−0.83, 0.12)	0.28 (0.10, 1.12)	1.86 (0.02)	(−1.62, 0.87)
IG(0.5, 0.11) prior	−0.33 (−0.66, −0.01)	0.08 (0.03, 0.39)	1.10 (0.03)	(−1.11, 0.35)
HN(0, 0.058) prior	−0.33 (−0.63, −0.04)	0.05 (0.00, 0.28)	0.60 (0.03)	(−0.98, 0.32)
Fixed effects	−0.28 (−0.46, −0.10)	—	—	(−0.46, 0.10)

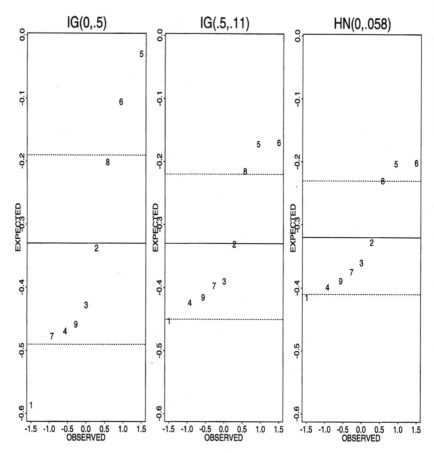

Figure 3 Normal scores plots for E $[\theta_i|y]$ ($i = 1, \ldots, 9$), for dentifrice example for three priors. The plotting symbol represents the study identifier. The solid line on each plot represents the posterior mean of μ, and the dotted lines a 50% posterior credible interval.

posterior expectations E$[\theta_i \mid y]$ ($i = 1, \ldots, 9$), for the three priors. In line with Table 2, there is greatest shrinkage with the half-normal prior. The posterior median and a 50% interval for $\mu \mid y$ are also displayed on the plot. As noted for Table 2, the location for all three priors is virtually identical, while the widths of the intervals are sensitive to the choice of prior. Note that, in all three plots, the estimated effects in studies 2 and 8 remain approximately equal to their observed values in Table 1. The reason for this behavior in study 2 is that the observed effect is close to the population mean, resulting in little opportunity for shrinkage, while in study 8 the precision is high. Although interpretation is difficult due to the small number of studies, visual inspection of Fig. 3 indicates that the normality assumption seems dubious, with a lighter-tailed alternative appearing more appropriate.

The plots in Fig. 3 do not convey the variability of estimates of individual random effects, information that is contained in samples from the posterior distribution. An alternative graphic that has been used by Wakefield (31) to reflect this variability consists of normal-score plots for *draws* from the posterior distribution $p(\theta_i \mid y)$, rather than the single plot using posterior means in Fig. 3. Sixteen such draws are shown in Fig. 4; the θ_i have been standardized by $E[\mu \mid y]$ and $E[\tau \mid y]$. The plots in which the points are almost horizontal correspond to draws in which τ^2 was close to zero. We see that, as before, a lighter-tailed alternative is suggested, although this conclusion is not as clear-cut as in Fig. 3. The ordering of the treatment effects across the plots indicates the width of the posterior $p(\theta_i \mid y)$. For example, study 1 has a small sample size and is ranked between 1 and 9 across the draws. Study 5, which produces the only positive observed treatment difference, is ranked highest most frequently.

B. Example 2: Anti-hypertensive

The next example considers a meta-analysis of 13 randomized trials of anti-hypertensive drugs presented in Ref. 32 and later re-analyzed in Ref. 2. The data are summarized in Table 1 and in Fig. 1(b) in terms of the log odds ratios of strokes for treatment versus controls. A negative Y_i indicates that fewer strokes occurred on treatment.

A subjective prior is not available for this meta-analysis, so we use the half-normal default prior (14). Based on the restricted maximum-likelihood estimate, $\hat{\tau}^2 = 0$, calculations similar to those in Example 1 yield a normal variance 0.132. The rejection algorithm for this example had an acceptance rate of only 0.0006. The density estimate in Fig. 5, which compares the prior and posterior distribution of τ^2, is based on 312 sample points, which still took under two minutes to generate. The posterior median of τ^2 is 0.029 with a 95% equal-tail credible interval given by (0.00, 0.20), which is concentrated closer to zero than the interval under the half-normal prior in Example 1 (see Table 2). Based on this

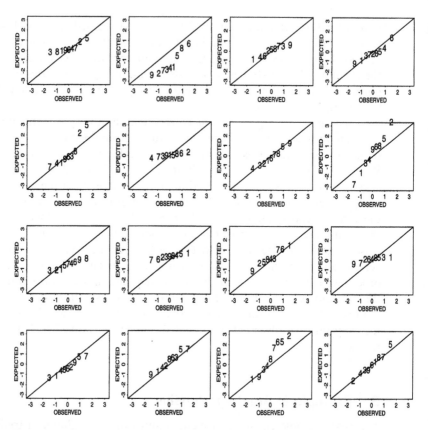

Figure 4   Normal scores plots for 16 random draws from the posterior distribution $p(\theta_i|y)$ $(i = 1, \ldots, 9)$ for the dentifrice example. The plotting symbol represents the study identifier.

prior, the posterior median and 95% credible interval for the overall treatment effect μ is given by -0.56 and $(-0.76, -0.38)$, respectively, indicating that fewer strokes occurred on treatment. Fig. 6 shows a normal scores plot of the posterior means of the θ_i. In this example, normality appears to be a reasonable assumption.

The Bayes factor in favor of homogeneity on the $\log_{10}$ scale was estimated to be 1.09 ± 0.20, which indicates strong evidence in this direction. In Ref. 2, the frequentist test for heterogeneity also produced insufficient evidence to reject the null hypothesis of homogeneity of treatment

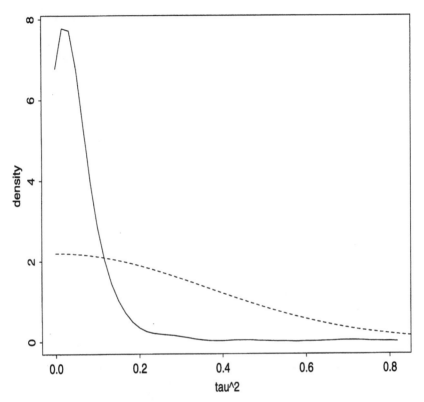

Figure 5 Kernel density estimates of marginal posterior density of τ^2 (solid line) and half-normal prior density (broken line) for anti-hypertensive example.

effect. Hence we also report posterior summaries for μ based on a fixed analysis for comparison. The posterior median for μ is -0.53, approximately the same as before, and the 95% credible interval is $(-0.68, -0.39)$, slightly more narrow than in the full model.

One method that has been advocated for detecting biases that may be present in small studies, including publication bias, is the funnel plot, in which the estimated study effects are plotted against precision; see DuMouchel and Normand (Chap. 6), Larose (Chap. 8), and Sutton et al. (Chap. 15). More formally, one may regress treatment differences against some function of study size to determine whether study size is correlated with outcome. For example, the original study estimates are

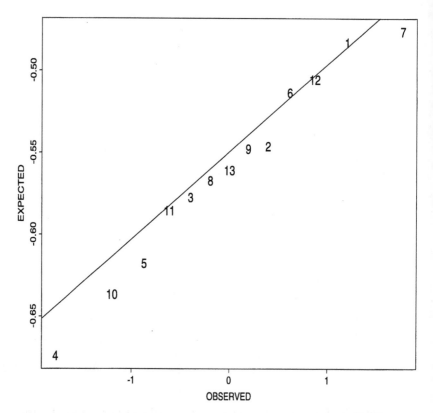

Figure 6 Normal scores plot for anti-hypertensive example. The plotting symbol represents the study identifier.

regressed on precision in Ref. 33 and on the log of the sample size in Ref. 13. Here, for illustration, we fit model (1) and (3), taking $x_i = w_i^{-1/2}$ to assess whether standard errors of the study-level treatment effects are correlated with reported outcome. For this model, the normal variance based on the restricted maximum-likelihood estimator is $V = 0.101$, which is slightly less than in the covariate-free model. This makes intuitive sense, since we would expect less between-study variability when a covariate is included in the model. A 95% equal-tail credible interval for β, the regression coefficient for this covariate, is $(-2.04, 0.75)$, indicating that smaller studies are slightly more likely to report outcomes in favor of the treatment. This can be seen in Fig. 1(b), where with the exception of study 9, studies with larger standard errors report negative estimates. The Bayes factor in favor of $\tau^2 = 0$ based on the covariate model is estimated as 2.40 ± 0.21, which indicates stronger evidence in favor of homogeneity, because the covariate explains some of the between-study variability.

C. Example 3: Pre-eclampsia

The final example is a meta-analysis of nine clinical trials investigating the risk of pre-eclampsia as a result of taking diuretics during pregnancy (34, 10). The data are summarized in Table 1 and displayed in Fig. 1(c) in terms of the log odds ratios of pre-eclampsia for treatment versus control. In three of the studies, there is a higher observed risk of pre-eclampsia for subjects in the treatment group.

A subjective prior is not available for this example, so we resort to the half-normal default prior given by (14); calculations yield a normal variance 7.28. The rejection algorithm for this example was very efficient, with an acceptance rate of 0.5. Figure 7 displays the half-normal prior and the marginal posterior density estimate for τ^2 based on this prior. The posterior mode lies away from zero, suggesting between-study variability. The prior density is relatively flat near the informative portion of the likelihood, alleviating any fears that a prior with mode at zero might be inappropriate when heterogeneity is present. The unit-information normal variance (13) appropriately adjusts the prior to be substantially less informative than the likelihood. The Bayes factor in favor of the null hypothesis of homogeneity on the $\log_{10}$ scale was estimated to be -1.09 ± 0.01, which corresponds to strong evidence in favor of the heterogeneity model by an inversion of Jeffreys' rule. Reference 10 reports that a likelihood ratio test for heterogeneity produces a highly significant result ($p = 0.006$). The median value of τ^2 is 0.52 with 95% credible interval (0.09, 2.68). The profile likelihood method used gave a point estimate of 0.24 with a 95% confidence interval of (0.03, 1.19), suggesting smaller values of the between-study variability.

In terms of substantive conclusions, a sampling-based approach allows inference for a range of parameters of interest. Here, for illustration and to allow comparison with Ref. 10, we consider the posterior distribution of $\phi = \exp(\mu)$, which corresponds to the *population* odds ratio comparing the odds of pre-eclampsia in the treatment group versus the odds in the control group. The parameter ϕ is the median of the distribution of odds ratios across studies. Figure 8 shows a histogram estimate of the posterior distribution of ϕ. The profile likelihood method

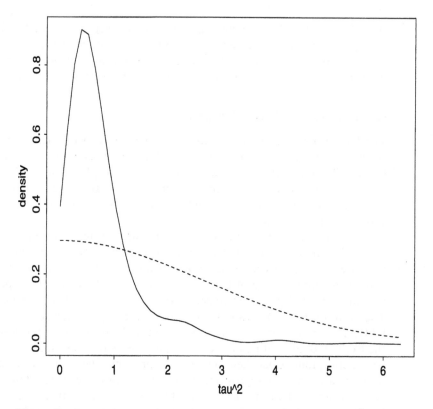

Figure 7 Kernel density estimates of marginal posterior density of τ^2 (solid line) and half-normal prior density (broken line) for pre-eclampsia example.

gave a point estimate of the population odds ratio as 0.60 with 95% confidence interval (0.37, 0.95). The Bayesian analysis gives a posterior median of 0.60 with a 95% credible interval (0.32, 1.08), which is wider than the likelihood interval (which only provides the nominal coverage asymptotically). Based on our posterior samples, $\Pr(\phi > 1 \mid Y) = 0.05$, showing strong evidence of an increased risk of pre-eclampsia in the control group. The flexibility to examine the treatment effect on meaningful scales helps to identify whether the treatment is *clinically* superior.

Informal visual examination of a normal-score plot for the random effects in this example (not shown) did not suggest that the normality assumption was inappropriate, although interpretation was again difficult with just nine studies.

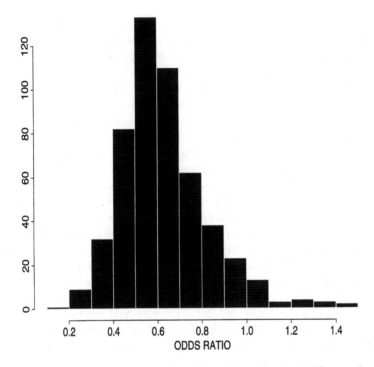

Figure 8 Posterior distribution of $\exp(\mu)$ under the half-normal prior, pre-eclampsia example.

VI. DISCUSSION

We have outlined an algorithm for performing a complete Bayesian meta-analysis, including estimation of fixed effects, evaluation of between-study heterogeneity, and assessment of goodness of fit. The rejection algorithm that we proposed for fitting our Bayesian hierarchical model can be easily programmed in statistical packages such as Splus or SAS, and because it generates independent samples, runs quickly under this software. We have discussed the delicate issue of sensitivity to choice of prior for the between-study variance component, and offered a default choice for qualitative assessment. Although our half-normal prior is motivated by its usefulness in evaluating homogeneity of treatment effect, example 3 illustrates that it is also suitable for estimation purposes when heterogeneity is present in the sample. This occurs because the prior is automatically adjusted to be less informative than the likelihood over the entire parameter space. Finally, we have stressed that an essential part of any meta-analysis is an assessment of the appropriateness of distributional assumptions and demonstrated simple diagnostics for normality.

All three of our studies had relatively small sample sizes, which is typical for meta-analyses. Larger numbers of studies and study-level covariates require more complicated models and diagnostics. If covariates are present then sample summaries of random effects such as quantiles may be plotted against available study-level covariates; such a technique is used in Ref. 35 for population pharmacokinetic models. In Ref. 7, the study-level raw data and their confidence intervals are plotted in this way. If normality at the second stage is found to be dubious, a more flexible mixture distribution may be necessary. In a parametric Bayesian framework this mixture may consist of a known number of elements (36) or an unknown number (37). Reference 38 describes a non-parametric maximum-likelihood approach. A Bayesian non-parametric approach based on mixtures of Dirichlet processes (39) provides greater flexibility and incorporation of uncertainty at the expense of increased computational overhead. For meta-analyses requiring complicated models such as these, efficient sampling algorithms such as those outlined in this chapter become increasingly valuable.

References

1. R Kass, L Wasserman. A reference Bayesian test for nested hypotheses and its relationship to the Schwarz criterion. Journal of the American Statistical Association 90:928–934, 1995.
2. A Whitehead, J Whitehead. A general parametric approach to the meta-analysis of randomised clinical trials. Statistics in Medicine 10:1665–1677, 1991.
3. D Spiegelhalter, L Freedman, M Parmar. Bayesian approaches to randomised clinical trials (with discussion). Journal of the Royal Statistical Society, Series A 157:357–415, 1994.
4. D Smith, G Givens, R Tweedie. Adjustment for publication bias and quality bias in Bayesian meta-analysis. In: D Stangl, DA Berry, eds. Meta-analysis in Medicine and Health Policy. New York: Marcel Dekker, 2000.
5. W DuMouchel, SL Normand. Computer-modeling and graphical strategies for meta-analysis. In: D Stangl, DA Berry, eds. Meta-analysis in Medicine and Health Policy. New York: Marcel Dekker, 2000.

6. S Thompson, S Pocock. Can meta-analyses be trusted? Lancet 338:1127–1130, 1991.

7. S Thompson. Why sources of heterogeneity in meta-analysis should be investigated. British Medical Journal 309:1351–1355, 1994.

8. D Pauler, J Wakefield, R Kass. Bayes factors for variance components models. Journal of the American Statistical Association, to appear.

9. R DerSimonian, N Laird. Meta-analysis in clinical trials. Controlled Clinical Trials 7:177–188, 1986.

10. R Hardy, S Thompson. A likelihood approach to meta-analysis with random effects. Statistics in Medicine 15:619–629, 1996.

11. K Abrams, B Sansó. Approximate Bayesian inference for random effects meta-analysis. Statistics in Medicine 17:201–218, 1998.

12. T Smith, D Spiegelhalter, M Parmar. Bayesian meta-analysis of randomized trials using graphical models and BUGS. In: D Berry, D Stangl, eds. Bayesian Biostatistics. New York: Marcel Dekker, 1996.

13. T Smith, D Spiegelhalter, A Thomas. Bayesian approaches to random-effects meta-analysis: a comparative study. Statistics in Medicine 14:2685–2699, 1995.

14. S Thompson, T Smith, S Sharp. Investigating underlying risk as a source of heterogeneity in meta-analysis. Statistics in Medicine 16:2741–2758, 1998.

15. D Spiegelhalter, A Thomas, N Best. BUGS: Bayesian inference using Gibbs Sampling. Medical Research Council Biostatistics Unit, 1996.

16. W Gilks, C Robert. Strategies for improving MCMC. In: W Gilks, S Richardson, D Spiegelhalter, eds. Markov Chain Monte Carlo in Practice. New York: Chapman and Hall, 1996, pp 89–114.

17. T DiCiccio, R Kass, A Raftery, L Wasserman. Computing Bayes factors by combining simulation and asymptotic approximations. Journal of the American Statistical Association 92:903–915, 1997.

18. J Hodges. Some algebra and geometry for hierarchical models, applied to diagnostics (with discussion). Joural of the Royal Statistical Society, Series B 60:497–536, 1998.

19. B Ripley. Stochastic Simulation. New York: Wiley, 1987.

20. H Jeffreys. Theory of Probability. 3rd ed. Oxford: Oxford University Press, 1961.

21. S Berry. Understanding and testing for homogeneity across 2×2 tables: application to meta-analysis. Statistics in Medicine 17:2353–2369, 1998.

22. C Morris, S-L Normand. Hierarchical models for combining information and for meta-analysis. In: J Bernardo, J Berger, D Lindley, S Smith, eds. Bayesian Statistics 4. Oxford: Oxford University Press, 1992.

23. J Kelsall, J Wakefield. Discussion of Bayesian models for spatially correlated disease and exposure data by N Best, L Waller, A Thomas, E Conlon, R Arnold. In: J Bernardo, J Berger, A Dawid, A Smith, eds.

Proceedings of the Sixth Valencia Meeting on Bayesian Statistics. New York: Wiley, 1999.

24. R Kass, SK Vaidyanathan. Approximate Bayes factors and orthogonal parameters, with application to testing equality of two binomial proportions. Journal of the Royal Statistical Society, Series B 54:129–144, 1992.

25. K Chaloner. Elicitation of prior distributions. In: D Berry, D Stangl, eds. Bayesian Biostatistics. New York: Marcel Dekker, 1996, pp 141–156.

26. DK Pauler. The Schwarz criterion and related methods for normal linear models. Biometrika 85:13–27, 1998.

27. M Johnson. Comparative efficacy of NaF and SMFP dentifrices in caries prevention: a meta-analysis overview. Caries Research 27:328–336, 1993.

28. N Silliman. Hierarchical selection models with applications in meta-analysis. Journal of the American Statistical Association 92:926–936, 1997.

29. A Dempster, L Ryan. Weighted normal plots. Journal of the American Statistical Association 80:845–850, 1985.

30. N Lange, L Ryan. Assessing normality in random effects models. Annals of Statistics 17:624–642, 1989.

31. J Wakefield. Discussion of: Some algebra and geometry for hierarchical models, applied to diagnostics, by JS Hodges. Journal of the Royal Statistical Society, Series B 60:497–536, 1998.

32. R Collins, R Peto, S MacMahon, P Herbert, N Fiebach, K Eberlein, J Godwin, N Qizilbash, J Taylor, C Hennekens. Blood pressure, stroke, and coronary heart disease. Part 2, short-term reductions in blood pressure: overview of randomized drug trials in their epidemiological context. Lancet 335:827–838, 1990.

33. M Egger, G Davey Smith, M Schneider, C Minder. Bias in meta-analysis detected by a simple graphical test. British Medical Journal 315:629–634, 1997.

34 R Collins, S Yusuf, R Peto. Overview of randomized clinical trials of diuretics in pregnancy. British Medical Journal 290:17–23, 1985.

35. J Wakefield. The Bayesian analysis of population pharmacokinetic models. Journal of the American Statistical Association 91:62–75, 1996.

36. C Robert. Mixtures of distributions: inference and estimation. In: W Gilks, S Richardson, D Spiegelhalter, eds. Markov Chain Monte Carlo in Practice. New York: Chapman and Hall, 1996.

37. S Richardson, P Green. On Bayesian analysis of mixtures with an unknown number of components. Journal of the Royal Statistical Society, Series B 59:731–792, 1997.

38. M Aitken. A general maximum likelihood analysis of overdispersion in generalised linear models. Statistics in Computing 6:251–262, 1996.

39. M Escobar, M West. Bayesian density estimation and inference using mixtures. Journal of the American Statistical Association 90:577–588, 1995.

10

Meta-analysis of Population Pharmacokinetic Data

Nargis Rahman
Imperial College of Science, Technology, and Medicine, London, England

Jon Wakefield
University of Washington, Seattle, Washington

Summary

Pharmacokinetic data are collected following the administration of known dosage regimens and consist of drug concentrations with associated sampling times. Population pharmacokinetic data contain such information on a number of individuals, possibly along with individual-specific characteristics. During drug development, a number of population pharmacokinetic studies are typically carried out, and combining such studies is of great importance, in particular for the design of future studies. In this chapter, we will describe a model which we have recently developed (Wakefield and Rahman, 2000) to carry out population pharmacokinetic meta-analysis, and illustrate its application using data from four phase I studies. The data from a set of population pharmacokinetic studies may be described by a four-stage hierarchical model with a nonlinear first stage for within-individual variability, a second stage for between-individual variability *within* a particular study, a third stage for between-study variability, and a fourth stage containing prior distributions. We can also describe the data by a model that does not acknowledge between-study differences. We will fit these two models to our data and compare conclusions. Often the most relevant and informative summary will be the predictive distribution. Such a distribution may be obtained for a new individual for key pharmacokinetic parameters such as volume or clearance, or for concentrations following the administration of a known dosage regimen. The last of these is particularly important for design. We emphasize the assessment of model adequacy, since it is an area that is infrequently addressed in hierarchical modeling, and the Bayesian approach is ideally suited to the consideration of informative diagnostics.

I. INTRODUCTION

Pharmacokinetics considers the absorption, distribution, and elimination over time of a drug and its metabolites. Pharmacokinetic data consist of drug concentrations along with known sampling times and dosage regimens. A dosage regimen is defined by the route of administration and the sizes and timings of the doses. Population data arise when pharmacokinetic data are measured on a group of individuals, along with subject-specific covariates such as age, sex, or weight. Pharmacokinetic parameters of interest include the terminal half-life which is the time for the amount of drug to be halved in the "terminal phase," and the apparent volume of distribution which relates the concentration to the total amount of absorbed drug. This is not a volume in the usual sense of the word, and it can be greater than body volume. The volume of blood (or plasma) which is cleared of drug per unit time is known as the clearance and is also a parameter of fundamental interest. The rate of elimination of the drug at any time is given by clearance multiplied by concentration. These parameters are useful for characterizing the drug and have traditionally been used to choose the dose and dosing interval that should be given to subjects in later phases of clinical development.

Population pharmacokinetic data can be modeled using a hierarchical structure; see, for example, Wakefield et al. (1998), Steimer et al. (1994), and Yuh et al. (1994). Pharmacokinetic models are nonlinear in the parameters. This makes inference difficult from both a classical and a Bayesian perspective. A variety of methods have been proposed; see Racine-Poon and Wakefield (1998) for a review of estimation methods for population pharmacokinetic models and Davidian and Giltinan (1995) for an analysis of nonlinear repeated-measures data.

The structure of this chapter is as follows. In Sec. II, we introduce the dataset to be used to illustrate the meta-analysis. In Sec. III, we introduce a general four-stage hierarchical model which may be used to combine population pharmacokinetic data. In Sec. IV, we analyze the data introduced in Sec. II, and Sec. V contains a concluding discussion. Our approach throughout is Bayesian. The application of Bayesian methods is described in Gelman et al. (1995), and a collection of case studies using Bayesian methods is provided by Berry and Stangl (1996). An Appendix contains details of the Markov chain Monte Carlo approach that we utilize.

II. THE DATASET

The datasets that we will use to demonstrate the meta-analysis came from a phase I pharmacokinetic study of the drug metropolol. The drug is a beta-blocker used for the treatment of hypertension. The data are measurements of metropolol concentrations in the blood plasma. There are four studies, each containing six subjects. In the first study, each subject received a single dose of approximately 43 μmol of metropolol by intravenous injection. In studies 2, 3, and 4, each subject received approximately 86 μmol of metropolol by intravenous injection on two occasions. The goal of each of the studies was to understand the pharmacokinetics of the oral formulation, which was either co-administered or administered in a different occasion. A carbon-labeled formulation was used in studies 2, 3, and 4, and each subject also received an oral tablet of approximately 120 mg at each dosing occasion. For illustration, we only consider the intravenous data. In all of the studies, plasma metropolol concentrations were determined via analysis of blood samples collected at times ranging between 0 and 31 hours post dose. Figure 1 shows plots of the log concentration against time for each subject and each study. A summary of the studies is contained in Table 1.

Pharmacokinetic data are modeled by viewing the body as divided into nominal compartments, and then modeling the flow between compartments and the outside environment (Gibaldi and Perrier, 1982). The models are known as *compartmental models*, and they often result in a model for drug concentration that consists of a sum of exponentials. An initial impression of the number of exponential terms may be obtained by plotting log concentrations versus time; each exponential term will correspond to a linear portion. From Fig. 1, there is a suggestion that the

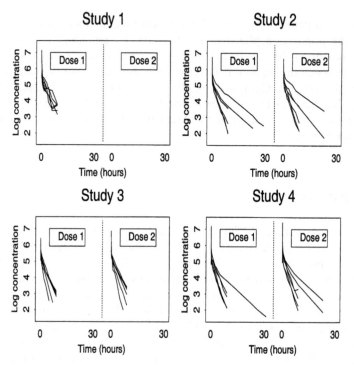

Study 1

Figure 1 Plots of log concentration versus time for each of six subjects in four studies.

data for all subjects in all studies can be characterized adequately by a two-compartment model, that is, by a model consisting of a sum of two negative exponentials. This is confirmed by analyses carried out for this drug by Lunn and Aarons (1997). Figure 2 shows a schema for a two-compartment model. The predicted concentration at time t is given by

Table 1 Design of Four Phase I Studies

Study no.	No. of subjects	Range of samples per subject	Dose size (μmol)
1	6	11	43
2	6	11–14	86/86
3	6	9–11	86/86
4	6	11–14	86/86

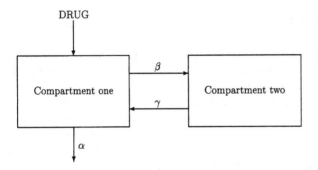

Figure 2 Graphical representation of a two-compartment model.

$$f(\theta, t) = \frac{D}{V_1}\left\{Ae^{-\lambda_1 t} + (1 - A)e^{-\lambda_2 t}\right\},$$

where

$$\lambda_1 = \tfrac{1}{2}\left\{\alpha + \beta + \gamma + [(\alpha + \beta + \gamma)^2 - 4\alpha\gamma]^{1/2}\right\},$$

$$\lambda_2 = (\alpha + \beta + \gamma) - \lambda_1, \qquad A = (\lambda_1 - \gamma)/(\lambda_1 - \lambda_2),$$

$$\alpha = \exp(\log \text{Cl} - \log V_1), \qquad \beta = \exp(\log \text{Cld} - \log V_1),$$

$$\gamma = \exp(\log \text{Cld} - \log V_2), \qquad \text{Cld} = \beta V_1 = \gamma V_2,$$

and $\theta = (\log \text{Cl}, \log \text{Cld}, \log V_1, \log V_2)$. V_1 is the volume of compartment 1 (litres, l) and V_2 is the volume of compartment 2 (l). The clearance ($l\,hr^{-1}$) associated with drug transfer between compartments 1 and 2 is denoted by Cld, while Cl is the clearance ($l\,hr^{-1}$) associated with the elimination of the drug from compartment one to the outside environment. The first-order disposition-rate constants are α, β, and γ, where α is associated with the elimination of the drug from the body, and β and γ are associated with drug transfer between compartments 1 and 2 (see Fig. 2).

III. THE MODEL

In this section, we will describe the four-stage hierarchical model which can be used to combine population pharmacokinetic studies (Wakefield and Rahman, 2000). Let y_{kij} denote the jth observed concentration for the ith subject in study k, collected at times t_{kij}, where $j = 1, \ldots, n_{ki}$, $i = 1, \ldots, n_k$, and $k = 1, \ldots, K$. We denote the pharmacokinetic parameters of individual i in study k by $\boldsymbol{\theta}_{ki}$, which is of dimension p, and the pharmacokinetic model predicting the concentration at time t_{kij} by $f_{ki}(\boldsymbol{\theta}_{ki}, t_{kij})$. The subscript on the function f indicates that different individuals may have different dosage regimens. A four-stage model to describe data of these kind is then given as follows.

A. Model for the Within-individual Variability of Individual i in Study k

The data $\log y_{kij}$ are distributed as $N\{\log f_{ki}(\boldsymbol{\theta}_{ki}, t_{kij}), \sigma_k^2\}$. We have assumed that the within-individual error is lognormal, which implies approximately a constant coefficient of variation. This model often reflects the assay precision. It is important to check the appropriateness of the pharmacokinetic model and the error distribution, because assay precision is not the only source of variability. When patients, rather than volunteers, are being studied, the environment in which the study is being carried out is not so controlled, and so there may be other substantial sources of within-subject variation.

We assume a different within-subject variance, σ_k^2, for each study. This allows for the possibility that, for example, the assay changed over time. The variances may decrease in size over time if the assay technique improved considerably over time and the assay precision makes a significant contribution to the within-subject error term.

B. Model for the Between-individual Variability in Study k

The $\boldsymbol{\theta}_{ki}$ are distributed as $N(\boldsymbol{\mu}_k, \boldsymbol{\Sigma})$, so the population-mean parameter for study k is denoted by $\boldsymbol{\mu}_k$, and $\boldsymbol{\Sigma}$ is the variability of the pharmacokinetic parameters in study k. We are assuming that the variability between subjects is constant across studies; hence $\boldsymbol{\Sigma}$ does not depend on k. We could allow a different $\boldsymbol{\Sigma}$ for each study, and then assume that these arose from a common distribution. However, the implementation of this model would not be straightforward, and unless there were a large number of exchangeable studies, then the prior for the parameters of the common distribution of the Σ_k will drive the analysis.

We could also assume that the σ_k^2 arose from a common distribution, although we would not do this if we expected a time effect, since this would invalidate the exchangeability assumption. With a common distribution, we would obtain the predictive concentration for a new subject using a common within-subject variability distribution. We could also impose the constraint $\sigma_1^2 > \ldots > \sigma_K^2$ if we believed that the assay was improving. We assume the prior distributions for the precisions σ_k^{-2} to be gamma distributions denoted by $\mathrm{Ga}(\frac{1}{2}\nu_0, \frac{1}{2}\nu_0\tau_0)$, for $k = 1, \ldots, K$.

C. Model for the Between-study Variability in Population Means

We assume that the $\boldsymbol{\mu}_k$ are distributed as $\mathrm{N}(\boldsymbol{\eta}, \boldsymbol{\Omega})$. The parameter $\boldsymbol{\eta}$ is the average (across-studies) population mean, and $\boldsymbol{\Omega}$ describes the variability in the population mean across studies. If $\boldsymbol{\Omega}$ is identically zero, then we obtain a model with no between-study differences, which corresponds to a fixed-effect homogeneity model. Keeping the K studies distinct corresponds to a fixed-effect analysis with K different unrelated population means. This is the common approach within the pharmaceutical industry, where data from a series of population pharmacokinetic trials are not formally combined. The disadvantage of this model is that we cannot carry out a prediction for a new study, since there is no common distribution of population means. The prior distribution for $\boldsymbol{\Sigma}^{-1}$ is assumed to be the inverse Wishart distribution $W\{(\rho_1 \boldsymbol{R})^{-1}, \rho_1\}$. Here ρ_1 represents the "strength of belief" in the prior estimate, $\boldsymbol{R}$, of the variance–covariance matrix $\boldsymbol{\Sigma}$, see Wakefield et al. (1994). We choose $\rho_1 \geq p$ to ensure a proper posterior distribution.

D. Prior Distributions

Here we specify prior distributions for $\boldsymbol{\eta}$ and $\boldsymbol{\Omega}$. The prior for $\boldsymbol{\eta}$ is taken to be normal: $\mathrm{N}(\boldsymbol{c}, \boldsymbol{C})$. The prior distributions for $\boldsymbol{\Omega}^{-1}$ is taken to be the inverse Wishart distribution $W\{(\rho_2 \boldsymbol{S})^{-1}, \rho_2\}$. The interpretation of ρ_2 and $\boldsymbol{S}$ is as with ρ_1 and $\boldsymbol{R}$ in stage 3.

We can therefore summarize the model as:

Stage 1: $\log y_{kij} \sim N\{\log f_{ki}(\theta_{ki}, t_{kij}), \sigma_k^2\}$

$(k = 1, \ldots, K; i = 1, \ldots, n_k; j = 1, \ldots, n_{ki}).$

Stage 2: $\theta_{ki} \sim N(\mu_k, \Sigma)$ $(k = 1, \ldots K; i = 1, \ldots, n_k),$

$\sigma_k^{-2} \sim Ga(\tfrac{1}{2} v_0, \tfrac{1}{2} v_0 \tau_0)$ $(k = 1, \ldots, K).$

Stage 3: $\mu_k \sim N(\eta, \Omega)$ $(k = 1, \ldots, K),$

$\Sigma^{-1} \sim W\{(\rho_1 R)^{-1}, \rho_1\}.$

Stage 4: $\eta \sim N(c, C),$ $\Omega^{-1} \sim W\{(\rho_2 S)^{-1}, \rho_2\}.$

This model has one more stage than the traditional meta-analysis hierarchical model, with the additional stage accounting for between-individual differences within studies. This model is most appropriate for combining phase I studies, since these studies are well controlled and are generally in healthy volunteers. The between-study exchangeability assumption is therefore more likely to be appropriate.

IV. ANALYSIS OF DATA

The predicted concentrations for the ith subject in study k at time t_{kij} is given by

$$f_{ki}(\theta_{ki}, t_{kij}) = \frac{D_{ki}}{V_{1ki}} \left\{ A_{ki} e^{-\lambda_{1ki} t_{kij}} + (1 - A_{ki}) e^{-\lambda_{2ki} t_{kij}} \right\},$$

where $\theta_{ki} = (\log Cl_{ki}, \log Cld_{ki}, \log V_{1ki}, \log V_{2ki})$. Here, the dimensionality of the pharmacokinetic parameters is $p = 4$. In studies 2, 3, and 4, each subject received two doses, but none of the drug from the first occasion is left when the second dose is administered. We assume the same pharmacokinetic parameters for both dosing occasions; that is, we do not allow between-occasion variability as studied by Lunn and Aarons (1997). In truth, we believe that the pharmacokinetic parameters are not constant, and that between-occasion variability occurs in many trials which take multiple measurements over long periods of time (particularly in phase III of drug development). However, in general, between-occasion variability is ignored. For example, consider a trial in which one patient provides 10 drug concentration measures over a long period (perhaps receiving multiple doses), and another just provides one measurement. We would assume that each patient had their own set of pharmacokinetic parameters, and we would not model between-occasion variability for the subject with 10 measurements.

We can write the joint posterior distribution of all the unknown parameters in terms of conditional distributions (see Appendix for full details), and then obtain samples from the posterior distribution using Markov chain Monte Carlo techniques implemented in the software BUGS (Spiegelhalter et al., 1996). We carried out an analysis from the model described in Section 3 (the exchangeability analysis) and an analysis in which all of the individuals were assumed to arise from a common population (the homogeneity analysis). The latter analysis does not allow for between-study differences. The meta-analysis was carried out sequentially, so that in the first analysis only the first two studies were included; then the third study was added and finally the fourth. The mean for the normal prior for η is $c = (0, \dots, 0)$ and the variance–covariance is $C = 100I_4$, where I_p represents a $p \times p$ identity matrix. For the within-subject precisions, σ_k^{-2} $(k = 1, \dots, K)$ we set $\nu_0 = \tau_0 = 0$ in the gamma prior. The priors for Σ and Ω were specified via $R = S = 0.04I_4$ which corresponds to a 20% coefficient of variation for the individual and study-level parameters. We set $\rho_1 = \rho_2 = p = 4$, to have the "flattest" proper prior. For the homogeneity analysis we took the priors to be $\mu \sim N(c, C)$, and $\Sigma^{-1} \sim W\{(\rho R)^{-1}, \rho\}$, with $\rho = 4$ and $\sigma^{-2} \sim Ga(\frac{1}{2}\nu_0, \frac{1}{2}\nu_0\tau_0)$.

In general, for this two-compartment model, the pharmacokinetic parameters are given by $V = Cl/\lambda_2$ and $t_{1/2} = (\log 2)/\lambda_2$, where V is the apparent volume of distribution, Cl is the clearance, and $t_{1/2}$ is the terminal half life—that is, corresponding to the smaller exponential term λ_2 which dominates as time increases. We can directly obtain the population level mean (for within a particular study or across the studies) for $\log Cl$, since $\log Cl$ is simply given by θ_1 under our parametrization; hence the between-study distribution for $\log Cl$ is given by $N(\eta_1, \Omega_{11})$. We cannot obtain directly the population-level distribution for the volume and half-life parameters, since we cannot express the log of these quantities as simple linear combinations of θ. But what we can do is generate samples of the quantity of interest, and then construct the empirical distribution (see Wakefield and Rahman, 2000, for details).

Figure 3 shows predictions for the clearance parameter (Cl); the 90% intervals are shown along with the median indicated by the middle symbol. Each number represents a prediction for a typical individual from the study denoted by that number, from the exchangeability ana-

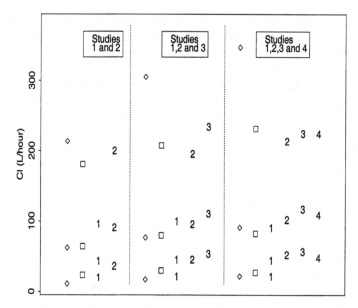

Figure 3   Predictions for a new individual for the clearance parameter. Each set of three points represent the 5%, 50%, and 95% quantiles of the predictive distribution. The diamonds represent predictions for a new individual from the exchangeability analysis, and the squares from the homogeneity analysis. The numbered analyses are predictions for individuals from the studies indicated. These intervals arise from the exchangeability model.

lysis. The intervals are smaller for study 1; indicating that there is less between-subject variability in study 1 for the clearance parameter. The squares represent an overall prediction from the homogeneity analysis, and the diamonds represent an overall prediction from the exchangeability analysis. These predictions are for a new individual that is from a study exchangeable with those already considered. The homogeneity model underestimates uncertainty, because it does not acknowledge between-study differences.

Figures 4 and 5 show the same type of plots for the half-life and volume parameters respectively. For the half-life predictions, the intervals for study 1 are not as narrow as in the clearance or volume plots. For study 1, the prediction for volume is much lower than the predictions for the other studies. This could be due to some characteristic specific to that study: for example, study 1 subjects may be in a different age or weight range.

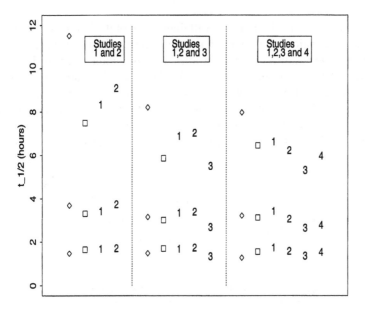

Figure 4   Predictions for a new individual for the half-life parameter. See the legend of Fig. 3 for further explanation.

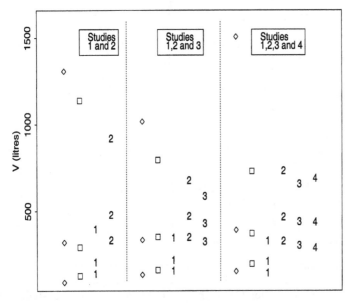

Figure 5   Predictions for a new individual for the volume parameter. See the legend of Fig. 3 for further explanation.

We now consider a number of diagnostic plots to assess whether the model assumptions are appropriate. Interpreting these diagnostics is hindered by the small number of studies and individuals per study. A discussion of Bayesian diagnostics can be found in, for example, the paper of Gelfand et al. (1992). We examine a number of residual plots. Each residual is a function of the parameters, and therefore we obtain a posterior distribution of each residual. We represent these posteriors using boxplots. Figure 6 plots the residual versus the normal scores, for the

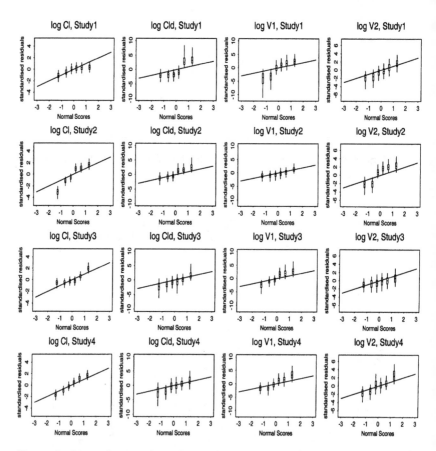

Figure 6 Normal-score plots of second-stage residuals from the exchangeability analysis. Each row represents the residual plot for the four elements (log Cl, log Cld, log V_1, log V_2) of the parameter θ_{ki} for the four studies. There are six boxplots in each subplot, since each study contains six subjects.

second stage of the model from the exchangeability analysis, where all studies were included. Row k (indexing studies) represents the residual plot for the four parameters of θ_{ki} ($k = 1, \ldots, 4$). There are six boxplots in each subplot, since each study contains six subjects. With small numbers, interpretation is difficult, but for all studies and all parameters the residuals appear close to a straight line, and so the second-stage normal assumption seems appropriate. Figure 7 plots the third-stage residual versus normal scores from the exchangeability analysis which included all four studies. There are four subplots, since the parameter μ_k is of dimension 4, and the four boxplots denote the four studies. There is a slight deviation from the straight line for the first and third elements of the parameter μ_k, but the other two elements do not appear to be incon-

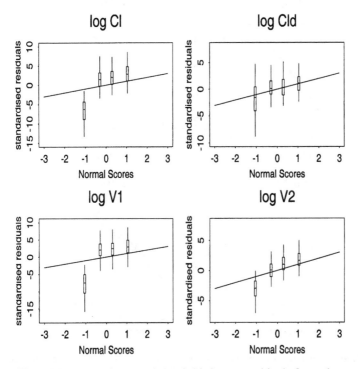

Figure 7 Normal-score plots of third-stage residuals from the exchangeability analysis for the four elements (log Cl, log Cld, log V_1, log V_2) of the parameter μ_k. There are four boxplots in each subplot, since there are four studies in total.

sistent with a straight line. This plot could be reflecting the fact that study 1 has a different dose size and also a different formulation.

Figure 8 plots the second-stage residual versus normal scores from the homogeneity analysis which included all four studies. Again there are four subplots, because the dimension of θ_i is 4. Shown are the medians of the distributions of the residuals, with the plot symbol indicating which study each subject came from; that is, there are four groups of six subjects with the groups representing the four studies. These residuals appear to lie on a straight line. So the second-stage normal assumption for the homogeneity model seems appropriate; but there is a clear grouping of individuals within particular studies, showing that the assumption that all 24 subjects are exchangeable is clearly incorrect. Any deviation from

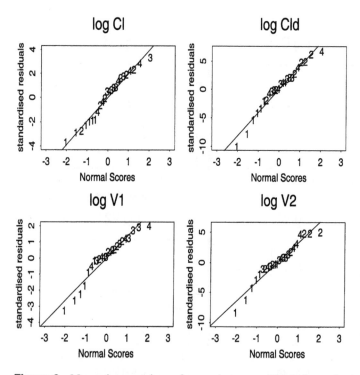

Figure 8   Normal-score plots of second-stage residuals from the homogeneity analysis for the four elements ($\log \mathrm{Cl}$, $\log \mathrm{Cld}$, $\log V_1$, $\log V_2$) of θ_i. Shown are the medians of the distributions of the residuals with the plot symbol indicating which study each subject came from. The line is the least-squares fit through the points.

the straight line seems to be mostly due to those subjects who belong to study 1.

The medians for the within-subject variability for the four studies are $\sigma_1^2 = 0.051$, $\sigma_2^2 = 0.026$, $\sigma_3^2 = 0.043$, and $\sigma_4^2 = 0.072$ under the exchangeability model. Under the homogeneity model, the median for the within-subject variability is $\sigma^2 = 0.049$. The within-subject variability is fairly similar across the four studies. We note that, for this drug, the intravenous formulation was only included during the phase I studies to aid in the understanding of the absorption characteristics of the oral tablet; but, for illustration of the meta-analysis model, we demonstrate how predictive concentrations may be constructed. Figure 9 shows con-

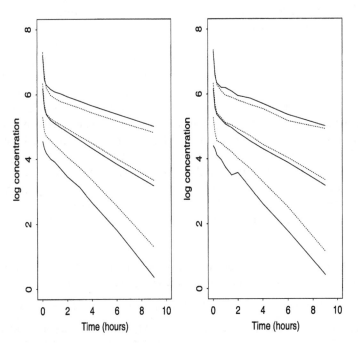

Figure 9 Predictive concentration intervals (on a logarithmic scale) following an intravenous dose of size 86 μmol. The left panel is a predicted distribution for the *modeled* concentration, the right for the *observed* concentration. The dotted lines indicate the 5%, 50%, and 95% points of the predictive distribution for the homogeneity analysis, the solid lines the same points for the exchangeability analysis.

centration predictions as a function of time, following a dose of size 86 μmol, under the exchangeability and homogeneity models. There are three sets of lines for each analysis, showing the 5%, 50%, and 95% points of the predictive distribution. The dotted lines are from the homogeneity model, and the solid lines are from the exchangeability model. The left plot shows predictions for $f(\theta, t)$, and the right plot for $f(\theta, t) + \varepsilon$, where $\varepsilon \sim N(0, \sigma^2)$. For the second set of predictions, we drew values of σ^2 from the posterior distribution of σ_4^2. For both sets of plots, the predicted intervals are wider under the exchangeability model. So the prediction from the exchangeability model would suggest a greater proportion of individuals who may experience toxicity or be under-dosed. The drug metropolol is not highly toxic, and so the differences obtained from the homogeneity and exchangeability models are not of clinical importance here. For highly toxic drugs, the differences seen in Fig. 9 could be of great importance.

Table 2 shows the medians of the coefficients of variation for the covariance matrices from the homogeneity and exchangeability analysis, in which all four studies were included. For the exchangeability analysis, the between-study variability (Ω) is, in general, greater than the between-individual variability. This suggests that a random-effect model may be appropriate to explain between-study differences. The between-individual variability from the homogeneity analysis is greater than the between-individual variability from the exchangeability analysis, implying that some of the between-study variability has been absorbed into Σ.

Recall that we assume a common within-study variability Σ. To assess the between-individual variability for each study alone, we fitted a population model to the data from each study separately. Table 3

Table 2 Posterior Medians of Coefficients of Variation of Pharmacokinetic Parameters Under the Homogeneity and Exchangeability Models[*]

Homogeneity		Exchangeability		Exchangeability	
Parameter	Median	Parameter	Median	Parameter	Median
$\sqrt{\Sigma_{11}}$	0.59	$\sqrt{\Sigma_{11}}$	0.40	$\sqrt{\Omega_{11}}$	0.47
$\sqrt{\Sigma_{22}}$	0.79	$\sqrt{\Sigma_{22}}$	0.73	$\sqrt{\Omega_{22}}$	0.24
$\sqrt{\Sigma_{33}}$	0.57	$\sqrt{\Sigma_{33}}$	0.35	$\sqrt{\Omega_{33}}$	0.93
$\sqrt{\Sigma_{44}}$	0.41	$\sqrt{\Sigma_{44}}$	0.22	$\sqrt{\Omega_{44}}$	0.24

[*]Σ is the between-individual variability; Ω is the between-study variability.

Table 3 Posterior Medians of the Between-individual Coefficients of Variation (Diagonal Elements) and Correlation (Off-diagonal Elements) for Each of the Four Studies

Parameter	Median
Study 1	$\begin{pmatrix} 0.21 & & & \\ -0.84 & 0.74 & & \\ 0.81 & -0.96 & 0.73 & \\ 0.55 & -0.61 & 0.62 & 0.20 \end{pmatrix}$
Study 2	$\begin{pmatrix} 0.58 & & & \\ 0.15 & 0.46 & & \\ 0.73 & -0.08 & 0.17 & \\ 0.09 & 0.87 & -0.05 & 0.30 \end{pmatrix}$
Study 3	$\begin{pmatrix} 0.34 & & & \\ -0.62 & 0.31 & & \\ 0.08 & -0.69 & 0.34 & \\ 0.38 & 0.03 & -0.48 & 0.16 \end{pmatrix}$
Study 4	$\begin{pmatrix} 0.44 & & & \\ 0.90 & 0.87 & & \\ -0.66 & -0.44 & 0.34 & \\ 0.68 & 0.82 & -0.21 & 0.24 \end{pmatrix}$

shows the posterior medians of the between-individual coefficients of variation (diagonal elements) and the correlations (off-diagonal elements) for each of the four studies. There is quite a lot of variability in the correlations, but the between-individual coefficients of variation are similar for the four studies. However, it is difficult to assess the assumption of a common Σ from this table alone.

We now assess the sensitivity to the prior distributions for the between-subject and between-study variability. For the analysis which included all four studies, we used five sets of priors for Σ and Ω. For each set of priors, we set $\rho_1 = \rho_2 = p = 4$. The first set of priors were $R = S = 0.04I_4$, which corresponds to a 20% coefficient of variation for the individual-level and study-level parameters. The other sets of priors were chosen to reflect a 50% and 10% coefficient of variation; so the second set of priors were $R = S = 0.25I_4$, the third set was $R = S = 0.01I_4$, the fourth set was $R = 0.04I_4$ and $S = 0.25I_4$, and the fifth set was $R = 0.01I_4$, and $S = 0.25I_4$.

Figures 10, 11, and 12 show plots of the overall and study-specific prediction intervals for the clearance, half-life, and volume parameters respectively under the five sets of priors. The results for the clearance parameter are reasonably robust to the choice of prior specification. For the half-life and volume parameters, the width of the intervals of the overall predictions do vary with the different sets of priors, but the intervals for the study-specific predictions remain relatively constant across the five sets of priors. For all pharmacokinetic parameters, the medians of the predictive distributions are robust to the choice of prior specification.

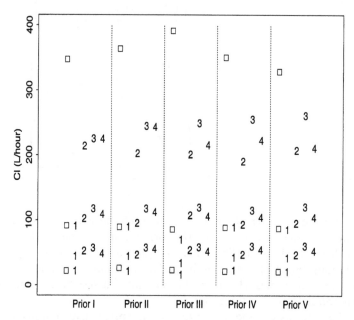

Figure 10 Comparison of prediction intervals for the clearance parameter from the exchangeability analysis under five sets of priors. The prior estimate for the between-individual variability is given by R; for the between-study variability, it is given by S. Each set of three points represent the 5%, 50%, and 95% quantiles of the predictive distribution. The squares represent an overall prediction for a new individual, and the numbered analyses are predictions for individuals from the studies indicated. In each case, we chose priors with diagonal matrices. The five priors chosen, with diagonal elements of R listed first, and those of S second, are I : (0.04,0.04), II : (0.25,0.25), III : (0.01,0.01), IV : (0.04,0.25), and V : (0.01,0.25).

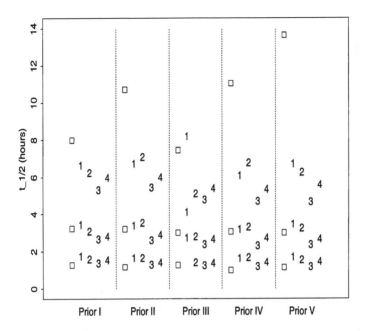

Figure 11 Comparison of prediction intervals for the half-life parameter from the exchangeability analysis under five sets of priors. See the legend of Fig. 10 for further explanation.

V. DISCUSSION

The meta-analysis model that has been presented in this chapter is useful for combining studies that are "similar." The four studies that we combined satisfied this requirement, since they were all phase I studies with similar experimental conditions and formulation. There were other population pharmacokinetic studies available for this drug, but these had different models corresponding to different formulations—for example, slow release. In our model, we have assumed that the individual pharmacokinetic parameters do not change over the two dosing occasions. However, the between-occasion variability within a subject can also be modeled (Lunn and Aarons, 1997). If covariate information is available, then we can include these easily in the model at the individual or study level.

The meta-analysis model from Sec. III allows us to obtain either study-level predictions or a general prediction for an individual who

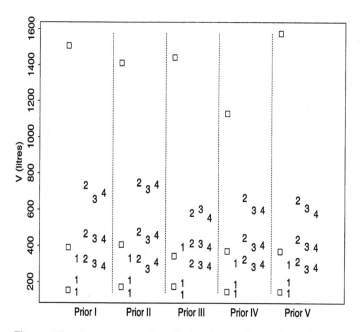

Figure 12 Comparison of prediction intervals for the volume parameter from the exchangeability analysis under five sets of priors. See the legend of Fig. 10 for further explanation.

arises from the population of exchangeable studies. There is very little work on design for population studies, but an obvious approach is to simulate data for new subjects under a number of different competing designs. We can then analyse these data and evaluate which design is best suited to provide the information that is needed from the new study. If studies are not combined and a study specific prediction is used, then the predictive interval will be narrower than if we obtain a general prediction from the meta-analysis model, and so the study-specific prediction will be overly precise and may not adequately represent the variability found across the studies.

Acknowledgments

We would like to thank Peter Lloyd of Clinical Pharmacology at Novartis Incorporated for allowing us to use the metropolol dataset, and Amy Racine-Poon for providing us with the data and for helpful comments. We would also like to thank Glaxo Wellcome for providing financial support for the first author.

Appendix: Implementation Details

We first describe the posterior distribution for all of the unknown parameters. Let $y = (y_1, \ldots, y_K)$, where $y_k = (y_{ki}, \ldots, y_{kn_k})$ denotes the totality of the data for study k, and y_{ki} represents the data of individual i in study k, for $k = 1, \ldots, K$. Then, using the conditional independence assumption implied by the model of Sec. III, we have

$$p(\theta_{11}, \ldots, \theta_{Kn_k}, \sigma_1^2, \ldots, \sigma_K^2, \mu_1, \ldots, \mu_K, \Sigma, \eta, \Omega | y)$$

$$= p(y)^{-1} \left\{ \prod_{k=1}^{K} \prod_{i=1}^{n_k} p(y_{ki} | \theta_{ki}, \sigma_k^2) p(\theta_{ki} | \mu_k, \Sigma) \right\}$$

$$\times \left\{ \prod_{k=1}^{K} p(\mu_k | \eta, \Omega) p(\sigma_k^2) \right\} p(\eta) p(\Sigma) p(\Omega).$$

To construct a Markov chain to sample from this posterior distribution, we require the conditional distributions of the individual parameters (or sets of parameters). Gelfand et al. (1990) describe the use of the Gibbs sampler for linear hierarchical models and Wakefield et al. (1994) for nonlinear hierarchical models. Gibbs sampling is not straightforward for the model in Sec. III, because the conditional distributions for the θ_{ki}, which are given by

$$p(\theta_{ki} | \cdot) \propto p(y_{ki} | \theta_{ki}, \sigma_k^2) p(\theta_{ki} | \mu_k, \Sigma),$$

are not of standard form. Here, and in what follows, we assume that we have conditioned upon all other parameters and data, so that

$$p(\theta_{ki} | \cdot) = p(\theta_{ki} | \theta_{-ki}, \sigma_1^2, \ldots, \sigma_K^2, \mu_1, \ldots, \mu_K, \Sigma, \eta, \Omega, y),$$

where θ_{-ki} represents the set of individual specific pharmacokinetic parameters $\theta_{k'i'}$ ($k' = 1, \ldots, K; i' = 1, \ldots, n_k$), without element θ_{ki}. We can use the Metropolis–Hastings algorithm (Smith and Roberts, 1993) to sample from the conditional distribution of θ_{ki}. The other conditional distributions are given by

$$\sigma_k^{-2}|\cdot \sim \mathrm{Ga}\left\{\frac{1}{2}\left(\nu_0 + \sum_{i=1}^{n_k} n_{ki}\right),\right.$$

$$\left.\frac{1}{2}\left(\nu_o\tau_0 + \sum_{i=1}^{n_k}\sum_{j=1}^{n_{ki}}[y_{kij} - \log f_{kij}(\theta_{ki}, t_{kij})]^2\right)\right\},$$

$$\mu_k|\cdot \sim \mathrm{N}\left\{(n_k\Sigma^{-1} + \Omega^{-1})^{-1}\left(\sum_{i=1}^{n_k}\Sigma^{-1}\theta_{ki} + \Omega^{-1}\eta\right), (n_k\Sigma^{-1} + \Omega^{-1})^{-1}\right\},$$

$$\Sigma^{-1}|\cdot \sim \mathrm{W}\left\{\left[\rho_1 R + \sum_{k=1}^{K}\sum_{i=1}^{n_k}(\theta_{ki} - \mu_k)(\theta_{ki} - \mu_k)^{\mathrm{T}}\right]^{-1}, \rho_1 + \sum_{k=1}^{K}n_k\right\},$$

$$\eta|\cdot \sim \mathrm{N}\left\{\left(\sum_{k=1}^{K}\Omega^{-1} + C^{-1}\right)^{-1}\left(\sum_{k=1}^{K}\Omega^{-1}\mu_k + C^{-1}c\right),\right.$$

$$\left.\left(\sum_{k=1}^{K}\Omega^{-1} + C^{-1}\right)^{-1}\right\},$$

$$\Omega^{-1}|\cdot \sim \mathrm{W}\left\{\left[\rho_2 S + \sum_{k=1}^{K}(\mu_k - \eta)(\mu_k - \eta)^{\mathrm{T}}\right]^{-1}, \rho_2 + K\right\}.$$

References

Berry DA, Stangl DK. Bayesian Biostatistics. New York: Marcel Dekker, 1996.

Davidian M, Giltinan DM. Nonlinear Models for Repeated Measures Data. Chapman and Hall, 1995.

Gelfand AE, Dey DK, Chang H. Model determination using predictive distributions with implementation via sampling-based methods. In: JM Bernardo, JO Berger, AP Dawid, AFM Smith, eds. Bayesian Statistics 4, pp 147–167. Oxford University Press, 1992.

Gelfand AE, Hills SE, Racine A, Smith AFM. Illustration of Bayesian inference in normal data models using Gibbs sampling. *J. Amer. Statist. Ass.* 85, 972–985, 1990.

Gelman A, Carlin JB, Stern HS, and Rubin DB. Bayesian Data Analysis. Chapman and Hall, 1995.

Gibaldi M, Perrier D. Drugs and the Pharmaceutical Sciences. Marcel Dekker, 1982.

Lunn DJ, Aarons L. Markov chain Monte Carlo techniques for studying inter-occasion and intersubject variability: application to pharmacokinetic data. Applied Statistics 46, 73–91, 1997.

Racine-Poon A, Wakefield JC. Statistical methods for population pharmacokinetic modelling. Statistical Methods in Medical Research 7, 63–84, 1998.

Smith AFM, Roberts GO. Bayesian computation via the Gibbs sampler and related Markov chain Monte Carlo methods. Journal of the Royal Statistical Society B 55, 3–23, 1993.

Spiegelhalter D, Thomas A, Best NG. BUGS: Bayesian Inference Using Gibbs Sampling. Medical Research Council Biostatistics Unit, 1996.

Steimer J, Vozeh S, Racine-Poon A, Holford N, O'Neill R. The population approach: rationale, methods and applications in clinical pharmacology and drug development. In: P Welling, H Balant, eds. Springer Verlag, 1994.

Wakefield JC, Aarons L, Racine-Poon A. The Bayesian approach to population pharmacokinetic/pharmacodynamic modelling. In: BP Carlin, AL Carriquiry, C Gatsonis, A Gelman, RE Kass, I Verdinelli, M West, eds. Case Studies in Bayesian Statistics, p. 205–265, Springer-Verlag, 1998.

Wakefield JC, Rahman N. The combination of population pharmacokinetic studies. To appear in Biometrics, 2000.

Wakefield JC, Smith AFM, Racine-Poon A, Gelfand AE. Bayesian analysis of linear and non-linear population models using the Gibbs sampler. Applied Statistics 41, 201–221, 1994.

Yuh L, Beal SL, Davidian M, Harrison F, Hester A, Kowalski K, Vonesh E, Wolfinger R. Population pharmacokinetic/pharmacodynamic methodology and applications: a bibliography. Biometrics 50, 566–675, 1994.

11

Meta-analysis of Individual-patient Survival Data Using Random-effect Models

Daniel J. Sargent
Mayo Clinic, Rochester, Minnesota

Benny C. Zee
Queen's University, Kingston, Ontario, Canada

Chantal Milan
Centre d'Epidemiologie et de Population de Bourgogne, Dijon, France

Valter Torri
Istituto di Ricerche Farmacologiche Mario Negri, Milano, Italy

Guido Francini
Siena University, Siena, Italy

Abstract

Models based on random effects are very popular in meta-analyses based on published data. One reason for their popularity is that they allow for inter-trial heterogeneity. In this chapter, we present a class of models that allow for random-effect analyses of individual-patient survival data. These random-effect survival models allow for between-trial heterogeneity in treatment effects and/or baseline hazard functions, and in addition, allow for multilevel modeling of the treatment effects. These models are fitted in a Bayesian setting using Markov chain Monte Carlo methods. We illustrate the models using a meta-analysis of individual patient survival data from five randomized trials comparing chemotherapy to non-treatment in patients with Dukes' stage B2 colon cancer.

I. INTRODUCTION

A question of current relevance in oncology is whether treatment with chemotherapy following the removal of a tumor is beneficial in patients with Dukes' stage B2 colon cancer. The patients considered here have had a complete resection of their tumor, and at the time of resection, had no lymph-node involvement or distant metastases. Previous randomized controlled clinical trials conducted by major groups in Europe and North America have demonstrated a beneficial effect on overall survival from a chemotherapeutic regimen containing two drugs, 5-fluorouracil (5-FU) and leucovorin (LV), in a broader group of patients that includes Dukes' B2 patients. However, these trials were not designed or adequately powered to address whether the treatment is beneficial in the subgroup of B2 patients. These patients comprise less than 50% of the clinical-trial population, so a pooling of several datasets is necessary in order to provide sufficient power to detect a benefit. In this chapter, we will consider the pooling of individual-patient survival times from five randomized clinical trials that included both a no-treatment arm and at least one arm containing those treated with 5-FU + LV. Further details of the dataset are presented in Sec. IV.

A key issue in analyzing this pooled dataset, and in many other meta-analyses, is inter-trial heterogeneity. As will be discussed in Sec. IV, slight variations in the protocol-specified treatments exist between the five trials that we will pool. In addition, these trials were carried out in different countries. These facts suggest a need for an approach to analysis that has the ability to account for the possibly differing risks that are present within each trial. These differing risks manifest in two ways. First, the baseline level of risk may differ between the trials. Second, the observed treatment effect may differ in the various trials. Any observed variation in baseline risk and/or in treatment effect may be simply random variability among studies, or it may imply that there is heterogeneity between studies. The possibility of differing risks in the different trials suggest that some degree of trial-specific analysis may be appropriate. The main focus of this chapter is to present and discuss a number of these trial-specific analysis techniques, using the Dukes' B2 dataset to demonstrate the methods. We label these techniques "random-effect survival analysis," or "hierarchical survival analysis."

The remainder of the paper is organized as follows. Section II introduces the topic of random-effect survival analysis, and discusses a number of ways in which the random-effect models may be useful in meta-analysis. Section III provides a brief overview of the computing methods that are used to fit these models. Section IV provides more information on the Dukes' B2 dataset, and then presents a number of different analyses of that dataset. Section V concludes with a discussion.

II. RANDOM-EFFECT SURVIVAL ANALYSIS

A. Survival Analysis

The analysis of time-to-event data is complicated by many factors. A successful survival model must account for the possibility of censoring, account for covariates (both baseline and time-varying), and make as few assumptions as possible about the underlying event process. The standard model used for survival analysis is the Cox proportional-hazards regression model (Cox, 1972). In the standard proportional-hazards regression framework, we model the relationship between covariates and the hazard of death at time t through a proportional-hazards model. Specifically, let individual i have event time T_i and covariate vector x_i. Define the hazard function for individual i, $\lambda(t; x_i)$, by

$$\lambda(t; x_i) = \lim_{\Delta t \to 0} \frac{\Pr[T_i \in (t, t + \Delta t) \mid T_i \geq t, x_i]}{\Delta t}. \tag{1}$$

Standard proportional-hazards regression relates the hazard of death with the covariates through the equation

$$\lambda(t; x_i) = \lambda_0(t) \exp(x_i\theta), \tag{2}$$

where $\lambda_0(t)$ is a baseline hazard function. In Cox proportional-hazards regression, $\lambda_0(t)$ is modeled nonparametrically, and estimates of the relative-risk parameter, θ, are obtained from the "partial likelihood" (Cox, 1975). Formally, from a frequentist viewpoint, the use of the partial likelihood as the basis for inference on θ can be justified through the counting-process theory of Aalen (1978); see Anderson and Gill (1982). A Bayesian justification is provided by Kalbfleisch (1978).

B. Random-effect Models

Models based on random effects are very popular in meta-analysis based on published data. The basis for random-effect models is the assumption that the observed values of the parameters of interest are drawn from a population of possible values, just as the patients that are enrolled in a trial are drawn from a population of possible patients. In this sense, random-effect models are a type of hierarchical model, hierarchical in the sense that the various effects (parameters) in the model are specified in a hierarchy. In a hierarchical model, a typical statistical model (e.g., regression, ANOVA, etc.) provides the base model, with additional levels of modeling relating the parameters to each other in a sensible fashion.

To motivate the random-effect modeling, consider first the simplest hierarchical model, the simple random-effect ANOVA. In this model we have a total of n observations y_{ij} $(i = 1, \ldots, K, j = 1; \ldots, J)$. The one-way variance-components model assumes that, for each i, the observations y_{ij} have a common mean θ_i, and then assumes that the θ_i are draws from a distribution with mean μ. Specifying the distributions typically used, the model is

$$y_{ij} \mid \theta_i, \sigma^2 \sim \mathrm{N}(\theta_i, \sigma^2) \quad (i = 1, \ldots, K; j = 1, \ldots, J),$$

$$\theta_i \mid \mu, v \sim \mathrm{N}(\mu, v) \quad (i = 1, \ldots, K).$$

In a Bayesian setting, the specification of the model is completed by the addition of prior distributions for μ, σ^2, and v. This model is commonly used in meta-analysis of continuous normally-distributed response data.

One reason for this model's popularity is that it allows for the possibility of inter-trial heterogeneity. If θ_i is a trial-specific treatment effect, treating the θ_i as draws from a distribution with an unknown variance allows for between-trial heterogeneity. If the treatment effect varies substantially between the trials, v will be estimated to be non-zero, and inter-trial heterogeneity will be allowed in the model. If, on the other hand, substantial variability does not exist between trials, then v will be estimated to be close to zero, and the individual treatment effects will be substantially shrunk towards the overall mean. In essence, v acts as a smoothing parameter. If v is large, there is little "smoothing," i.e., the random effects are not substantially shrunk toward the overall mean. For small values of v, there is substantial smoothing, with the individual random effects being constrained to be close to the overall mean.

C. Random-effect Survival Models

The theory and application of the one-way random-effect model to normal data is well known. In this subsection we apply the model to the survival-data setting, and discuss several generalizations.

Several authors have considered proportional-hazards survival models with random effects. The popular "frailty" model (Clayton, 1978, 1991) for multiple event-time data is one example of a survival model with random effects. Gustafson (1995), Stangl (1995), and Stangl and Greenhouse (1995) considered more general hierarchical survival models, but these authors assumed an exponential baseline hazard function. Sargent (1998) presented a general framework through which random-effect can be introduced into the Cox proportional-hazards model. This framework is based on using the Cox partial likelihood as the base for the hierarchical model, and allows very general random-effect structures for the model parameters.

To illustrate the general framework of Sargent (1998), consider first the case of the one way random-effect model applied to survival data. We proceed under the assumption that a proportional-hazards model of the form (2) will be used to relate the covariates of interest to the time to death. Specifically, suppose that there are m covariates of interest that we wish to model as fixed effects, and for the moment, a single study-specific treatment effect that we wish to model as random with r levels. For patient i, define $x_i = (x_{1i}, \ldots, x_{mi})$ to be a row vector of covariates whose coefficients $(\beta_1, \ldots, \beta_m)$ are modeled as fixed effects. Define γ_j to be the random treatment effect for trial j, and $y_i = (y_{i1}, \ldots, y_{ir})$ to be a row vector representing the treatment covariate for patient i. In the random-treatment-effect model, y_{ij} will be 1 if patient i is in the treatment group in trial j, and 0 otherwise. Using this notation, the proportional-hazards model can be written

$$\lambda(t; x_i, y_i) = \lambda_0(t) \exp(x_i\beta + y_i\gamma), \tag{3}$$

where $\beta = (\beta_1, \ldots \beta_m)^\mathsf{T}$ and $\gamma = (\gamma_1, \ldots \gamma_r)^\mathsf{T}$. We define $C(\beta, \gamma \mid x, y)$ to be the Cox partial likelihood based on the model (3).

The second layer of the hierarchical model is the distribution for the random effects. Typically, the γ_i are modeled as draws from a distribution g with mean μ and variance v. Frequently this distribution is taken to be Gaussian, but the use of Markov chain Monte Carlo (MCMC) computing allows a more flexible choice of distributions (see Sec. III). The final layer of this model is the prior distributions for the model parameters. Throughout this paper, we will use flat (improper) prior distributions for the β_i, the γ_i, and μ, introducing a proper prior distribution only on $h_\nu = 1/\nu$. The use of a proper prior for h_ν is required for computational stability of the MCMC algorithm. Let $q(\nu \mid \omega)$ represent this prior distribution, where ω is the vector of hyperparameters governing q.

Given this model specification, we can write the joint posterior distribution for the model parameters, $p(\beta, \gamma, \nu \mid x, y, \omega)$, as the product of the three components $C(\beta, \gamma \mid x, y)$, $g(\gamma \mid \mu, \nu)$, and $q(\nu \mid \omega)$ (Carlin and Louis, 1996). Explicitly,

$$p(\beta, \gamma, \nu \mid x, y, \omega) \propto C(\beta, \gamma \mid x, y)g(\gamma \mid \mu, \nu)q(\nu \mid \omega). \tag{4}$$

This is a typical framework for a hierarchical model, where in the nomenclature of Hodges (1998), the three model components are labeled the data component, the constraint component, and the prior component, respectively.

We now turn to the consideration of more general hierarchical survival models. The model (3) uses a random mechanism to model the treatment effect in each study. This random treatment effect can be included in either a stratified or a non-stratified Cox model. A second use for a model with random effects is to bridge the dichotomy between the usual stratified and non-stratified models. A non-stratified model assumes that all patients have the same baseline hazard function. A stratified model allows arbitrary baseline hazard functions for each strata, with no relationship between the functions in the different strata. An intermediate model assumes that the baseline hazard functions for each strata have the same shape, but allows them to have different magnitudes. To implement this model, we use the usual (unstratified) Cox partial likelihood, but allow a study-specific random effect. In a meta-analysis, this model may be useful in cases where there are many studies (strata), with limited numbers of patients in each study.

The general framework of Sargent (1998) also allows models with more than one level of random effects. In many cases, there are natural groupings of random effects, be they random treatment effects or random study effects. For instance, some of the interventions (treatments) may be more similar in some of the studies than in others. In these cases, it makes sense to group those studies that used the most similar interventions. As another example, in an international meta-analysis, several of the studies may be from the same country. There may be reason to believe that the baseline level of risk differs more between countries than within countries, suggesting that a multilevel model of random study effects within random country effects may be appropriate.

III. IMPLEMENTATION

The modeling approach outlined in Sec. II requires a method of parameter estimation with the flexibility to handle a wide range of possible posterior distributions. We will use Markov chain Monte Carlo methods to perform the parameter estimation. MCMC methods are designed to obtain samples from a density $\pi(\tau)$, which is either known only up to a constant of proportionality or is known but is difficult to sample directly. MCMC methods generate samples from $\pi(\tau)$ by creating a Markov chain on the state space of τ which has as its equilibrium distribution $\pi(\tau)$. For theoretical details on Markov chain Monte Carlo methods, see Tierney (1994) or Smith and Roberts (1993); Carlin and Louis (1996) provide many implementational details.

For the models presented here, we use a mixture of a univariate Metropolis algorithm (for model parameters with support on the real line) and Gibbs sampling for variance components when conjugacy is available. For the Metropolis algorithm, we use a Gaussian candidate distribution with the current realization as the mean, and variance chosen to give an acceptance rate of 25–50% (Gelman et al. 1995). In all of the algorithm runs discussed in Sec. IV, three parallel chains were started at overdispersed values, and monitored for convergence using the Gelman and Rubin (1992) diagnostic and by monitoring within-chain autocorrelation.

IV. APPLICATION TO THE B2 DATASET

A. Further Details on the Dataset

As a demonstration of the methods described in Sec. II, we apply several hierarchical Cox models to the dataset described in Sec. I. First we provide further details on the data for this meta-analysis.

The data in this meta-analysis is from five clinical trials that randomized patients with completely resected Dukes' stage B2 and C colon cancer to either treatment with 5-FU + LV or non-treatment. Here we focus on only the patients with stage B2 tumors. Specifically, the trials are from the following groups: NCCTG Intergroup (O'Connell et al., 1997); NCIC-CTG, G.I.V.I.O., and FFCD (IMPACT, 1995); and Siena (Francini et al., 1994). Table 1 gives characteristics of each trial. Further details of the individual trial designs, power calculations, and a meta-analysis using standard techniques (a stratified log-rank test) were presented by the IMPACT B2 group (IMPACT B2, 1999).

B. Simple Random-effect Models

The primary endpoint for this meta-analysis will be overall survival. All 1016 eligible patients are included in the analysis on an intent-to-treat basis. Table 2 presents results from four different Cox proportional-hazards models, all with stratification by study in the usual manner, some with fixed effects and some with random effects. The first model is a standard Cox model, stratified by study. This model gives an esti-mated relative risk of 0.80, comparing treatment to control. This value is not significantly different from 1 (standard log-rank two-sided p-value = 0.12). The second model in Table 2 is a stratified Cox model with study-specific fixed treatment effects. This model uses only data from each individual study to produce that study's estimated treatment effect. Based on this model, it appears that there may be heterogeneity between trials, since the estimated γ_i vary from -1.10 to 0.24. However, a like-lihood-ratio test comparing the study-specific fixed-effect model to the single-treatment-effect model does not produce a significant p-value ($p = 0.55$).

The third column in Table 2 presents the results from model (3)— the model with random study-specific treatment effects—stratified by study in the usual manner. Here we have assumed a Gaussian random-

Table 1 Trial Characteristics

	FFCD	NCIC–CTG	SIENA	GIVIO	NCCTG
Total randomized	268	370	239	888	317
B2s randomized	168	221	121	449	57
FU dose (mg/m^2)	400	370	400	370	425
LV dose (mg/m^2)	200	200	200	200	20
Duration of therapy (months)	6	6	12	6	6
Treatment effect	γ_1	γ_2	γ_3	γ_4	γ_5

Table 2 Treatment Effects with Standard Errors for Various Fixed- and Random-effect Models—Usual Stratification

Parameter	Single fixed treatment effect	Group-specific fixed treatment effect	Group-specific Gaussian random treatment effect	Group-specific t random treatment effect
μ	−0.22 (0.14)	—	−0.21 (0.20)	−0.21 (0.18)
γ_1	—	−0.25 (0.34)	−0.21 (0.20)	−0.22 (0.20)
γ_2	—	0.24 (0.35)	−0.11 (0.22)	−0.11 (0.22)
γ_3	—	−0.28 (0.33)	−0.22 (0.20)	−0.22 (0.20)
γ_4	—	−0.25 (0.20)	−0.22 (0.17)	−0.22 (0.16)
γ_5	—	−1.10 (0.68)	−0.29 (0.28)	−0.31 (0.30)

effect distribution, with a prior for h_v having mean 50 and standard deviation 100. We selected this prior by considering what values for the random effects were reasonable, as follows. A conservative belief regarding the amount of heterogeneity present between the studies would be that the true γ_i vary around μ by ±0.10. Using 0.10 as an estimate of the standard deviation of g leads to a prior estimate of h_v of 100. A prior belief that represents a feeling of substantial heterogeneity might be that the γ_i vary around μ by ±1.0. Using 1.0 as the prior standard deviation of g leads to a prior estimate of h_v of 1. The prior chosen, with mean 50 and standard deviation of 100, contains these extremes.

As can be seen from Table 2, the random-treatment-effect model presents a compromise between the single fixed-effect model and the model with fixed study-specific treatment effects. In the random-effect model, the study-specific treatment effects are shrunk from the model with study-specific treatment effects towards the overall estimate from the single-treatment-effect model. The point estimate of μ, the overall treatment effect, changes little from the model with a single fixed treatment effect; however μ's standard error increases substantially from 0.14 to 0.20. Thus, in the model allowing for heterogeneity, the certainty regarding the overall treatment effect is somewhat diminished. The posterior mean for h_v in this model is 106.2, a substantial increase from the prior mean of 50, indicating that, in this case, the data do not support a heterogeneous treatment effect. This is clear from the estimated γ_i, which are largely shrunk to the overall estimate and away from their estimates in the model with study-specific treatment effect. In addition, the large value of h_v indicates that substantial information can be shared in estimating the γ_i. Thus the standard errors for the γ_i in the model with group-specific random treatment effects are smaller than those in the model with group-specific fixed treatment effects.

The final column in Table 2 presents results from model (3), but with a Student's t distribution for the random treatment effects. The use of the t distribution requires a full Metropolis computing algorithm, as the normal–gamma conjugacy present in the Gaussian random-effect model does not extend to the t distribution. As we have no means to select a value for the degrees of freedom (ρ) of the t distribution, we place a prior distribution on ρ and add it to the sampling algorithm. We use a prior for ρ with mean and standard deviation 10, and use the same prior for the random-effect distribution as in the normal treatment-effect model. For this dataset, the Student's t random effects has virtually no effect on any of the parameter estimates. The posterior mean of ρ is 8.4, with posterior standard deviation 9.8: almost identical to the prior. This indicates that the data contain little information regarding ρ, which is not surprising given that only five observations are contributing to ρ's estimation. Runs of the MCMC algorithm using other prior distributions for ρ gave similar results—that is, that the posterior moments for ρ were virtually identical to the prior moments. Therefore, in this case, the data do not support the Student's t model, since they are not sufficient to move the degrees-of-freedom parameter from its prior distribution. The Student's t model may be more appropriate in other situations; we postpone further discussion to Sec. V.

One advantage of the MCMC-based computational algorithms used here is that, once the sampler has generated draws from the joint posterior distribution, virtually any desired summary statistic can be computed. In this case, one particularly interesting summary is the posterior probability that μ is less than zero, recalling that a negative μ indicates a beneficial treatment effect. Based on the draws generated from the posterior distributions for the three relevant models in Table 2 (the study-specific fixed-effect model does not estimate an overall treatment effect), the posterior probability that $\mu < 0$ under each model is 0.946, 0.898, and 0.916, respectively. While these values are suggestive of a lower relative risk on treatment, the 95% highest posterior density regions for μ include zero in each model, indicating that a "statistically significant" benefit for the treatment has not been demonstrated.

C. Restricted Stratification Models

The models in Table 2 are based on a usual stratified Cox model (henceforth the US model). As discussed in Sec. II, the random-effects Cox model allows for a more restricted version of a stratified model than standardly used. For example, consider a model having both a study-specific random treatment effect and a study-specific random effect. In the notation of Sec. II, $y_{1i} = (y_{1i1}, \ldots y_{1ir})$ is a row vector where $y_{1ij} = 1$ if patient i is in trial j, and 0 otherwise; $y_{2i} = (y_{2i1}, \ldots y_{2ir})$ is a row vector, where $y_{2ij} = 1$ indicates that patient i is on the treatment group in trial j; and $y_i = (y_{1i}, y_{2i})$. Then the model is

$$\lambda(t; x_i, y_i) = \lambda_0(t) \exp(y_i(\xi \mid \gamma)), \tag{5}$$

where $\xi = (\xi_1, \ldots \xi_r)^{\mathsf{T}}$ are random study effects and $(\xi \mid \gamma)$ denotes a column vector where the first r elements are ξ and the last r elements are γ. For identifiability we must fix one of the ξ_i so that there is a reference group. We set ξ_4 (the study effect for the GIVIO study) to zero, since that is the study with the largest number of patients. The remaining ξ_i are then interpreted as log relative risks comparing each trial's baseline hazard function to the baseline hazard function in the GIVIO trial. To fit this model, we assume that the ξ_i are normally distributed with mean zero and variance σ^2, that is,

$$\xi_i \sim N(0, \sigma^2) \quad (i = 1, 2, 3, 5). \tag{6}$$

As in the random-treatment-effect model, we used a gamma prior for $h_{\sigma^2} = \sigma^{-2}$ with mean 50 and standard deviation 100.

Table 3 presents results from the first three models in Table 2 for the treatment effects (deleting the model with the Student's t random-effect distribution), confined to models with restricted stratification (henceforth RS) which allows a study-specific effect to multiply a common baseline hazard function. As is clear from the table, the RS models had no effect on the estimate of the overall treatment effect μ, and little effect on most of the individual γ_i. However, the standard errors of the treatment effects were consistently smaller in the RS models. This can be explained by the fact that, in the RS models, fewer degrees of freedom are used in estimating the baseline hazard function in each group. Thus more precision is available for the estimation of the treatment-effect parameters.

One interesting observation from Table 3 is that, for four of the five studies, the estimated study-specific treatment effect changed very little

Table 3 Treatment Effects with Standard Errors for Various Fixed- and Random-effect Models—Restricted Stratification

Parameter	Single fixed treatment effect	Group-specific fixed treatment effect	Group-specific random treatment effect
μ	−0.22 (0.13)	—	−0.24 (0.17)
γ_1	—	−0.27 (0.29)	−0.24 (0.18)
γ_2	—	−0.14 (0.33)	−0.22 (0.19)
γ_3	—	−0.28 (0.31)	−0.23 (0.19)
γ_4	—	−0.12 (0.20)	−0.18 (0.16)
γ_5	—	−1.15 (0.65)	−0.31 (0.24)
ξ_1	−0.07 (0.15)	−0.06 (0.19)	−0.06 (0.15)
ξ_2	−0.31 (0.18)	−0.34 (0.26)	−0.27 (0.21)
ξ_3	−0.01 (0.15)	−0.01 (0.18)	−0.01 (0.15)
ξ_4	0	0	0
ξ_5	−0.09 (0.20)	0.02 (0.21)	−0.07 (0.19)

between the US and the RS models. However, γ_2, the treatment effect in the NCIC-CTG trial, changed rather substantially. In the model with the fixed study-specific treatment effect, the posterior mean of γ_2 dropped from 0.24 in the US model to −0.14 in the RS model, while in the model with the random study-specific treatment effect, γ_2's posterior mean dropped from −0.11 to −0.22. An explanation for this fact can be gained from Figs 1 and 2. In Fig. 1, the Kaplan–Meier estimates of the survival functions for each trial's control group are plotted, while Fig. 2 plots the same for the treatment group in each trial. From Fig. 2, the survival experience in the treatment groups was very similar across the five trials. However, from Fig. 1, the survival experience in the control group in the NCIC–CTG trial was superior to that of the control groups in the other four trials. An investigation of important prognostic factors (performance status, age, tumor grade, sex) showed that this group was balanced with the NCIC–CTG treatment group on all of these factors; thus this difference cannot be explained by the available covariates.

This observation explains why γ_2 differs so substantially between the US and RS models. In the US model, the treatment effect for each trial is estimated by comparing the treatment group to the control group in that trial alone. In the US model, the treatment effect for the NCIC–CTG trial is based on comparing the treatment group to a

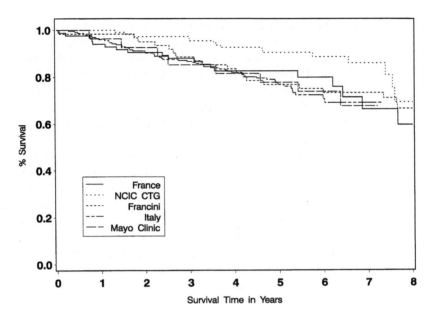

Figure 1 Kaplan–Meier estimate: survival by study, control patients.

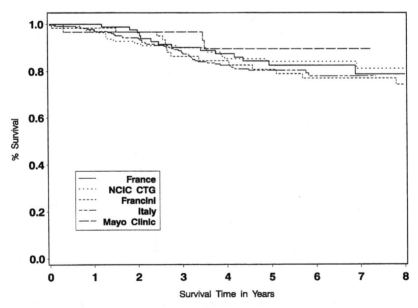

Figure 2 Kaplan–Meier estimate: survival by study, 5-FU patients.

control group that had a very favorable survival profile. In contrast, in the RS model, the survival experience of each control group is shrunk towards the survival experience of the control group in the other four trials. In the RS model, the treatment effect for the NCIC–CTG trial is estimated by comparing the treatment group to a control group that has been shrunk towards the other four control groups. Therefore, the treatment effect in the NCIC–CTG trial looks better in RS model than in the US model.

Further insight into the difference between the fixed- versus random-effect models and the US versus RS models can be gained by looking at the fitted survival curves for each trial under each model. Figure 3 plots the fitted survival curves for two models with a single fixed treatment effect, while Fig. 4 plots the fitted survival curves for two models with a study-specific random treatment effect. In each of these figures, the top row of plots is from a US model, while the bottom row is from a RS model. The first column of plots in each figure represents the control group, while the second column represents the treatment group. The

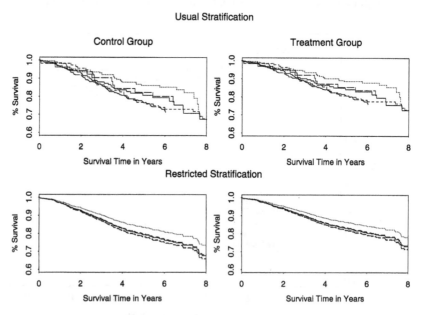

Figure 3 Estimated survival curves, models with a single fixed treatment effect.

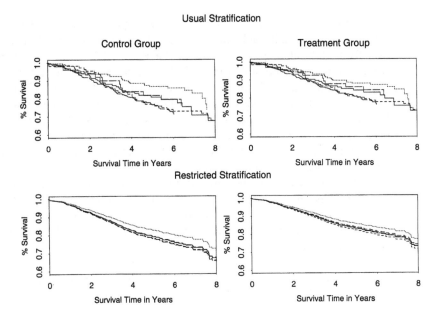

Figure 4 Estimated survival curves, models with a random treatment effect.

individual curves within each plot are purposefully not labeled: the goal of these plots is to demonstrate trends, not to make inference on the individual trials.

Looking first at the plots for the control group in Fig. 3, it is immediately obvious that, in the US model (the top row of plots), the fitted survival functions bear no relationship to each other. In contrast, in the RS model (the bottom row of plots), the fitted survival functions are proportionally related. What is not as clear immediately—but proves true upon inspection—is that, in the control group plot in the RS model (the plot in the lower left), the dotted survival curve (which represents the NCIC–CTG trial) is closer to the average of the rest of the fitted curves than it is in the US model (the plot in the upper left). As discussed earlier, this is because, in the RS model, information is shared between the control groups, resulting in fitted curves being shrunk towards each other. Comparing the two columns in Fig. 3, it is clear that, in the model with a single fixed treatment effect, there is an estimated beneficial treatment effect (the estimated survival curves for the treatment group are higher than the estimated curves for the control group). In the fixed-effect model, the treatment effect is manifested by proportionally increasing the fitted survival curve in each study by a constant amount. A final observation from this set of plots is that, in the RS model, estimation of the survival function is possible for the full range of the data in all trials, as opposed to the survival-function estimation ending at the end of data in each particular trial in the US model. This results from the fact that the RS model pools data from across the trials, and it uses this pooling to obtain survival-function estimates beyond the range of the data for trials with more limited follow-up. Whether this feature of the RS model is desirable depends on each investigator's preference.

Turning to the models with random treatment effects illustrated in Fig. 4, it is again clear that there is an overall treatment effect (the fitted survival curves are higher in the treatment group than in the control group). However, in the random-treatment-effect model, the treatment effect is not consistent across the different studies. For this reason, in Fig. 4, the fitted survival curves in the second column are not a scale multiple of the curves in the first column. In Fig. 4, the relationship between the curves in the first and second column of plots is specific to each study.

D. Hierarchical Treatment-effect Models

The models considered in Secs IV.B–C have a relatively simple hierarchical structure, that is, they have a single level of random effects. An additional level of modeling may be beneficial in grouping studies that have similar characteristics. For example, in the B2 meta-analysis, four of the five studies had treatment durations of 6 months, while the remaining study (the study from Siena) had a 12-month treatment duration. For a second example, four of the five studies used a LV dose of 200 mg/m^2, while the remaining study (the NCCTG study) used a LV dose of 20 mg/m^2. We might expect that studies that share important treatment characteristics would have results that are more similar than results from studies that differ on these characteristics.

To include this type of grouping into the analysis, we used stratified Cox models with a hierarchical random-treatment-effect structure. For the model that grouped the four trials with a treatment duration of 6 months together, we fit a Cox model with treatment effects $\gamma = (\gamma_1, \ldots, \gamma_5)$, with the hierarchical structure

$$\gamma_1, \gamma_2, \gamma_4, \gamma_5 \sim f(\gamma_{6mo}, \sigma^2_{6mo}), \qquad \gamma_{6mo}, \gamma_3 \sim f(\mu, \sigma^2_{5-FU+LV}). \qquad (7)$$

The model that groups the four trials based on LV dose is defined by obvious analogy. We used the same prior distributions for σ^2_{6mo}, $\sigma^2_{5-FU+LV}$, and σ^2_{200mg/m^2} as in Sec. IV.B.

Table 4 compares the results of the simple random-treatment-effect model, the model with treatment effects grouped by duration of treatment, and the model with treatment effects grouped by LV dose. As shown in Table 4, the model with a hierarchical treatment-effect structure based on treatment duration had little effect on any of the individual parameter estimates. This is due to the fact that the survival experience in the Siena trial (represented by γ_3) was very similar to the survival experience in three of the four remaining trials. Thus the fact that the Siena trial was given more freedom in the model than the other treatment effects (by putting fewer restrictions on γ_3 than the other parameters) had little effect. The standard error of the overall treatment effect did increase in this model, probably because the extra level of modeling moved the estimation of the overall treatment effect farther from the data, increasing the uncertainty about its value.

The model with the hierarchical treatment structure based on LV dose did produce results that were somewhat different that the simple random-treatment-effect model. The estimate of the overall treatment effect dropped from -0.21 to -0.39, and its standard error increased from 0.20 to 0.74. This results from the overall treatment effect μ in essence becoming an average of two parameters: γ_{200mg/m^2} and γ_5. In

Table 4 Treatment Effects with Standard Errors for Random-effect Models with Hierarchical Treatment-effect Structure

Parameter	Simple random treatment effect	Duration-specific random treatment effect	LV dose-specific random treatment effect
μ	-0.21 (0.20)	-0.21 (0.31)	-0.39 (0.74)
γ_1	-0.21 (0.20)	-0.20 (0.21)	-0.20 (0.21)
γ_2	-0.11 (0.22)	-0.08 (0.22)	-0.10 (0.22)
γ_3	-0.22 (0.20)	-0.22 (0.22)	-0.21 (0.20)
γ_4	-0.22 (0.17)	-0.20 (0.16)	-0.21 (0.16)
γ_5	-0.29 (0.28)	-0.27 (0.27)	-0.54 (0.51)
γ_{6mo}	—	-0.18 (0.17)	—
γ_{200mg/m^2}	—	—	-0.20 (0.19)

this model, the estimate of γ_5 differs from the other γ_i by a large absolute magnitude, and has a large uncertainty. The model that we have fitted assumes that the overall treatment effect μ is drawn from a distribution for which $\gamma_{200\text{mg/m}^2}$ and γ_5 are the only two observations, resulting in a great deal of uncertainty about its true value. In this case, where there are few data to estimate γ_5, this model may be less appropriate than others considered previously.

Throughout this section, we have considered a number of different models for the Dukes B2 dataset. Several of the models have provided additional insight into the individual trials—in particular, the restricted stratified model which suggests that the control group in the NCIC–CTG study seems to have particularly favorable survival. Comfortingly, all of the analyses have delivered a consistent message: that, while the data suggest some benefit from treatment with 5-FU + LV on overall survival, the magnitude of benefit does not reach the level of statistical significance.

V. DISCUSSION

The goal of this chapter has been to suggest and demonstrate a variety of approaches to modeling individual-patient survival data in the setting of a meta-analysis. These approaches are based on the random-effect Cox model. Some of the analyses presented above, in particular the models with the Student's t random-effect distribution, and the models with a hierarchical random-treatment-effect structure, may not have been best illustrated on the Dukes' B2 dataset, where the meta-analysis was based on only five trials. Models with hierarchical treatment-effect structures may best be suited for meta-analyses of many trials, so that each of the hierarchical parameters will have several different "observations" to aid in its estimation. The same is true of the Student's t models, where many observations (treatment effects) are necessary in order to estimate the distribution's degrees of freedom.

Throughout this work, we have made an explicit decision not to compare the fit of the fixed-effect versus the random-effect models. We consider this choice to be a modeling decision, which should be specified prior to the analysis. Just as in standard analysis of variance, or meta-analysis of continuously distributed Gaussian data, the choice between a fixed- or random-effect analysis should be based on how the data were collected, and what type of inferences are desired.

We have presented models and analyses in a Bayesian setting. The main reason for the Bayesian methodology is the ability to fit a wide array of models, with varying distributional assumptions, using the same estimation procedures. All of the models discussed above were fitted in FORTRAN using the same base code, and, with the exception of the model with the Student's t random-effect distribution, only perhaps 10 lines of code needed to be altered to fit any of the models considered. Computing time for these models was very reasonable, with the most complex model requiring approximately 30 minutes of CPU time using an Ultra Sparc-ii 300 MHz processor, with most models requiring on the order of 5 CPU minutes.

The only restriction imposed by the Bayesian methodology is the requirement of prior distributions on the highest-level variance components. These prior distributions are required for computational stablility; without these priors, the MCMC algorithm does not converge. In most cases, considerations such as in Sec. IV.B can be used to specify the prior distributions based on reasonable ranges for the model parameters. In the case of the Dukes' B2 analysis, the results shown are not sensitive to reasonable perturbations in the prior distributions.

In summary, we feel that the models presented here may be very useful in the meta-analysis setting. Random-effect models are very popular in the linear model, normal errors setting, and here we have demonstrated how the random-effect Cox models can be applied and interpreted in the survival analysis setting. The main advantage of the random-effect analysis is the ability to account for, and investigate, inter-trial heterogeneity, thus gaining insight into the individual trials.

Acknowledgments

The work of the first author was supported in part by Mayo Cancer Center Core Grant. The members of the Collaborating Committee for this project were: C. Erlichman (Mayo Clinic, Rochester, Minnesota), S. Marsoni (Groupo Italiano di Valutazione Interventi in Oncologia), J.F. Seitz (Fondation Francaise Cancerologie Digestive), J. Skillings (NCI Canada Clinical Trial Group), L. Shepard (NCI Canada Clinical Trial Group), C. Milan (Fondation Francaise Cancerologie Digestive), M.

O'Connell (Mayo Clinic, Rochester, Minnesota), R. Petrioli (University of Siena), L. Bedenne (Fondation Francaise Cancerologie Digestive), M. Giovannini (Fondation Francaise Cancerologie Digestive), Y.P. Letruet (Fondation Francaise Cancerologie Digestive), B. Tardio (Groupo Italiano di Valutazione Interventi in Oncologia), A. Zaniboni (Groupo Italiano di Valutazione Interventi in Oncologia), G. Pancera (Groupo Italiano di Valutazione Interventi in Oncologia), G. Martignoni (Groupo Italiano di Valutazione Interventi in Oncologia), M. Kahn (Mayo Clinic, Rochester, Minnesota), R. Labianca (Groupo Italiano di Valutazione Interventi in Oncologia), and A. Barni (Groupo Italiano di Valutazione Interventi in Oncologia).

References

Aalen OO. Nonparametric inference for a family of counting processes. Annals of Statistics 6:701–726, 1978.

Anderson PK, Gill RD. Cox's regression model for counting processes: a large sample study. Annals of Statistics 10:1100–1120, 1982.

Carlin BP, Louis TA. Bayes and Empirical Bayes Methods for Data Analysis. London: Chapman and Hall, 1996.

Clayton DG. A model for association in bivariate life tables and its applications in epidemiological studies of familial tendency in chronic disease incidence. Biometrika 65:141–151, 1978.

Clayton DG. A Monte Carlo method for Bayesian inference in frailty models. Biometrics 47:467–485, 1991.

Cox DR. Regression models and life tables (with discussion). Journal of the Royal Statistical Society, Series B 34:187–220, 1972.

Cox DR. Partial likelihood. Biometrika 62:269–275, 1975.

Francini G, Petrioli R, Lorenzini L, et al. Folinic acid and 5-fluorouracil as adjuvant chemotherapy in colon cancer. Gastroenterology 106:899–906, 1994.

Gelman A, Roberts G, Gilks W. Efficient Metropolis jumping rules. In: JO Berger, JM Bernardo, AP Dawid, AFM Smith, eds. Bayesian Statistics 5. Oxford: University Press, 1995, pp 599–9608.

Gelman A, Rubin DB. Inference from iterative simulation using multiple sequences (with discussion). Statistical Science 7:457–511, 1992.

Gray RJ. A Bayesian analysis of institutional effects in a multivariate cancer clinical trial. Biometrics 50:244–254, 1994.

Gustafson P. A Bayesian analysis of bivariate survival data from a multicenter cancer clinical trial. Statistics in Medicine 14:2523–2535, 1995.

Hodges JS. Some algebra and geometry for hierarchical models, applied to diagnostics (with discussion). Journal of the Royal Statistical Society, Series B 60:497–536, 1998.

International multicentre pooled analysis of colon-cancer trials (IMPACT) investigators. Efficacy of adjuvant fluorouracil and folinic acid in colon cancer. Lancet 345:939–944, 1995.

International multicentre pooled analysis of B2 colon-cancer trials (IMPACT B2) investigators. Efficacy of adjuvant fluorouracil and folinic acid in B2 colon cancer. Journal of Clinical Oncology, 17:1356-1363, 1999.

Kalbfleisch JD. Nonparametric Bayesian analysis of survival time data. Journal of the Royal Statistical Society, Series B 40:214–221, 1978.

O'Connell MJ, Mailliard JA, Kahn MJ, et al. Controlled trial of fluorouracil and low-dose leucovorin given for six months as postoperative adjuvant therapy for colon cancer. J Clin Oncol 15:246–250, 1997.

Sargent DJ. A general framework for random effects survival analysis in the Cox proportional hazards setting. Biometrics 54:1486–1497, 1998.

Smith AFM, Roberts GO. Bayesian computation via the Gibbs sampler and related Markov chain Monte Carlo methods (with discussion). Journal of the Royal Statistical Society, Series B 55:3–23, 1995.

Stangl D. Prediction and decision making using Bayesian hierarchical models. Statistics in Medicine 14:2173–2190, 1995.

Stangl D, Greenhouse J. Assessing placebo response using Bayesian hierarchical survival models. Lifetime Data Analysis 4:5–28, 1998.

Tierney L. Markov chains for exploring posterior distributions (with discussion). Annals of Statistics 22:1701–1762, 1994.

12
Adjustment for Publication Bias and Quality Bias in Bayesian Meta-analysis

David D. Smith
U.S. Food and Drug Administration, Rockville, Maryland

Geof H. Givens
Colorado State University, Fort Collins, Colorado

Richard L. Tweedie
University of Minnesota, Minneapolis, Minnesota

Abstract

Meta-analysis reviews, collects, and synthesizes individual sample surveys to estimate an overall effect size. If the studies for a meta-analysis are chosen through a literature review, an inherent selection bias may arise, since in particular, studies may tend to be published more readily if they are statistically significant, give a desirable answer, or are deemed to be of higher quality. Here, "quality" depends on sample characteristics, study-design elements such as blinding and control, and many other objective and subjective factors.

Within a Bayesian hierarchical model allowing stratification on quality, we develop a data-augmentation technique to estimate and adjust for the numbers and outcomes of studies that may be missing in each stratum. This permits inferences to account for potential publication bias. We apply this method to a meta-analysis of studies of cervical cancer rates associated with use of oral contraceptives.

I. INTRODUCTION

There has been an enormous recent increase in the use of meta-analysis as a statistical technique for combining the results of many individual analyses (1–3). While the combined analysis may have increased inferential power over any individual study, there are several drawbacks to meta-analysis (4–6). One well-documented concern is the need to collect all studies, both published and unpublished, relevant to the meta-analysis if the subsequent inferences are to be valid (7–9). A meta-analysis based on only a subset of all relevant studies may result in biased conclusions.

Publication of any new study in a scientific journal depends on several factors (10). Statistical significance may be a major determining factor of publication (11). Some researchers (e.g. students with Masters' or Ph.D. theses) may not submit a nonsignificant result for publication, and editors may not publish nonsignificant results even if they are submitted (12). Therefore, there may be a non-representative proportion of significant studies in the scientific literature. Moreover, the quality of studies may affect whether it will be published; a "high quality" paper, even if it does not exhibit statistical significance, may well be accepted where a "low quality" and insignificant result will fare less well (as perhaps it should).

Publication bias is problematic for a meta-analysis whose data come solely from the published scientific literature. A non-representative proportion of significant studies in the literature will lead to a non-representative proportion of significant studies in the meta-analysis data set. A standard meta-analysis model will then result in a conclusion biased toward significance. This phenomenon is known as "publication bias," or the "file-drawer problem" (7).

There are several methods which may be used to detect and estimate the impact of publication bias on a meta-analysis. Funnel plots and related graphical methods are useful for determining visually the existence of missing studies (13, 14). There are also several quantitative methods which estimate the number of missing studies and explicitly model the probability of publication (8, 9, 15). Many of these methods are limited to models where publication bias is based on a single factor, such as significance level of the study suppressed. They often fail to account for other factors, such as study quality, which may be an important consideration in determining publication.

In this chapter, we extend a Bayesian method described in Ref. 16, which covers the situation where publication probability is due solely to significance levels, to a stratified model which allows for other aspects to be taken into account. Estimation uses the data-augmentation principle (17). Specifically, we construct an algorithm which imputes latent sets of missing studies into a meta-analysis, in accordance with the probabilities that they are missing in given significance ranges, or quality ranges, or the like.

We then apply this model to a set of studies collected in Ref. 18, which examines studies on the association between the use of oral contraceptives and cervical cancer. We investigated the effect of publication bias due to significance levels alone in Ref. 19. Here we are able to consider the effect of the quality assessments of Ref. 18, and show that this makes a considerable difference in the final evaluation of the data set. Specifically, after allowance for these biases, the estimated relative risk of cervical cancer from oral contraceptive use is reduced considerably, indicating that a suppression bias against studies of insignificance or poor quality may seriously distort the results of an ordinary meta-analysis of these data.

II. ADJUSTING FOR PUBLICATION BIAS

A. Hierarchical Bayes Models for Observed and Latent Data

In what follows, we shall consider situations as in Ref. 16, where the measure of association in the meta-analysis is the relative risk (RR). The RR is commonly used to measure the association between a potentially toxic agent and a disease endpoint (3), although our work could equally apply to other measures such as risk differences or mortality ratios. For distributional reasons, we work on a log scale, and so let $\Delta = \log RR$. If $\Delta = 0$, then exposure to the agent is associated with no change in health risk; $\Delta > 0$ implies that exposure is associated with an increased health risk, and $\Delta < 0$ implies that exposure is associated with a health benefit.

The augmentation algorithm we shall use is an extension of that of Ref. 16 to a multi-tier situation; frequentist models in the single-tier case have also been considered (8, 9, 15).

We assume that n observed studies belong to s different classes, strata, or groups, which we shall call "tiers," with n_i observed studies belonging to each tier ($i = 1, \ldots, s$). For concreteness, tiers here will be thought of as referring to the quality classifications of the studies (however defined), although clearly they could refer to some other study characteristic, such as national origin of the study, or to clusters of data identified on the basis of multivariate data. Furthermore, our models are easily extended to additional hierarchical levels when more structured tier assumptions are appropriate.

The individual studies produce estimates of Δ, say Y_{ij}, for $i = 1, \ldots, s$ and $j = 1, \ldots, n_i$. We let p_{ij} equal the one-sided p-value of the (i, j)th study for testing the null hypothesis that $\Delta \le 0$.

Without considering the possibility of publication bias, the hierarchical model for the studies is

$$Y_{ij} = \Delta + \alpha_i + \beta_{j(i)} + \epsilon_{ij}, \tag{1}$$

where $\alpha_i \sim N(0, \eta^2)$ represents heterogeneity between tiers, $\beta_{j(i)} \sim N(0, \tau_i^2)$ represents heterogeneity between studies within tier i, and $\epsilon_{ij} \sim N(0, \sigma_{ij}^2)$ represents within-study variability of study (i, j). The α_i, $\beta_{j(i)}$, and ϵ_{ij} are assumed to be mutually independent. We write $\boldsymbol{\sigma^2} = \{\sigma_{ij}^2\}$ and $\boldsymbol{\tau^2} = \{\tau_i^2\}$.

Using the hierarchical model in (1), the likelihood of the data $\mathbf{Y} = (Y_{11}, \ldots, Y_{sn_s})$ is

$$p(\mathbf{Y} \mid \Delta, \eta^2, \tau^2, \sigma^2) \propto \frac{\exp\left(-\frac{1}{2}(\mathbf{Y} - \mathbf{1}\Delta)'\boldsymbol{\Sigma}^{-1}(\mathbf{Y} - \mathbf{1}\Delta)\right)}{|\boldsymbol{\Sigma}|^{1/2}}, \tag{2}$$

where $\boldsymbol{\Sigma}$ is the $n \times n$ covariance matrix of $\mathbf{Y}$ determined by the above assumptions. Although the number of tiers, s, must be fixed in advance in our model, we allow for tiers that contain no observed data: that is, for tiers with $n_i = 0$. However, to fully implement such a situation it is necessary to make specific assumptions about the characteristics of the studies within any tier for which $n_i = 0$ and about the nature of publication bias within such a tier.

Now we modify the model above to account for publication bias. We assume that, in addition to the n observed studies from the s groups, there are an additional m studies from these same groups which were not observed, due to publication bias. The number m and the relative risks which might have been found from these m studies are unknown and must be estimated. Uncertainty about these estimates must be reflected in the final meta-analysis inference, and we do this by treating them as parameters in a Bayesian analysis.

Let the estimated log relative risks from these missing studies be denoted as Z_{ij} for $i = 1, \ldots, s$ and $j = n_i + 1, \ldots, n_i + m_i$, where m_i is the number of missing studies in tier i and $m = \sum_i m_i$, and let $\mathbf{Z} = \{Z_{ij}\}$. For notational convenience, we will also denote the complete set of estimated log relative risks for all studies, both observed and missing, by $\mathbf{X} = \{X_{ij}\}$ for all i and j, where $X_{ij} = Y_{ij}$ when (i, j) indexes an observed study and $X_{ij} = Z_{ij}$ when (i, j) indexes a missing study.

We assume that the same random-effect model in (1) holds for the outcomes of the missing studies, namely

$$Z_{ij} = \Delta + \alpha_i + \beta_{j(i)} + \epsilon_{ij}, \tag{3}$$

where $\alpha_i \sim N(0, \eta^2)$, $\beta_{j(i)} \sim N(0, \tau_i^2)$, and $\epsilon_{ij} \sim N(0, \sigma_{ij}^2)$ are mutually independent. Now σ^2 includes the variances of the latent studies as well as those of the observed studies.

B. A Model for Selection Bias

There are various selection criteria that one might consider when trying to model publication bias. Following Refs 8, 9, and 16, we assume that, within each tier, the selection mechanism for a study is based on the study's p-value for rejecting the null hypothesis that $\Delta \leq 0$ in favor of the alternative hypothesis $\Delta > 0$. This mechanism is compatible with the widely believed possibility that statistically significant studies are more likely to be published than insignificant studies; but here we will also assume that the selection might be based on the tier in which the study falls, so that for example, a non-significant high-quality study may be accepted with a higher probability than a non-significant low-quality study.

To make this dependence explicit, we consider a partition of the unit interval into c interval segments, say $I_1, \ldots, I_c$. A p-value from any study must fall into exactly one of these intervals, and we assume that publication is governed by the probabilities

$$w_i^{(k)} = \Pr[\text{a tier } i \text{ study with } p\text{-value in } I_k \text{ is published}].$$

We let $\mathbf{w} = \{w_i^{(k)}; i = 1, \ldots, s; k = 1, \ldots, c\}$, $n_i^{(k)}$ be the number of tier i studies observed with p-values in I_k, and $m_i^{(k)}$ be the number of missing tier i studies with (unobserved) p-values in I_k. Henceforth, superscripts refer to p-value interval, and subscripts refer to tiers. Then $n_i = \sum_k n_i^{(k)}$ and $m_i = \sum_k m_i^{(k)}$, where the $n_i^{(k)}$ are known and the $m_i^{(k)}$ are unknown. Let $\mathbf{m} = \{m_i^{(k)}; i = 1, \ldots, s; k = 1, \ldots, c\}$.

We adopt the following model for the number of tier i missing studies with p-values in I_k:

$$m_i^{(k)} \mid \mathbf{w} \sim \begin{cases} \text{negative-binomial}(n_i^{(k)}, w_i^{(k)}) & \text{if } n_i^{(k)} > 0, \\ \text{logarithmic}(w_i^{(k)}) & \text{if } n_i^{(k)} = 0, \end{cases} \tag{4}$$

where U has a logarithmic(λ) distribution if the probability mass function of U is $p(U = u \mid \lambda) = (1 - \lambda)^u / (-u \log \lambda)$ on $u = 1, 2, \ldots$ (20). Each tier of studies is treated independently.

Finding the conditional distribution in (4) requires knowledge of the weight vector $\mathbf{w}$. In the single-tier context, Refs 8 and 9 present a maximum-likelihood method for estimating the $w_i^{(k)}$ from a meta-analysis dataset. We pursue a Bayesian analog in this chapter, following Ref. 16.

We do not include here a systematic investigation of the case when an entire tier is unobserved ($n_i = 0$ for some i).

C. The Complete Data Likelihood and Conditional Posterior Distributions

The observed data are the outcomes, $\mathbf{Y}$, of the observed studies, and we condition on the numbers of observed studies in each tier, n_i. Using (2), we write the likelihood for the observed data under this conditioning as

$$p(\mathbf{Y} \mid \Delta, \eta^2, \tau^2, \sigma^2) \propto \frac{\exp\left(-\frac{1}{2}(\mathbf{Y} - \mathbf{1}\Delta)'\Sigma^{-1}(\mathbf{Y} - \mathbf{1}\Delta)\right)}{|\Sigma|^{1/2}} \tag{5}$$

$$\prod_{i=1}^{s} \prod_{j=1}^{n_i} \prod_{k=1}^{c} \mathbf{1}_{\{p_{ij} \in I_k\}}.$$

The latent data are the outcomes, $\mathbf{Z}$, of the unobserved studies, and the numbers of such studies in each tier, $\mathbf{m}$. Together, $\mathbf{Y}$, $\mathbf{Z}$, and $\mathbf{m}$ comprise the complete data $(\mathbf{X}, \mathbf{m})$. At times, it is convenient to consider the latent data $(\mathbf{Z}, \mathbf{m})$ as nuisance parameters to be marginalized out of final inference about Δ.

Under models (1) and (3) for the observed and missing studies, the partial conditional likelihood for the study outcomes is

$$p(\mathbf{X} \mid \Delta, \eta^2, \tau^2, \sigma^2, \mathbf{m}) \propto \frac{\exp\left(-\tfrac{1}{2}(\mathbf{X} - \mathbf{1}\Delta)'\Omega^{-1}(\mathbf{X} - \mathbf{1}\Delta)\right)}{|\Omega|^{1/2}}$$

$$\prod_{i=1}^{s} \prod_{j=1}^{n_i+m_i} \prod_{k=1}^{c} \mathbf{1}_{\{p_{ij} \in I_k\}}, \tag{6}$$

where Ω represents the covariance structure of $\mathbf{X}$ and, hence, has dimension depending on $\mathbf{m}$ and elements depending on τ^2, σ^2, and η^2. We stress that (6) is conditional on knowing $\mathbf{m}$. Treating $\mathbf{m}$ as an unknown latent vector and conditioning instead on the parameter $\mathbf{w}$, we may write the complete data likelihood as

$$p(\mathbf{X}, \mathbf{m} \mid \Delta, \eta^2, \tau^2, \sigma^2, \mathbf{w}) \propto p(\mathbf{X} \mid \Delta, \eta^2, \tau^2, \sigma^2, \mathbf{m}) \prod_{i=1}^{s} Q_i(\mathbf{w}, \mathbf{m}), \tag{7}$$

where

$$Q_i(\mathbf{w}, \mathbf{m}) = \prod_{k=1}^{c} \left[\left[\binom{n_i^{(k)} + m_i^{(k)} - 1}{m_i^{(k)}} (w_i^{(k)})^{n_i^{(k)}} (1 - w_i^{(k)})^{m_i^{(k)}} \right. \right.$$

$$\left. \times \mathbf{1}_{\{m_i^{(k)} \in \{0,1,2,\ldots\}\}} \right]^{\mathbf{1}_{\{n_i^{(k)} > 0\}}}$$

$$\left. \times \left[\frac{-(1 - w_i^{(k)})^{m_i^{(k)}}}{(m_i^{(k)} \log w_i^{(k)})} \mathbf{1}_{\{m_i^{(k)} \in \{1,2,\ldots\}\}} \right]^{\mathbf{1}_{\{n_i^{(k)} = 0\}}} \right]$$

(cf. Ref. 16). In our Bayesian analysis, we adopt independent prior distributions $p(\Delta)$, $p(\eta^2)$, $p(\tau^2)$, $p(\sigma^2)$, $p(\mathbf{w})$, and $p(\mathbf{Z})$. Since $\mathbf{m}$ and $\mathbf{w}$ are related through (4), no separate prior for $\mathbf{m}$ is needed since its conditional distribution is known once $\mathbf{w}$ is known. Degenerate priors are allowed, and for example, we take σ_{ij}^2 to be known for individual observed studies.

Likelihood (7) is an extension of (2) but now includes parameters $\mathbf{w}$ which can be used to model publication bias. Reference 9 considers only the observed data and uses an observed data likelihood of a form analogous to (7). For identifiability, it assumes that the probability of publication equalled 1.0 in the most significant p-value interval and considered maximum-likelihood estimation only up to a multiplicative constant. Following this approach, we also scale the $w_i^{(k)}$, as indicated below, although we do not assume that the maximum publication probability corresponds to the most significant p-value interval. However such a monotonicity constraint is straightforward to enforce, and in Sec. IV we discuss the effect on our inferences of constraining the $w_i^{(k)}$ to be monotonically increasing as the p-value decreases.

Using prior distributions and the complete data likelihood, we derive the univariate conditional posterior distributions. We use $p(q \mid \cdot)$ to represent the conditional posterior distribution of q given all other parameters. The univariate conditionals for Δ, η^2, and τ^2 are then easily found from (7) as

$$p(\Delta \mid \cdot) \propto \frac{p(\Delta)}{A(\Delta)} \exp\left(-\frac{1}{2}(X - 1\Delta)'\Omega^{-1}(X - 1\Delta)\right), \tag{8}$$

$$p(\eta^2 \mid \cdot) \propto \frac{p(\eta^2)}{A(\eta^2)} \left[\frac{\exp\left(-\frac{1}{2}(X - 1\Delta)'\Omega^{-1}(X - 1\Delta)\right)}{|\Omega|^{1/2}}\right], \tag{9}$$

$$p(\tau^2 \mid \cdot) \propto \frac{p(\tau^2)}{A(\tau^2)} \left[\frac{\exp\left(-\frac{1}{2}(X - 1\Delta)'\Omega^{-1}(X - 1\Delta)\right)}{|\Omega|^{1/2}}\right], \tag{10}$$

where A is a normalizing function, $A(\Delta, \eta^2, \tau^2, \sigma^2, w)$, which we write in varying notation to emphasize its dependence on the particular parameter of interest.

The conditional density for the pair (Z, σ^2) is also straightforward:

$$p(Z, \sigma^2 \mid \cdot) \propto \frac{p(Z, \sigma^2)}{A(\sigma^2)} \left[\frac{\exp\left(-\frac{1}{2}(X - 1\Delta)'\Omega^{-1}(X - 1\Delta)\right)}{|\Omega|^{1/2}}\right]$$

$$\times \prod_{i=1}^{s} \prod_{j=1}^{n_i+m_i} \prod_{k=1}^{c} 1_{\{p_{ij} \in I_k\}}, \tag{11}$$

where the Y variables in X are set equal to the values of the observed data.

We consider Z and σ^2 in a bivariate form since, for any new study, the values of Z_{ij} and σ_{ij}^2 must be chosen to ensure that the constraint $p_{ij} \in I_k$ is satisfied.

In practice, for each (i, k), it is efficient to draw $m_i^{(k)}$ missing study variances, σ_{ij}^2, from $p(\sigma^2 \mid \cdot)$ with no constraint on the outcomes or p-values of the missing studies; then simulate the $m_i^{(k)}$ missing study p-values, p_{ij}, uniformly on I_k; and finally calculate the corresponding $Z_{ij} = \sigma_{ij}\Phi^{-1}(p_{ij})$.

Since, as discussed above, we have no prior on m, its conditional distribution is simply $p(m \mid w) \propto \prod_{i=1}^{s} Q_i(w, m)$.

Finally, because of the scaling we impose on the weights **w**, the posterior conditional distribution of **w** is given by

$$p(\mathbf{w} \mid \cdot) \propto \frac{p(\mathbf{w})}{A(\mathbf{w})} \prod_{i=1}^{s} p_1(\mathbf{w}_i \mid \cdot), \tag{12}$$

where for any fixed i, $\mathbf{w}_i = (w_i^{(1)}, \ldots, w_i^{(c)})$ and $p_1(\mathbf{w}_i \mid \cdot)$ is the conditional probability density function that results when $\mathbf{w}_i \times \max_k w_i^{(k)}$ has conditional density proportional to $Q_i(\mathbf{w}, \mathbf{m})$.

The posterior for Δ is not amenable to an analytical solution. Numerical techniques must be used, and we use a Gibbs sampling (21, 22) strategy to obtain approximate samples from the desired posterior distribution. Gibbs techniques can be used to obtain a sample from a desired target distribution by simulating realizations from a Markov chain whose stationary distribution is equal to the target.

The target distribution is the joint posterior distribution implied by the priors and complete data likelihood for our model. This target is marginalized to obtain the observed data posterior, from which inference is drawn. By sequentially sampling from the univariate conditional posterior distributions of the parameters, we can simulate approximate realizations from the joint posterior. The distribution of sampled points converges to the posterior distribution because the conditionals in equations (8)–(12) have the entire parameter space as its support, thus producing an aperiodic irreducible Markov chain (22). We use the quantile method to obtain posterior interval estimates. The details of the implementation, sampling, burn-in, simulation length, and subsampling are similar to those of Ref. 16, and we do not repeat them here.

III. PERFORMANCE ON SIMULATED DATA SETS

A. Simulated Data Sets

Reference 16 used a number of simulation scenarios to show that the performance of the single-tier model is satisfactory. Here we investigate the performance of the method on data sets with more than one tier (i.e., where $s > 1$).

We consider studies that fall into $s = 3$ tiers, and evaluate performance when (a) there is no overall association ($\Delta = 0$, RR $= 1.0$), and (b) when there is a particular positive association ($\Delta = 0.4$, RR ≈ 1.5). We are also interested in the effect that heterogeneity within tiers has on the results. To that end, we simulated data sets where within-tier studies were homogeneous ($\tau_1^2 = \tau_2^2 = \tau_3^2 = 0$) and where there were differences in the heterogeneity of each tier ($\tau_1^2 = 0$, $\tau_2^2 = 0.3$, $\tau_3^2 = 0.6$). In tabled results, we denote these as "zero" and "nonzero" respectively. We drew the individual study variances, $\hat{\sigma}_{ij}^2$, from (i) $\hat{\sigma}_{ij}^2 \sim$ gamma (shape = 3, mean = 0.33), for all i,j, and (ii) $\hat{\sigma}_{1j}^2 \sim$ gamma (shape = 3, mean = 0.1), $\hat{\sigma}_{2j}^2 \sim$ gamma (shape = 3, mean = 0.33), and $\hat{\sigma}_{3j}^2 \sim$ gamma (shape = 3, mean = 0.6). We denote these as "ident" and "diff" respectively. In all simulated data sets, we set η^2 equal to zero for simplicity.

We simulated 60 studies for each of the three tiers and each of the eight parameter combinations above. We split the studies' p-values into three intervals ($c = 3$), which were the same for all tiers:

$$I_1 = [0, 0.10], \quad I_2 = (0.10, 0.50], \quad I_3 = (0.50, 1].$$

Recall that the subscript i is the index for tier and k is the index for p-value interval.

One-sided p-values were calculated using only X_{ij} and $\hat{\sigma}_{ij}^2$. In practice, one would not have knowledge of the values of the other variance components when calculating p-values.

In each tier, we applied one of three different schemes of suppression ($1 - w_i^{(k)}$) to the studies within the tier. These were: no suppression (N), light suppression (L) and heavy suppression (H), given respectively for any i by

$$\text{N: } w_i^{(1)} = w_i^{(2)} = w_i^{(3)} = 1.00;$$

$$\text{L: } w_i^{(1)} = 1.00, \quad w_i^{(2)} = 0.75, \quad w_i^{(3)} = 0.50;$$

$$\text{H: } w_i^{(1)} = 1.00, \quad w_i^{(2)} = 0.50, \quad w_i^{(3)} = 0.25.$$

To create overall suppression schemes, we combined these choices in three ways: NNN, LLH and LHH, where, for example, LLH is

$$\text{Tier 1: } w_1^{(1)} = 1.00, \quad w_1^{(2)} = 0.75, \quad w_1^{(3)} = 0.50;$$

$$\text{Tier 2: } w_2^{(1)} = 1.00, \quad w_2^{(2)} = 0.75, \quad w_2^{(3)} = 0.50;$$

$$\text{Tier 3: } w_3^{(1)} = 1.00, \quad w_3^{(2)} = 0.75, \quad w_3^{(3)} = 0.25.$$

We performed two Bayesian meta-analyses for each simulated data set. The first was a standard Bayesian meta-analysis which does not adjust for publication bias. The second meta-analysis used the augmentation method in Sec. II. The two meta-analyses had the same priors. The prior on Δ was flat; the priors on τ_i^2 for all i were exponentials with mean

3.0; the prior on η^2 was an exponential with mean 0.3; the priors on the within-study population variances, σ_{ij}^2, were degenerate at $\hat{\sigma}_{ij}^2$, the observed study variances; the priors on the missing within-study population variances were empirical priors based on the observed study variances. The priors on the missing studies' log relative risks were uniform on the real line.

We consider two methods of comparing a standard meta-analysis with a publication-bias-adjusted meta-analysis. The first is to compare the posterior credibility intervals for Δ for both methods. The second compares the square root of expected squared-error loss, or risk, given by

$$\mathcal{R}(\Delta^*) = \sqrt{\int_{-\infty}^{\infty} p(\Delta \mid \cdot)(\Delta - \Delta^*)^2 \, d\Delta},$$

where $p(\Delta \mid \cdot)$ is the marginal posterior of Δ, and Δ^* is the true value, which is either 0 or 0.4 in our simulations.

We define $\mathcal{R}_0(\Delta^*)$ as the square root of risk of a standard Bayesian meta-analysis, and $\mathcal{R}_A(\Delta^*)$ as the square root of risk of the augmentation method described in Sec. II. For each simulated data set, we calculate the risk ratio $\mathcal{R}_A(\Delta^*)/\mathcal{R}_0(\Delta^*)$. Therefore, risk ratios less than 1.0 favor the augmentation method, and ratios greater than 1.0 favor the standard Bayesian meta-analysis.

B. Results of the Simulations

Table 1 shows the results of the risk calculations. Table 2 shows the posterior mean and a 95% credible interval (CI) for Δ in each parameter/suppression combination when $\Delta = 0$. Table 3 gives these results for $\Delta = 0.4$.

Table 1 indicates that the posterior of Δ resulting from a standard Bayes meta-analysis is slightly more concentrated near the true Δ when there are no studies missing as compared with the data-augmentation technique. This is partly due to the augmentation algorithm generally underestimating the true $\log RR$ in these cases, and partly due to the fact that the credible intervals for the standard Bayesian meta-analysis are more narrow than those of the augmentation algorithm. In those cases where publication bias is present, the meta-analysis adjusted for publication bias generally decreases the ratio of the risks and favors the augmentation method. In addition, Table 1 shows that $\mathcal{R}_A(\Delta^*)$ is more consistent across the suppression schemes than $\mathcal{R}_0(\Delta^*)$. The $\mathcal{R}_0$

Table 1 Risk Calculations of a Meta-analysis Which Fails to Adjust for Publication Bias, $\mathcal{R}_0(\Delta^*)$, and the Augmentation Method, $\mathcal{R}_A(\Delta^*)$, on Simulated Data in Three Tiers

Δ^*	Suppr.	τ^2	σ^2	$\mathcal{R}_A(\Delta^*)$	$\mathcal{R}_0(\Delta^*)$	$\mathcal{R}_A(\Delta^*)/\mathcal{R}_0(\Delta^*)$
0	NNN	Zero	Ident.	0.15	0.12	1.18
0	NNN	Zero	Diff.	0.13	0.13	0.99
0	NNN	Nonzero	Ident.	0.18	0.15	1.20
0	NNN	Nonzero	Diff.	0.13	0.16	0.80
0	LLH	Zero	Ident.	0.25	0.22	1.11
0	LLH	Zero	Diff.	0.23	0.21	1.09
0	LLH	Nonzero	Ident.	0.28	0.33	0.86
0	LLH	Nonzero	Diff.	0.19	0.27	0.70
0	LHH	Zero	Ident.	0.26	0.24	1.07
0	LHH	Zero	Diff.	0.22	0.23	0.96
0	LHH	Nonzero	Ident.	0.31	0.37	0.85
0	LHH	Nonzero	Diff.	0.28	0.34	0.83
0.4	NNN	Zero	Ident.	0.26	0.25	1.07
0.4	NNN	Zero	Diff.	0.25	0.20	1.25
0.4	NNN	Nonzero	Ident.	0.29	0.22	1.32
0.4	NNN	Nonzero	Diff.	0.27	0.24	1.14
0.4	LLH	Zero	Ident.	0.29	0.29	1.04
0.4	LLH	Zero	Diff.	0.29	0.31	0.95
0.4	LLH	Nonzero	Ident.	0.28	0.37	0.76
0.4	LLH	Nonzero	Diff.	0.33	0.48	0.68
0.4	LHH	Zero	Ident.	0.29	0.34	0.86
0.4	LHH	Zero	Diff.	0.26	0.35	0.74
0.4	LHH	Nonzero	Ident.	0.30	0.45	0.66
0.4	LHH	Nonzero	Diff.	0.30	0.57	0.52

*Risk ratios (far right column) less than 1.0 favor the augmentation method, and ratios greater than 1.0 favor the standard Bayesian meta-analysis.

(Δ^*) posterior risk increases as more studies are suppressed, which is the expected effect of publication bias.

In those data sets where $\Delta = 0$ (Table 2), suppressing studies in a standard meta-analysis results in an elevated point estimate of Δ. As more studies are suppressed, the bias of the standard meta-analysis increases. However, in all cases in which studies were suppressed, the augmentation algorithm adjusted the mean log RR toward 0. All stan-

Table 2 Results of a Meta-analysis Which Fails to Adjust for Publication Bias ("Std"), and the Augmentation Algorithm ("Aug") on Simulated Data in Three Tiers*

	Suppr.	τ^2	σ^2	Post. mean Δ	Std. err.	95% bounds Lower	Upper
Aug	NNN	Zero	Ident.	−0.01	0.13	−0.26	0.24
Std	NNN	Zero	Ident.	−0.02	0.12	−0.26	0.23
Aug	NNN	Zero	Diff.	0.02	0.12	−0.26	0.21
Std	NNN	Zero	Diff.	−0.02	0.12	−0.26	0.23
Aug	NNN	Nonzero	Ident.	−0.02	0.17	−0.34	0.33
Std	NNN	Nonzero	Ident.	0.01	0.14	−0.27	0.28
Aug	NNN	Nonzero	Diff.	−0.04	0.11	−0.28	0.15
Std	NNN	Nonzero	Diff.	0.01	0.14	−0.27	0.28
Aug	LLH	Zero	Ident.	0.10	0.19	−0.30	0.44
Std	LLH	Zero	Ident.	0.10	0.16	−0.23	0.40
Aug	LLH	Zero	Diff.	0.08	0.17	−0.28	0.41
Std	LLH	Zero	Diff.	0.10	0.15	−0.21	0.38
Aug	LLH	Nonzero	Ident.	0.15	0.16	−0.16	0.48
Std	LLH	Nonzero	Ident.	0.20	0.17	−0.16	0.51
Aug	LLH	Nonzero	Diff.	0.09	0.12	−0.11	0.33
Std	LLH	Nonzero	Diff.	0.14	0.17	−0.20	0.46
Aug	LHH	Zero	Ident.	0.09	0.19	−0.24	0.47
Std	LHH	Zero	Ident.	0.12	0.16	−0.21	0.43
Aug	LHH	Zero	Diff.	0.10	0.16	−0.22	0.39
Std	LHH	Zero	Diff.	0.12	0.15	−0.20	0.40
Aug	LHH	Nonzero	Ident.	0.16	0.17	−0.17	0.51
Std	LHH	Nonzero	Ident.	0.23	0.18	−0.14	0.55
Aug	LHH	Nonzero	Diff.	0.11	0.24	−0.38	0.56
Std	LHH	Nonzero	Diff.	0.20	0.18	−0.16	0.54

*The mean and 95% credible limits come from the marginal posterior of Δ. The true Δ is 0.

Table 3 Results of a Meta-analysis Which Fails to Adjust for Publication Bias ("Std"), and the Augmentation Algorithm ("Aug") on Simulated Data in Three Tiers*

	Suppr.	τ^2	σ^2	Post. mean Δ	Std. err.	95% bounds	
						Lower	Upper
Aug	NNN	Zero	Ident.	0.28	0.15	−0.04	0.56
Std	NNN	Zero	Ident.	0.34	0.17	−0.04	0.60
Aug	NNN	Zero	Diff.	0.27	0.13	−0.01	0.51
Std	NNN	Zero	Diff.	0.35	0.15	0.04	0.58
Aug	NNN	Nonzero	Ident.	0.25	0.16	−0.07	0.56
Std	NNN	Nonzero	Ident.	0.35	0.15	0.03	0.61
Aug	NNN	Nonzero	Diff.	0.26	0.16	−0.03	0.50
Std	NNN	Nonzero	Diff.	0.36	0.16	0.03	0.62
Aug	LLH	Zero	Ident.	0.28	0.20	−0.17	0.60
Std	LLH	Zero	Ident.	0.43	0.18	−0.01	0.67
Aug	LLH	Zero	Diff.	0.30	0.18	−0.02	0.66
Std	LLH	Zero	Diff.	0.47	0.16	0.10	0.77
Aug	LLH	Nonzero	Ident.	0.32	0.18	−0.05	0.64
Std	LLH	Nonzero	Ident.	0.52	0.18	0.13	0.83
Aug	LLH	Nonzero	Diff.	0.24	0.18	−0.13	0.58
Std	LLH	Nonzero	Diff.	0.54	0.21	0.08	0.90
Aug	LHH	Zero	Ident.	0.41	0.19	0.03	0.75
Std	LHH	Zero	Ident.	0.51	0.17	0.13	0.78
Aug	LHH	Zero	Diff.	0.36	0.17	0.00	0.62
Std	LHH	Zero	Diff.	0.50	0.18	0.11	0.81
Aug	LHH	Nonzero	Ident.	0.32	0.18	−0.09	0.60
Std	LHH	Nonzero	Ident.	0.56	0.19	0.16	0.89
Aug	LHH	Nonzero	Diff.	0.38	0.18	0.00	0.70
Std	LHH	Nonzero	Diff.	0.61	0.21	0.15	0.98

*The mean and 95% credible limits come from the marginal posterior of Δ. The true Δ is 0.40.

dard and data augmention meta-analyses included 0 in their CIs under this choice of parameters. The augmentation algorithm's credible intervals are wider than those of the standard meta-analysis, presumably because that analysis accounts for an additional source of uncertainty.

In those data sets where $\Delta = 0.4$ (Table 3), the overall number of studies suppressed was smaller compared to those where $\Delta = 0$, since there were fewer studies in I_2 and I_3. The standard Bayesian meta-analyses and the augmentation method covered the true Δ in all cases. The widths of the standard Bayesian meta-analyses' credible intervals were generally slightly more narrow than those of the augmentation method. However, these intervals were always positively biased, which is the expected effect in the presence of publication bias. In cases of heavy suppression (LHH), the point estimates of the augmentation method were very near the actual value of 0.40.

The results in Tables 1 to 3 are derived from a simulation design that is biased against the data augmentation approach for the following reason. To avoid a confounding influence on the results, we retained the same priors on suppression rate in all cases. This prior for **w** covered the same wide range of suppression rates in all trials, with the understandable result of degraded performance in the NNN trials. In real applications, we would reflect our *a priori* belief of low suppression (based on diagnostics such as funnel plots) through narrower priors on **w**. If we had designed simulations to mimic this behavior, thus varying the **w** prior across simulations, the performance of the augmentation approach would have improved.

IV. CERVICAL CANCER AND ORAL CONTRACEPTIVE USE

A. Data and One-tier Meta-analyses

In this section, we apply the the meta-analysis model described in Sec. II to a data set with tier structure and potential publication bias.

The studies that we will consider were collected in Ref. 18 using Medline and Indice Medico Español searches, and relate to the association between the use of oral contraceptives and cervical cancer. There has been an abundance of research on this topic around the world, but the studies' conclusions range from very strong associations to negative associations, making meta-analysis appropriate.

Reference 18 evaluates 62 published relative risks from 51 published papers. The data set was thought to be complete up to 1990. The study classifies the results into two groups: group I is the set of all 62 results, and Group II is a set of 26 "methodologically acceptable" results or higher-quality studies, although the rationale for being acceptable is not spelled out in any detail. It also splits outcomes into three categories corresponding to indicators of cervical cancer (dysplasia, carcinoma *in situ*, and invasive cancer). The study's references are given in the Appendix, and Table 4 summarizes its findings: these are based on fixed-effects meta-analyses on each of the three outcomes, despite finding evidence of heterogeneity among the three populations. The exclusion of "low quality" studies in group II substantially increases all estimates of RR.

When the results are stratified by type of cervical cancer, inference based on such small samples becomes questionable. An advantage of meta-analysis is lost when we subdivide: inferential power decreases. However, it may be biologically inappropriate to combine studies done on specific cancer types if we attempt to estimate the association of "cervical cancer in general" with oral contraceptive use. Nonetheless, we will assume that meta-analysis can validly address the question "is there an association between cervical cancer and oral contraceptive use?", even if we do not specify cancer types, or other covariates such as brand of oral contraceptive, explicitly. We must, however, interpret any biological relevance of these meta-analyses with caution, even though the random-effects and Bayesian methods are designed to allow for variation due to heterogeneous populations.

Table 4 Relative Risks with 95% CIs of the Delgado-Rodriguez et al. (18) Fixed-effects Meta-analyses of Cervical Neoplasia among Ever-users of Oral Contraceptives*

	Group I	Group II
Dysplasia	1.31 (1.24, 1.38)	1.52 (1.27, 1.82)
	(20 studies)	(6 studies)
Carcinoma *in situ*	1.29 (1.18, 1.41)	1.52 (1.31, 1.76)
	28 studies	(10 studies)
Invasive cancer	1.13 (0.99, 1.27)	1.21 (1.06, 1.37)
	(16 studies)	(10 studies)

*Two group I studies failed to report measures of variability.

In Ref. 19, we present fixed-effects, random-effects, and Bayesian random-effects meta-analyses on this data set, grouping all three outcomes so that inferential power is increased. These results appear in Table 5. Again, the inclusion of all group I studies decreases the estimate of RR, and the use of random-effect or Bayesian models also lowers the estimate in Group I, which one might take as an indication of the impact of serious heterogeneity in this group.

We first apply the publication bias method using only one tier to the group II or "high-quality" studies, and find that there appears to be publication bias present, as is suggested by the funnel plot in Fig. 1, where group II is depicted by solid circles. The posterior RR mean is 1.29 and the 95% CI is (1.07, 1.53). In comparison, a standard Bayesian meta-analysis on the same studies and same priors results in a RR of 1.46 and 95% CI of (1.08, 1.94). Hence, the publication-bias adjustment reduces the estimated excess risk while narrowing the posterior CI.

This analysis uses the following p-value intervals for a one-tier analysis:

$$I_1 = [0, 0.01], \quad I_2 = (0.01, 0.05], \quad I_3 = (0.05, 1].$$

We take the priors on Δ and τ_1^2 as diffuse, and the priors on $\{\sigma_{1j}^2\}$ as degenerate at their corresponding study estimates, $\{\hat{\sigma}_{1j}^2\}$, following Ref. 16. The priors on $w_1^{(2)}$ and $w_1^{(3)}$ are uniform on [0.5, 1.0]; we also enforce the monotonicity constraint $1 = w_1^{(1)} \geq w_1^{(2)} \geq w_1^{(3)} \geq 0$.

The number of observed studies in each of the p-value intervals are $n_1^{(1)} = 4, n_1^{(2)} = 7$, and $n_1^{(3)} = 15$, and the posterior means of $m_1^{(1)}, m_1^{(2)}$, and $m_1^{(3)}$ are 0, 1.76, and 8.81 respectively.

Table 5 Relative Risks with 95% CIs of the LaFleur et al. (19) Meta-analyses of Cervical Neoplasia Among Ever-users of Oral Contraceptives

Type of analysis	Group I	Group II
Overall fixed effects	1.30 (1.24, 1.35)	1.37 (1.26, 1.49)
Overall random effects	1.15 (1.10, 1.30)	1.38 (1.17, 1.63)
Overall Bayesian model	1.13 (0.95, 1.34)	1.46 (1.08, 1.94)

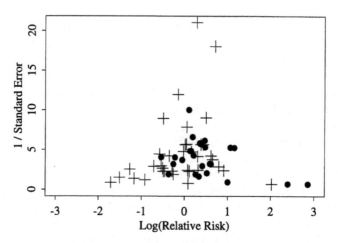

Figure 1 Funnel plot of all 62 studies in Delgado-Rodriguez et al. (18) that measure the association between oral contraceptive use and cervical cancer. The 26 Group II ("high-quality" tier 2) studies are denoted by filled circles.

B. Multi-tier Meta-analysis

In Ref. 18, it is claimed that there is a distinction in the quality of studies included in group II. We now analyze these data using the multi-tier approach described in Sec. II. There is information in the low-quality studies, and we should not lightly omit them from a meta-analysis. However, the low-quality studies may have a different chance of being published compared to the high-quality group II studies. Grouping the studies into quality tiers allows us to make publication-bias adjustments within each tier, with the hierarchical structure of the model reflecting these adjustments in the overall RR.

We group the 62 studies into two tiers: tier 1 (low-quality studies) will refer to those studies which are in group I but are excluded from group II; tier 2 (high-quality) studies will refer to those studies in group II. The high-quality studies appear in Fig. 1 as dark circles; the low-quality studies appear as crosses. Note that the tier 1 studies appear to cluster around a lower within-tier $\log RR$ than the tier 2 studies.

We first performed a standard Bayesian meta-analysis on all the studies, treated as a two-tier but complete data set. The estimated RR of this meta-analysis is 1.16, with a 95% CI of (0.97, 1.34). The posterior means of τ_1^2 and τ_2^2 are 0.11 and 0.09 respectively, and the posterior mean of η^2 is 0.14. Thus, although this meta-analysis failed to adjust for publication bias, the results suggested a weak nonsignificant association between cervical cancer and oral contraceptive use. The two-tier analysis result is rather similar to the non-tier augmentation analysis in Table 5, perhaps indicating that the missing studies allocated by the algorithm were similar to the actual group I studies suppressed in using only group II studies.

In our two-tier analysis, we used broad, roughly flat, priors for Δ, η^2, and τ^2. Rather than adopt improper priors, we used a normal distribution for Δ and exponential distributions for η^2 and τ^2—in each case with prior variances so large as to provide very flat prior support well beyond the range of reasonable values.

We use the p-value intervals $I_1 = [0, 0.10]$, $I_2 = (0.10, 0.50]$, $I_3 = (0.50, 1]$, and we place the same priors on the publication weights in both tiers: the prior distribution of weights in I_1 and I_2 are uniform(0.5, 1.0), and the prior distribution of weights in I_3 are uniform(0.2, 1.0). We again take the priors on the variances for the observed studies to be degenerate at the observed values. The priors that we place on the variances for the unobserved studies are broader than the empirical distribution of the observed study variances.

The RR of the augmented meta-analysis that accounts for publication bias was 1.06, with a 95% CI of (0.78, 1.42). Figure 2 compares the posterior distribution of this meta-analysis with the posterior of a Bayesian meta-analysis that does not adjust for publication bias. We also performed a meta-analysis with the monotonicity constraint enforced within both tiers, and this resulted in a RR of 1.01 and 95% CI of (0.80, 1.27). These results appear in Table 6. The posterior means of the τ_1^2, τ_2^2, and η^2 variance components are similar to the standard Bayesian meta-analysis; the meta-analysis with augmentation but with no monotonicity constraint gives the posterior means of τ_1^2, τ_2^2, and η^2 as 0.14, 0.11, and 0.06, respectively.

Figures 3 and 4 show relative-frequency histograms of the number of missing studies in each interval, with and without monotonicity restrictions. In the former case, the algorithm placed only about half as many missing studies into the insignificant range. This agrees with the message in the funnel plot of Fig. 1.

Clearly one should not dismiss the results of Ref. 18 on the basis of this augmentation approach. However, we conclude that publication bias may have a considerable impact on the inference in this data set. In

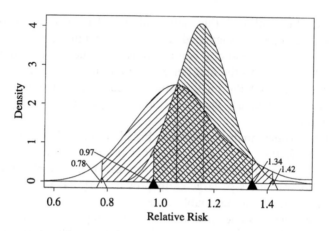

Figure 2 Relative-risk posteriors for a two-tier analysis of the cervical-cancer data. The leftmost posterior was calculated using data augmentation, and the right one assumes no publication bias. The 95% credible intervals correspond to the shaded areas under the curves.

particular, almost all the observed elevated risk estimates may be due to publication bias.

V. DISCUSSION

There have been many schemes proposed to judge the quality of studies for meta-analysis (see, for example, Ref. 23). A quality-adjusted meta-analysis has been performed considering quality as a covariate (24). Other studies group results into quality tiers and then use a weighting scheme to combine them: for example, in the EPA report on environ-

Table 6 Results of Two-tier Meta-analyses of the Delgado-Rodriguez et al. Cervical Cancer Data Set (18)

	Post. RR mean	95% cred. int.
Bayes, no augmentation	1.16	(0.98, 1.37)
Pub. bias adj.	1.06	(0.78, 1.42)
Pub. bias adj., monotone wts.	1.01	(0.80, 1.27)

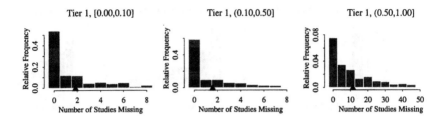

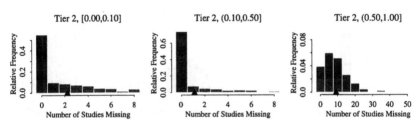

Figure 3 Posterior distributions of the number of missing studies, $m_i^{(k)}$, in each of three p-value divisions. The posterior mean is marked by a dark triangle.

mental tobacco smoke (25), whole tiers were either included or excluded from some meta-analyses.

We have found that the multi-tier version of the augmentation method in Ref. 16 provides an alternative quantitative approach which has utility in analyzing meta-analysis data sets with inherent tiers. Our simulation results suggest that the technique performs well in a variety of data sets which suffer from publication bias—although, in cases where there is no publication bias, the standard Bayesian analysis with no adjustments yields more satisfactory results.

Care must be taken when either meta-analytic method is used. A standard Bayesian analysis is of course preferable when there are no studies missing in the data set, since the augmentation algorithm presented tends to bias relative risk estimates downward. However, when there is suppression, the augmentation method performs well, adjusting the relative-risk credible intervals to recover from the bias caused by

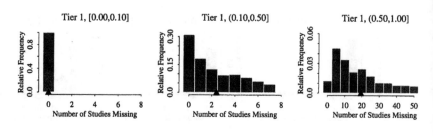

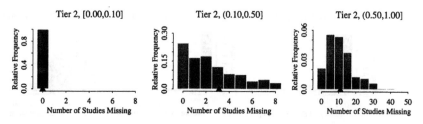

Figure 4 Posterior distributions of the number of missing studies, $m_i^{(k)}$, in each of three p-value divisions. This meta-analysis was performed with a monotonicity constraint on the publication probabilities within both tiers. The posterior mean is marked by a dark triangle.

missing studies. It is not true that our method always results in wider intervals and/or lower relative-risk estimates compared with approaches that do not account for publication bias. The example in Sec. IV.A, and others we have examined (26), shows that the opposite can occur when supported by the data. Further sensitivity work can be found in Ref. 26.

Overall, it is clear that the issues of publication bias are real and serious. While no method will completely overcome the problem of identifying intrinsically unavailable information, in examples such as that of the association between cervical cancer and oral contraceptives, we can make a reasonable effort to assess and account for the impact of such studies.

Table 7 Cervical Studies Considered in Meta-analyses by LaFleur et al. (1996), Collected by Delgado-Rodriguez et al. (1992)*

High-quality stratum				Low-quality stratum			
Study	Cancer	RR	95% CI	Study	Cancer	RR	95% CI
1	Dysplasia	0.77	(0.42,1.42)	27	Dysplasia	7.55	(0.45, 145)
2	Dysplasia	1.40	(0.43, 4.53)	28	Dysplasia	0.31	(0.15, 1.27)
3	Dysplasia	1.14	(0.76, 1.71)	29	Dysplasia	1.55	(1.10, 2.19)
4	Dysplasia	1.44	(1.03, 2.02)	30	Dysplasia	2.04	(1.83, 2.27)
5	Dysplasia	1.24	(0.79, 1.96)	31	Dysplasia	1.08	(0.48, 2.43)
6	Dysplasia	3.18	(2.19, 4.61)	32	Dysplasia	0.75	(0.33, 1.74)
7	Carc. *in situ*	2.72	(0.34, 21.8)	33	Dysplasia	0.61	(0.49, 0.75)
8	Carc. *in situ*	1.30	(0.47, 3.62)	34	Dysplasia	1.33	(1.22, 1.46)
9	Carc. *in situ*	1.17	(0.78, 1.74)	35	Dysplasia	1.85	(1.17, 2.94)
10	Carc. *in situ*	1.62	(1.13, 2.34)	36	Dysplasia	1.36	(0.85, 2.18)
11	Carc. *in situ*	2.90	(2.00, 4.10)	37	Dysplasia	0.86	(0.73, 1.01)
12	Carc. *in situ*	0.58	(0.36, 0.94)	38	Dysplasia	1.06	(0.84, 1.36)
13	Carc. *in situ*	1.68	(0.65, 4.31)	39	Dysplasia	0.70	(0.44, 1.11)
14	Carc. *in situ*	1.60	(1.20, 2.20)	40	Dysplasia	2.49	(1.10, 5.65)
15	Carc. *in situ*	0.95	(0.56, 1.59)	41	Carc. *in situ*	1.92	(1.14, 2.60)
16	Carc. *in situ*	1.43	(1.03, 2.00)	42	Carc. *in situ*	0.63	(0.27, 1.44)
17	Invasive	17.5	(0.99, 311)	43	Carc. *in situ*	0.60	(0.29, 1.24)
18	Invasive	11.0	(0.65, 184)	44	Carc. *in situ*	1.65	(1.34, 2.05)
19	Invasive	1.80	(1.00, 3.30)	45	Carc. *in situ*	0.40	(0.08, 2.03)
20	Invasive	1.11	(0.92, 1.35)	46	Carc. *in situ*	1.02	(0.72, 1.44)
21	Invasive	1.49	(1.10, 2.20)	47	Carc. *in situ*	0.49	(0.25, 0.94)
22	Invasive	0.69	(0.25, 1.89)	48	Carc. *in situ*	0.98	(0.65, 1.47)
23	Invasive	1.51	(0.78, 2.92)	49	Carc. *in situ*	1.12	(0.48, 2.62)
24	Invasive	0.80	(0.50, 1.30)	50	Carc. *in situ*	2.22	(1.12, 4.39)
25	Invasive	2.21	(0.90, 1.60)	51	Carc. *in situ*	1.47	(1.04, 2.07)
26	Invasive	1.85	(1.00, 3.14)	52	Carc. *in situ*	0.58	(0.30, 1.09)
				53	Carc. *in situ*	1.05	(0.75, 1.48)
				54	Carc. *in situ*	1.18	(0.78, 1.78)
				55	Carc. *in situ*	0.56	(0.36, 0.87)
				56	Carc. *in situ*	0.21	(0.15, 0.98)
				57	Invasive	0.28	(0.13, 0.60)
				58	Invasive	1.09	(0.09, 12.1)
				59	Invasive	0.61	(0.26, 1.40)
				60	Invasive	0.18	(0.02, 1.35)
				61	Invasive	0.22	(0.06, 0.80)
				62	Invasive	1.37	(0.60, 3.10)

*Group I studies refer to all 62 studies, and group II studies refer to the 26 high-quality studies in the left column.

Acknowledgments

The authors wish to thank Bonnie LaFleur and Sue Taylor of the University of Colorado Health Sciences Center for bringing the Delgado-Rodriguez et al. (18) collection to our attention, and for discussions on the epidemiological questions in that application of our methodology.

No official support or endorsement by the Food and Drug Administration of the content of this paper is intended or should be inferred.

Appendix: Cervical Cancer Study Bibliography

The following references were taken directly from Appendix 1 of Delgado-Rodgriguez et al. (18). Table 7 shows the data set used in Sec. IV. The study numbers below correspond to the Study column in Table 7.

1   MR Melamed, BJ Flehinger. Gynecol Oncol 1:290-298, 1973.
2   E Peritz, S Ramcharan, J Frank, et al. Am J Epidemiol 106:462–469, 1977.
3   L Andolsec, J Kovacic, M Kozuh, et al. Contraception 28:505–519, 1983.
4   MP Vessey, M Lawless, K McPherson, et al. Lancet 2:930–934, 1983.
5   DB Thomas. Obstet Gynecol 40:508–518, 1972.
6   EA Clarke, J Hatcher, GE McKeown, et al. Am J Obstet Gynecol 151:612–616, 1985.
7   MR Melamed, BJ Flehinger. Gynecol Oncol 1:290-298, 1973.
8   E Peritz, FA Pellegrin, RM Ray, et al. J Reprod Med 25 (suppl):346–372, 1980.
9   L Andolsec, J Kovacic, M Kozuh, et al. Contraception 28:505–519, 1983.
10 NH Wright, MP Vessey, B Kenward, et al. Br J Cancer 38:273–279, 1978.
11 V Beral, P Hannaford, C Kay, et al. Lancet ii:1331–1335, 1988.
12 DB Thomas. Obstet Gynecol 40:508–518, 1972.
13 SH Swan, WL Brown. Am J Obstet Gynecol 139:52–57, 1981.
14 KL Irwin, L Rosero-Bixby, MW Oberle, et al. JAMA 259:59–64, 1988.
15 R Molina, DB Thomas, A Dabancens, et al. Cancer Res 48:1011–1015, 1988.

16 CJ Jones, LA Brinton, RF Hamman, et al. Cancer Res 50:3657–3662, 1990.
17 L Andolsec, J Kovacic, M Kozuh, et al. Contraception 28:505–519, 1983.
18 MP Vessey, R Doll, R Peto, et al. J Biosoc Sci 8:373–427, 1976.
19 V Beral, P Hannaford, C Kay, et al. Lancet ii:1331–1335, 1988.
20 WHO. Br Med J 290:961–965, 1985.
21 LA Brinton, GR Huggins, HF Lehman, et al. Int J Cancer 38:339–344, 1985.
22 CJ Jones, LA Brinton, RF Hamman, et al. Cancer Res 50:3657–3662, 1990.
23 K Ebeling, P Nischan, C Schindler, et al. Int J Cancer 39:427–430, 1987.
24 KL Irwin, L Rosero-Bixby, MW Oberle, et al. JAMA 259:59–64, 1988.
25 F Parazzini, C La Vecchia, E Negri, et al. Int J Epidemiol 19:259–263, 1990.
26 LA Brinton, WC Reeves, MM Brenes, et al. Int J Epidemiol 19:4–11, 1990.
27 W Liu, L Koebel, J Shipp, et al. Obstet Gynecol 30:228–232, 1967.
28 MS Chai, WD Johnson, V Tricomi. N Y Stat J Med 70:2663–2666, 1970.
29 TS Kline, M Holland, D Wemple. Am J Clin Pathol 53:215–222, 1970.
30 M Bibbo, CM Keebler, GL Wied. J Reprod Med 6:79–83, 1971.
31 DF Miller. Am J Obstet Gynecol 115:978–982, 1973.
32 A Berget, T Weber. Dan Med Bull 21:172–176, 1974.
33 J Brux. Sem Hop Paris 50:1491–1495, 1974.
34 A Meisels, R Begin, V Schneider. Cancer 40:3076–3081, 1977.
35 CM Dougherty. Obstet Gynecol 36:741–744, 1970.
36 AW Diddle, WMH Gardner, PJ Williamson, et al. J Tenn Med Assoc 71:725–740, 1978.
37 V Vonka, J Kanka, J Jelinek, et al. Int J Cancer 33:49–60, 1984.
38 VU Geissler. Zentralbl Gynakol 110:267–276, 1988.
39 M Hren, J Kovacic, L Andolsek. Quaderni Clin Obstet Gynecol 33:1–12, 1978.
40 RWC Harris, LA Brinton, RH Cowdell, et al. Br J Cancer 42:359–369, 1980.
41 MR Melamed, LG Koss, BJ Flehinger, et al. Br Med J iii:195–200, 1969.
42 MS Chai, WD Johnson, V Tricomi. N Y Stat J Med 70:2663–2666, 1970.
43 JA Kirkland, MA Stanley. Canad Cytol 10:9–17, 1970.

44 M Bibbo, CM Keebler, GL Wied. J Reprod Med 6:79–83, 1971.
45 DF Miller. Am J Obstet Gynecol 115:978–982, 1973.
46 A Berget, T Weber. Dan Med Bull 21:172–176, 1974.
47 J Brux. Sem Hop Paris 50:1491–1495, 1974.
48 HJA Collette, G Linthorst, F Waard, et al. Lancet 1:441–442, 1978.
49 ME Attwood. J Obstet Gynaecol Br Commonw 73:662–665, 1966.
50 CM Dougherty. Obstet Gynecol 36:741–744, 1970.
51 AW Diddle, WMH Gardner, PJ Williamson, et al. J Tenn Med Assoc 71:725–740, 1978.
52 V Vonka, J Kanka, J Jelinek, et al. Int J Cancer 33:49–60, 1984.
53 VU Geissler. Zentralbl Gynakol 110:267–276, 1988.
54 AJ Worth, DA Boyes. J Obstet Gynaecol Br Commonw 79:673–679, 1972.
55 M Hren, J Kovacic, L Andolsek. Quaderni Clin Obstet Gynecol 33:1–12, 1978.
56 M Bibbo, CM Keebler, GL Wied. J Reprod Med 6:79–83, 1971.
57 A Berget, T Weber. Dan Med Bull 21:172–176, 1974.
58 J Brux. Sem Hop Paris 50:1491–1495, 1974.
59 HJA Collette, G Linthorst, F Waard, et al. Lancet 1:441–442, 1978.
60 V Vonka, J Kanka, J Jelinek, et al. Int J Cancer 33:49–60, 1984.
61 M Hren, J Kovacic, L Andolsek. Quaderni Clin Obstet Gynecol 33:1–12, 1978.
62 MB Andelman, J Zackler, HL Slutsky, et al. Int J Fertil 13:405–414, 1968.

References

1. L Hedges, I Olkin. Statistical Methods for Meta-analysis. New York: Academic Press, 1985.
2. I Olkin. Meta-analysis: methods for combining independent studies. Stat Sci 7:226, 1992.
3. H Cooper, LV Hedges, eds. The Handbook of Research Synthesis. New York: Russell Sage Foundation, 1994.
4. S Thompson, S Pocock. Can meta-analyses be trusted? Lancet 338:1127–1130, 1991.
5. NRC Committee on Applied and Theoretical Statistics. Combining Information: Statistical Issues and Opportunities for Research. Washington: National Academy Press, 1992.
6. K Mengersen, R Tweedie, B Biggerstaff. The impact of method choice in meta-analysis. Australian J Stat 37:19–44, 1995.
7. S Iyengar, JB Greenhouse. Selection models and the file drawer problem. Stat Sci 3:109–135, 1988.

8. K Dear, C Begg. An approach for assessing publication bias prior to performing a meta-analysis. Stat Sci 7:237–245, 1992.

9. L Hedges. Modeling publication selection effects in meta-analysis. Stat Sci 7:227–236, 1992.

10. K Dickersin, Y Min, C Meinert. Factors influencing publication of research results. JAMA 267:374–378, 1992.

11. P Easterbrook, J Berlin, R Gopalan, D Matthews. Publication bias in clinical research. Lancet 337:867–872, 1991.

12. British Medical Journal Editorial Staff. The editor regrets . . . (editorial). Brit Med J 280:508, 1983.

13. R Light, D Pillemer. Summing Up: the Science of Reviewing Research. Cambridge, MA: Harvard Univ. Press, 1984.

14. J Vandenbroucke. Passive smoking and lung cancer: a publication bias? Brit Med J 296:391–392, 1988.

15. NL Paul. Non-parametric classes of weight functions to model publication bias. Technical Report 622, Department of Statistics, Carnegie-Mellon Univ, Pittsburgh, PA, 1995.

16. GH Givens, DD Smith, RL Tweedie. Publication bias in meta-analysis: A Bayesian data-augmentation approach to account for issues exemplified in the passive smoking debate (with discussion). Stat Sci 12:221–150, 1997.

17. MA Tanner. Tools for statistical inference: observed data and data augmentation methods. In: J Berger, S Feinberg, J Gani, K Krickeberg, I Olkin, B Singer, eds. Lecture Notes in Statistics 67. New York: Springer-Verlag, 1991.

18. M Delgado-Rodriguez, M Sillero-Arenas, J Martin-Moreno, R Galvez-Vargas. Oral contraceptives and cancer of the cervix uteri. Acta Obstetricia et Gynecologica Scandinavica 71:368–376, 1992.

19. B LaFleur, S Taylor, D Smith, R Tweedie. Bayesian assessment of publication bias in meta-analyses of cervical cancer and oral contraceptives. Proceedings of the Joint Statistical Meetings, Chicago, 1996.

20. A Mood, F Graybill, D Boes. Introduction to the Theory of Statistics. 3rd ed. New York: McGraw-Hill, 1963.

21. S Geman, D Geman. Stochastic relaxation, Gibbs distributions and the Bayesian restoration of images. IEEE Trans Pattn Anal Mach Intell 6:721–741, 1984.

22. AFM Smith, GO Roberts. Bayesian computation via the Gibbs sampler and related Markov chain Monte Carlo methods (with discussion). JRSS Ser B 55:3–23, 1993.

23. D Moher, A Jadad, G Nichol, M Penman, P Tugwell, S Walsh. Assessing the quality of randomized controlled trials: an annotated bibliography of scales and checklist. Contr Clin Trials 16:62–73, 1995.

24. MP Longnecker, JA Berlin, MJ Orza, TC Chalmers. A meta-analysis of alcohol consumption in relation to risk of breast cancer. JAMA 260:652–656, 1988.
25. United States EPA Review. Health Effects of Passive Smoking: Assessment of Lung Cancer in Adults and Respiratory Disorders in Children. Washington: National Academy Press, 1992.
26. DD Smith. Adjusting for publication bias and quality effects in Bayesian random effects meta-analysis. PhD dissertation, Colorado State University, Fort Collins, CO, 1997.

13
Meta-analysis of Clinical Trials: Opportunities and Limitations

Richard Simon
Biometric Research Branch, National Cancer Institute, Bethesda, Maryland

Abstract

As meta-analysis has grown in use, the quality of meta-analyses has suffered. The tools of meta-analysis are sometimes developed and applied by individuals with little subject-matter knowledge and this rarely produces insightful results. Complex analysis methodologies are sometimes naively expected to serve as substitutes for adequate individual patient data, an adequate set of clinical trials and an adequate sampling plan. We review here the important basics for conducting meta-analyses of clinical trials. We try to describe frequent objectives of meta-analyses, why meta-analyses are important, and why they are frequently flawed and unreliable. We offer important methodologic features that should be observed in meta-analyses and comment on the importance of good clinical trials to achieving the objectives of meta-analyses.

I. WHY META-ANALYSIS?

Physicians refer to the medical literature for the most current information about how to treat their patients. How useful and reliable is the information they find? One difficulty is that the number of clinical trials, and the wide range of journals in which they are published, means that most physicians will be seeing only a portion of the reports—and that portion is usually a biased selection, because the most prominent journals often prefer the earliest or most statistically significant results. The individual clinical trials and their reports are also often defective in various ways, such as inadequate sample size, exclusion of randomized patients, and multiple testing. Conclusions of individual trials may conflict, making it unclear what to believe.

Another reason why reports of individual trials may be misleading is the fact that such reports generally avoid specification of a reasonable prior distribution for treatment effects. There are many clinical trials being conducted. For example, the National Cancer Institute sponsors about 150 randomized multi-center phase III treatment trials open for accrual in their cooperative oncology groups at any time, and there are many additional trials in the follow-up phase. Historically, the vast majority of the treatments studied have not represented therapeutic improvements. Consequently, for a first phase III clinical trial of most new treatments, the prior probability, π, that it represents a medically important improvement is often small, although there are exceptions for treatments with special scientific credentials. Consider a simple two-point model in which the treatment is viewed as either having no effect, $\Delta = 0$, or an effect of some medically important size: $\Delta = \delta$. Let the trial be planned to have power $1 - \beta$ with one-sided significance level α for detecting the treatment effect δ. Let the standardized normal deviate z represent the difference in outcome for the two treatments, where positive values represent advantage for the new treatment. We have, from Bayes theorem, that

$$\Pr(\Delta = \delta \,|\, z)/\Pr(\Delta = 0 \,|\, z) = \left(\frac{\pi}{1 - \pi}\right)\left(\frac{f(z \,|\, \Delta = \delta)}{f(z \,|\, \Delta = 0)}\right),$$

where f denotes the probability density function for z. For this model, we have

$$f(z \,|\, \Delta) \propto \exp(-(z - \Delta)^2/2\sigma^2),$$

where σ denotes the standard error of the maximum-likelihood estimate of Δ. Since the trial is designed to have power $1 - \beta$ for detecting a treatment effect of size $\Delta = \delta$ at significance level α, we have

$$z_{1-\alpha} + z_{1-\beta} = \delta/\sigma.$$

Thus, the posterior odds of an effective treatment is given by

$$\Pr(\Delta = \delta \mid z)/\Pr(\Delta = 0 \mid z) = \left(\frac{\pi}{1 - \pi}\right) \frac{\exp(-(z - z_{1-\alpha} - z_{1-\beta})^2/2)}{\exp(-z^2/2)}.$$

The posterior probability that $\Delta = 0$ is

$$\Pr(\Delta = 0 \mid z) = \left\{ 1 + \left(\frac{\pi}{1 - \pi}\right) \frac{\exp(-(z - z_{1-\alpha} - z_{1-\beta})^2/2)}{\exp(-z^2/2)} \right\}^{-1}.$$

If we plan the trial with one-sided α set at 0.025 and β set at 0.2, and obtain a result marginally significant at the one-sided 0.025 level, i.e., $z = 2$, then the posterior odds of treatment effectiveness becomes $5.2\pi/(1 - \pi)$. For many first phase III trials of a new treatment, π probably would not exceed 0.1. Hence, the posterior probability of the null hypothesis is 0.63 even when the trial results in $z = 2$. In this situation, it is likely that the treatment is not effective, although the trial was "statistically significant." This is a strong reason why confirmatory trials and meta-analyses of all relevant trials are crucial.

Treatment review papers have traditionally offered physicians an overview of clinical trials in a particular area. Meta-analysis is a term that denotes a more quantitative and systematic approach to reviewing research. Because it is quantitative, however, it is potentially much more powerful, influential, informative, and either more useful or more misleading than a traditional review. Meta-analysis has evoked strong reactions, both for and against it. There is much confusion about its validity and its role in medical therapeutics. In this chapter, we will explore the issue of whether meta-analysis is a tool that can help physicians care for their patients.

II. ORIGINS OF META-ANALYSIS

Meta-analysis originated in social science, where the need for policy makers to derive coherent summaries of the state of research was at least as great as the need for physicians to have sensible overviews of therapeutic questions today. Meta-analysis became popular in social sciences, but also controversial. The greatest concern was that of mixing apples and oranges—that is, combining studies that are too different in the intervention, subject selection, measurement of outcomes, research design, and quality. One critic said: "Although no single study was well enough done to prove that psychotherapy is effective, when you put all these bad studies together, they show beyond doubt that therapy works" (1). Many of the meta-analyses in social sciences are crude compared to the best meta-analyses in medicine. Randomized studies were combined with non-randomized studies, studies with grossly different endpoints were combined, and even the interventions combined in some cases bore only crude resemblance to each other. Some of the meta-analyses in medicine are much more refined. But the apples-and-oranges issue remains a concern in addressing the validity of meta-analyses.

III. THE PURPOSES OF A META-ANALYSIS

Some people believe that the purpose of a meta-analysis is "to obtain statistically significant results." They already know the answer, and want to convince others through the power of statistical significance. This is not always the goal. It is true that very small treatment effects based on very large numbers of patients may be statistically significant. Hence, finding a statistically significant difference in a meta-analysis may not in itself help a physician in deciding how to treat his or her patients. The size of the treatment effect is important, particularly when the treatment is toxic or expensive. In many cases, the class of treatments encompassed by a meta-analysis is broad, and the main outcome of the meta-analysis is a demonstration that the treatments have some positive effect on the endpoint of interest. The average effect estimated from the composite of treatments may not have much medical meaning, because it is an average of a broad class of treatments. If the average represents a substantial effect, then the physician may have been aided by the meta-analysis in deciding to treat his/her patient, although the meta-analysis will not tell the physician which specific treatment to use. If the average represents a small effect, however, one would not know whether this was because effective treatments were mixed with ineffective treatments, or because the treatments were generally only weakly effective. It may be difficult to learn this from the data; hence the meta-analysis, in this circumstance, would not be particularly helpful to the physician. In general, however, the purpose of a statistical analysis in a clinical trial, or a meta-analysis, is to estimate the size of treatment effects in order to aid clinical decision making.

Some purposes of meta-analyses are listed in Table 1. The first, overcoming publication bias and presenting results consistently, is

Table 1 Purposes of Meta-analysis

To overcome publication bias and consistently and objectively present all relevant trials

To resolve apparent conflicts in results of several trials

To overcome inadequate size of individual trials

To examine endpoints requiring larger samples than individual trials gathered

To evaluate subset effects

To examine generalizability of conclusions

To determine whether new trials are needed and to estimate probable treatment effects

To generate hypotheses for new trials

---

probably the most common and the most important. Our ability to retrospectively rationalize why one study is positive and others are negative is very refined. We have been taught that details matter—and it is always possible to find explanations for differences in study results. This approach to interpreting the literature is unsound. It may make us feel comfortable, because we can explain everything, but it leads to ascribing effectiveness to the treatment variant that by chance gives the largest observed effect. Our traditional guard against this trap is to require independent confirmation of results. This is well established in science and important in medicine. When we do a confirmatory study, we try to replicate the conditions of the original study exactly. When such a confirmatory study has not been done, then it is important to consider the results of all similar trials before interpreting the "good" result in isolation.

Meta-analyses are sometimes motivated by apparent conflicts in the results of similar clinical trials. The focus of such a meta-analysis may be to determine whether the difference is due to differences in the distribution of patients who entered the trials, the endpoints reported, or the methods of analysis used. If individual-patient data are available, the results can be re-analyzed to address these questions. In some cases, the differences that physicians see in the results of clinical trials are based on over-reliance on statistical significance. Just the fact that one of two results is statistically significant, while the other is not, does not mean that the two are inconsistent. Calculating an estimate of treatment effect for each study, and a confidence interval for each estimate, may quickly resolve such uncertainty and clarify that an average of the two estimates may be the best available summary of the expected treatment benefit for future patients. The two main components of data summarization in a meta-analysis are (a) displaying an estimate of treatment effect and its confidence limits for each study, and (b) averaging the study-specific estimates to obtain an overall estimate of treatment effect.

Table 1 lists three areas where obtaining overall average effects are of particular interest. The first is where most of the clinical trials are of inadequate size for detecting moderate but medically important treatment effects. This has been a very common situation in the past. Our society and academic system rewards independence, and hence everyone wants to do their own thing. This results in a mass of similar but non-identical clinical trials most of which are large enough to detect a big effect, but not to detect more moderate treatment benefits still of public-health importance. Meta-analysis is sometimes used to summarize results of trials done in this environment. Although the sample sizes of the individual trials may have been adequate to address the endpoint for which they were designed, meta-analysis may be used to analyze an alternative endpoint. For example, most U.S. clinical trials of primary breast cancer have been designed to detect differences in *disease-free* survival. To assess the effect of treatment on *survival* requires more patients and longer follow-up periods, because the effects are often smaller and manifest over a longer period. Hence, the meta-analysis of primary breast cancer organized by Peto et al. (2) was motivated by the wish to address survival; the effect of treatment on disease-free survival was already fairly clear based on individual studies.

Subset analysis is also an important motivation for meta-analyses. Subset analysis is one of the most controversial aspects in the reporting of clinical trials. Many investigators view subset analysis as unreliable, because they have seen so many apparently significant effects completely disappear in subsequent studies. Most clinical investigators and statisticians view subset analysis as generating hypotheses to be tested in other similar studies. It is often difficult to determine whether subset hypotheses generated in one study were confirmed or refuted in another, because the publications do not address it. Also, individual studies rarely have adequate size to reliably test subset hypotheses generated by another study. Consequently, a meta-analysis is the most appropriate setting for evaluating subset hypotheses. Even meta-analyses may have limited power for subset analysis, but it represents our best opportunity for examining the homogeneity of treatment effects based on randomized data.

Meta-analysis may also permit us to examine the generalizability of conclusions reached in the individual clinical trials. One of the difficulties physicians face in applying the results of clinical trials is trying to determine whether the patients treated in the trial are representative of the patients (s)he sees. The proportion of adult oncology patients treated in clinical trials is minuscule, and clinical trials sometimes have restrictive eligibility criteria. Randomization of treatment assignment among a highly selected group of patients does not ensure that the results of the trial are broadly applicable. A meta-analysis gives some opportunity to evaluate generalizability. Some proponents of meta-analysis view as an advantage the fact that clinical trials included in a meta-analysis are usually not identical.

Meta-analyses do not always come to the conclusion that a treatment does or does not work. Often the conclusion is that additional and better clinical trials are needed. The meta-analysis may provide some information about the likely size of the treatment effect to be realistically expected, and hence contribute to the development of a truly adequate clinical trial. The meta-analysis may also provide hypotheses to be tested in future clinical trials. For example, the average treatment effect for trials with longer-duration treatment may be substantially greater than the average for trials with shorter-duration treatment. This does not establish that longer-term treatment is better, because patient selection, treatment delivery, and other factors may be different for the two sets of trials. But it does provide a hypothesis to be prospectively tested in new clinical trials.

IV. THE STRUCTURE OF A META-ANALYSIS

Meta-analyses come in all shapes, sizes, and qualities. The best have the structure shown in Table 2 (3). Clinical trials are designed to be objective, with pre-specified patient-selection factors, treatments, and endpoints. When we address pre-specified hypotheses in a focused way, we can have more confidence in our findings than when we browse through a set of data searching for interesting patterns. A meta-analysis is a retrospective observational study and as such is inherently more subjective than a clinical trial (4). It is important to limit this subjectivity as much as possible, however, by prospectively structuring the process. This is best accomplished by writing a protocol for the meta-analysis.

Table 2 Structure of a Meta-analysis

Define specific objectives
Specify eligibility criteria for trials
Identify all eligible trials
Collect and validate data rigorously
Display trial-specific treatment effects with measures of precision
Examine variability of trial-specific treatment effects
Compute average treatment effects
Draw conclusions

The first step is to indicate the specific objectives to be achieved. This should be a written description of the questions to be answered and the hypotheses to be tested. The investigators should also describe the clinical conditions, therapies, and outcomes that will be grouped and compared. There should also be a written specification of eligibility criteria for clinical trials; for example all worldwide clinical trials of patients with histologically diagnosed FIGO stage I or II breast carcinoma in which there is truly randomized treatment assignment between surgery only and surgery followed within 30 days by tamoxifen administered for at least 1 year with no cytotoxic chemotherapy permitted. The meta-analysts must then try to identify all clinical trials that satisfy the eligibility criteria specified. If the eligibility criteria restricts consideration to published clinical trials, then the value of the meta-analysis may be severely compromised by publication bias (5). Most meta-analyses restrict themselves to published clinical trials. This is probably for convenience, although some investigators have expressed concern about using unpublished results because it has not passed peer review and may be of inferior quality. One of the primary motivations for doing a meta-analysis, however, is to avoid publication bias. This cannot be accomplished if the meta-analysis is limited to the published literature. Publication bias is real. It is not necessary, however, to include every clinical trial. One could restrict attention to trials sponsored by the NIH or trials conducted in the U.S. or western Europe. It is important to ensure that the likelihood of including a trial is not related to the results of the trial. It is difficult to identify all of the trials that should be included, and takes substantial effort. Many meta-analysts have called for the generation of a comprehensive registry of clinical trials to facilitate unbiased meta-analyses.

Meta-analyses in medicine should almost always be limited to randomized clinical trials. When treatment effects are large, meta-analyses are generally not needed. When treatment effects are moderate, they can only be studied if systemic biases of moderate degree are avoided. Meta-analyses attempt to avoid moderate-size random biases by accumulating large numbers, avoiding publication bias, and avoiding exclusion of randomized patients. Non-randomized studies, however, often have patient selection biases as large as the treatment effects.

Many meta-analyses are limited to published reports and abstract summary data from the publication. Most of the potential value of a meta-analysis is lost with this approach (3). As noted previously, published reports of clinical trials often have serious limitations. It is impossible to re-analyze without exclusions, check for valid randomization, examine subsets, and ensure uniform and unbiased approaches to analysis without the original individual-patient data. Not many data are needed for each patient—just a few baseline variables, the endpoints, and randomization information. In the discussion of their meta-analysis of adjuvant chemotherapy in head and neck cancer, Stell and Rawson commented: "This review has many sources of bias, some of which have already been referred to, but the main source of inaccuracy is the difficulty of extracting complete data from the various reports" (6). In collecting individual-patient data from investigators, sometimes additional follow-up is available, and this is also of value.

A major part of any meta-analysis is graphical display of the summary data from each study. For each study, a measure of treatment effectiveness compared to the control group is computed. A measure of uncertainty is also calculated for each study-specific treatment effect, and these estimates are graphically displayed. This is done for each endpoint of major interest. For survival, the measure of treatment effect is often the reduction in hazard rates for the treated group compared to the control group. Graphical display of the data often shows how imprecise the individual estimates of treatment effect are, and how much variability there is across studies.

The second part of the data summarization is the calculation of average treatment effects. These are computed as weighted averages of the study-specific treatment effects. There are two kinds of weighting used for computing the weighted averages. The most common approach is to assume study homogeneity: that the size of the true treatment effect is the same for each study, and that the observed results vary only because of random variation. This is a fixed-effect approach. For this approach, the weights are made proportional to the size of the individual studies. For survival data, the "size" of a study that determines the precision of the study-specific estimate of treatment effect is actually the number of deaths observed, not the number of patients entered. With this approach, one can compute a 95% confidence interval for the weighted-average treatment effect, and can test the null hypothesis that this average effect is really zero. Under the assumption of study homogeneity, this null hypothesis is equivalent to the true treatment effect being equal to zero in every study.

Testing the null hypothesis described above can be useful for determining whether there is evidence for any beneficial effect of therapy. The hypothesis test is valid even if the true treatment effects vary among studies. But the average treatment effect described in the previous paragraph and its measure of precision are based on the fiction that the true treatment effects are the same for each study. There are two ways one can analytically avoid making this assumption. First we can perform a statistical test of homogeneity of the study-specific treatment effects. It is important to do this, although the statistical power for detecting lack of homogeneity is often limited. The second approach is to use a random-effects model, which avoids the assumption of homogeneity of treatment effects (7); see also Sargent et al. (Chap. 11). In this case, a weighted-average treatment effect is computed using different weights from those described above. Less weight is given to very large studies than in the fixed-effect approach, because no one study by itself provides information about how treatment effect varies among studies. The confidence interval for this weighted average will be wider than for the fixed-effect average, because it incorporates across-study as well as within-study variability. There is some controversy in the statistical literature as to which model is more relevant. My view is that both are relevant, and results of both types of analysis should be presented. A measure of average treatment effectiveness is less meaningful if there is substantial inter-study variability.

These types of analysis are repeated for each endpoint and patient subset specified as an objective of the meta-analysis. Although more complex methods of analysis may be used to account for heterogeneity of patients or clinical trials, the basic structural features recommended here remain important. The reporting of the meta-analysis should not be a black-box process. The data for individual trials should be clearly presented, and the individual trials described and critically reviewed methodologically. The conclusions should be objectively based on the data but interpretations and limitations should be discussed from a knowledgeable medical perspective. There should be no rush to force strong conclusions or to tie too tidy a package.

V. IMPORTANT METHODOLOGIC FEATURES

Table 3 summarizes some of the important methodologic features of a meta-analysis. Most of the items have been already discussed; however several require additional comment. A central feature of meta-analysis is to "compare like with like." That is, we compute a study-specific estimate of treatment effect for each trial independently, and then average these study-specific estimates. We should not compare the performance of a treatment arm on one trial with a control arm on another trial. This key feature of meta-analyses of randomized clinical trials distinguishes such studies from much less reliable analyses of observational databases.

A corollary to the principle of comparing like with like is avoiding indirect comparisons, such as concluding that treatment A is superior to treatment B because the average size of effect of A versus control in one set of trials was greater than the average size of effect of B versus control in another set of trials, although A was never directly compared to B. This has been a source of considerable controversy. Such indirect comparisons are usually not sufficient, because of differences among the studies with regard to numerous factors such as patient-selection criteria.

It is also important to avoid combining incomplete data. Some studies report results for a secondary endpoint or particular subset of patients, because those results were interesting and suggestive. Combining data reported only by some studies is not reliable, and should be avoided. Obtaining individual-patient data often avoids this risk.

Table 3 Important Methodologic Features of a Meta-analysis

Has a narrow enough focus to be clinically meaningful
Avoids publication bias
Restricts studies to randomized trials
Obtains individual-patient data
Presents data from each individual trial
Avoids indirect comparisons

VI. THE ROLE OF META-ANALYSIS IN MEDICAL DECISION MAKING

The role of meta-analysis in medical decision-making, and the implications of the use of meta-analysis on the design of clinical trials, has been controversial. The National Cancer Institute and The National Heart, Lung and Blood Institute held a workshop in 1988 to discuss these issues (3). Much of the concern that physicians feel is based on what has been called the apples-and-oranges problem. That is, the estimates of treatment effect derived from meta-analyses are averages among studies that differ with regard to patient selection, treatments used, and quality control. How can such an average be used meaningfully to select a particular treatment for a particular patient? If the class of treatments included in the meta-analysis has no statistically significant benefit for a broad selection of patients in a set of clinical trials of diverse quality, then we cannot really conclude that none of the treatments are good. If the average treatment effect is positive and statistically significant, then the meta-analysis may still be of limited value in selecting a particular treatment for a particular patient, even if the range of studies is diverse with regard to treatment and quality. The average value itself may have little meaning in that circumstance. Some proponents of meta-analysis have argued that differences in treatment effect among drugs, dosage, or patient subsets are likely to be quantitative rather than qualitative, and hence overall pooled analyses are more likely to be reliable than more specifically focused analyses based on smaller numbers. In clinical decision-making with toxic or expensive treatments, however, quantitative differences are important. Determining that the average effect of a class of treatments on a class of patients is positive may be helpful in clinical decision-making, but it is usually not as specific as the information we hope to obtain from a well-designed randomized clinical trial.

Experience suggests that meta-analysis is generally not successful as a way of trying to make sense out of a very diverse set of studies most of which are grossly inadequate on their own. The variation in regimens, patients, designs, and study qualities will severely limit the strength of conclusions that can be reached. Marsoni et al. (8) decided not to attempt a meta-analysis in advanced ovarian cancer for this reason:

> We conclude that the internal coherence and development of randomized clinical trials in advanced ovarian cancer, and their methodologic soundness are quite poor. In this situation meta-analysis cannot

go beyond a systematic attempt to answer a very general treatment effectiveness question . . . (i.e., do any combination chemotherapies contribute to patient survival).

Nicolucci et al. (9) rejected a plan to do a meta-analysis of the treatment of advanced lung cancer for the same reason:

> To make more efficient use of sparse data from small trials, meta-analysis is becoming increasingly popular. To be valid, however, it must be based on combinations of trials in which similar enough therapies were tested and quality was at least roughly comparable . . . Our review of more than 20 years of research in the treatment of lung cancer also strongly suggests the lack of any apparent coordinated development of hypotheses. Given the systematic qualitative inadequacy of almost all trials, one may conclude that this way of conducting research may have led either to premature abandonment of potentially promising combination or to the unjustified testing of excessively complex or toxic combinations... While clinical research in lung cancer has contributed little to defining the best standard care, we conclude that its heterogeneity makes it unlikely that quantitative meta-analysis of existing trials will be constructive.

Meta-analysis was developed for integrating the results of small diverse studies in the social sciences. Glass et al. (1) state:

> Of course, it is unclear what meta-analysis will contribute to the progress of empirical research. One can imagine a future for research in the social and behavioral sciences in which questions are so sharply put and techniques so well-standardized that studies would hardly need to be integrated by merit of their consistent findings. But that future seems unlikely. Research will probably continue to be an unorganized, decentralized, non-standardized activity pursued simultaneously in dozens of places without thought to how it will all fit together in the end.

In medical therapeutics we have lived in this disorganized world and have hopefully left it. Innovation and independence are crucial in basic research and in pilot investigations. But when the time comes to do a phase III clinical trial, the do-your-own-thing approach often inhibits the ability to find useful treatments of value to public health. Many of the do-your-own-thing phase III trials are uninterpretable and add only confusion to the medical literature. Fortunately, many areas of medical therapeutics have in recent years become more aware of the need for doing more adequately sized phase III clinical trials. Because of the large number of clinical trials being conducted, it is still important to confirm initial positive findings in independent clinical trials, whenever possible. This is a basic feature of science, and is very important in medicine if we want reliable information (10). Physicians sometimes over-interpret the results of an individual clinical trial that gives favorable results, and dismiss others of similar design that did not. Given the number of clinical trials being conducted, this phenomenon leads to errors that can be avoided by confirmatory studies. Also, even a large well-done clinical trial is generally inadequate for reliably addressing subset questions and for evaluating the effect of treatment on long-term survival in a chronic disease like breast cancer. Hence there is a role for meta-analyses of well-designed clinical trials that can focus on rather specific treatments in well-defined patient populations. To the extent that we expect meta-analyses to be alternatives to good large clinical trials, we will be disappointed, however, and we will disappoint our patients. As Wittes (3) commented: "Clinicians will continue to rely chiefly on the results of individual clinical trials for the most reliable estimates of the impact of particular regimens on well-defined patient groups." To the extent that many trials have in the past been undersized, there is, however, no reason to assume that the pooling of randomized clinical trials can bail us out of the fundamental uncertainty that will accompany any situation where excellent clinical trials, the cornerstone of therapy evaluation, do not exist. The more pressing problem facing clinical-trial investigators, therefore, is to make these trials as excellent, pertinent, and informative as they can be.

References

1. GV Glass, B McGraw, ML Smith. Meta-analysis in clinical trials. Beverly Hills, CA: Sage Publications, 1981.
2. Group EBCTC. Effects of adjuvant tamoxifen and of cytotoxic therapy on mortality in early breast cancer. New England Journal of Medicine 319:1681–1692, 1988.
3. S Yusuf, R Simon, S Ellenberg. Proceedings of the workshop on the methodologic issues in overview of randomized clinical trials. Statistical Medicine 6:217–409, 1987.
4. R Simon. Overviews of randomized clinical trials. Cancer Treatment Reports 71:3–5, 1987.

5. CB Begg, JA Berlin. Publication bias and dissemination of clinical research. Journal of the National Cancer Institute 81:107–115, 1989.
6. PM Stell, NSB Rawson. Adjuvant chemotherapy in head and neck cancer. British Journal of Cancer 61:779–787, 1990.
7. R DerSimonian, N Laird. Meta-analysis in clinical trials. Controlled Clinical Trials 7:177–188, 1986.
8. S Marsoni, V Torri, A Taiana. Critical review of the quality and development of randomized clinical trials and their influence on the treatment of advanced epithelial ovarian cancer. Annals of Oncology 1:343–350, 1990.
9. A Nicolucci, R Grilli, AA Alexanian. Quality evolution and clinical implications of randomized controlled trials on the treatment of lung cancer. Journal of the American Medical Association 202:2101–2107, 1989.
10. MKB Parmar, RS Ungerleider, R Simon. Assessing whether to perform a confirmatory randomised clinical trial. Journal of the National Cancer Institute 88:1645–1651, 1996.

14

Research Synthesis for Public Health Policy: Experience of the Institute of Medicine

Michael A. Stoto
George Washington University School of Public Health and Health Services, Washington, D.C.

I. INTRODUCTION

In two recent studies, the Institute of Medicine (IOM) of the National Academy of Sciences has developed and applied new techniques of research synthesis, including meta-analysis, to help clarify politically charged policy issues. The Congress requested these studies because of concern that people who were suffering adverse health outcomes deserved compensation. The health outcomes were thought to be due to exposure to chemicals used in Vietnam in one case, and to the use of childhood vaccines in the other. As a non-governmental agency, the IOM cannot make compensation policy, but it can assess and summarize the existing scientific evidence to inform the federal agencies charged with these decisions. In both projects, the IOM used meta-analysis for combining statistical information wherever possible. The IOM's response to these two congressional mandates is contained in four reports (1–4), from which this paper is largely drawn. The purpose of this paper is to illustrate the approach that the IOM has taken, with a focus on statistical aspects of the analysis, and to discuss the potential applicability of this approach in other settings.*

The National Academy of Sciences (NAS) is a unique private institution charged with advising the U. S. government on matters of science and public policy. It was chartered by Congress in 1863 to provide a mechanism for eminent scientists and engineers to "investigate, examine, experiment, and report upon any subject of science or art" requested by the government. The Institute of Medicine was established in 1970 by the NAS to enlist distinguished medical and other professionals to study issues that affect public health. Like the NAS, the IOM conducts studies on particular topics and has a distinguished elected membership.

All IOM studies are carried out by expert volunteer committees appointed specifically for that purpose. Committee members are chosen to avoid conflicts of interest relative to the subjects of particular reports. These committees generally meet in closed sessions to evaluate the evidence and prepare findings and recommendations. All findings must be fully documented in written reports, and an appointed panel of expert reviewers ensures that the findings are consistent with the data in the reports.

---

*Much of this paper is drawn directly from the cited reports, and the conclusions are those of the authoring IOM committees. Members of the committees and staff responsible for these reports include Cynthia Abel, Russell Alexander, Michael Aminoff, Alan Aron, Arthur Asbury, John Bailey, Jesse Berlin, Karen Bolla, Norman Breslow, Kelley Brix, David Butler, Sanjay Baliga, Charles Carpenter, Lynn Cates, Gail Charnley, Mary Luz Coady, Graham Colditz, Kay Dickersin, Harvey Fineberg, Harold Fallon, Christopher Goetz, Marie Griffin, Seymour Grufferman, Katharine Hammond, Cynthia Howe, Christopher Howson, Richard B. Johnston Jr., Richard T. Johnston, Norbert Kaminski, Michael Katz, Michael Kramer, David Kriebel, Darwin Labarthe, David Lane, Tamar Lasky, Bryan Langholz, Catharyn Liverman, Carol Maczka, Kenneth McIntosh, Frederick Mosteller, Karle Mottet, Diane Mundt, Alfred Neugut, William Nicholson, Peter Nowell, Andrew Olshan, Malcolm Pike, Ken Ramos, Kathleen Rogers, Susan Rogers, Catherine Rose, Noel Rose, Penelope Shackelford, Bennett Shaywitz, Nancy Sprince, Kathleen Stratton, Paul Stolley, Michael Stoto, David Tollerud, and Clifford Weisel. The author is responsible for the way that the reports are summarized, and for the conclusions at the end of this paper.

II. AGENT ORANGE

A. Background

Between 1962 and 1971, U.S. military forces sprayed nearly 72 million litres of herbicide over approximately 3.6 million acres in Vietnam. The preparation known as Agent Orange accounted for approximately 42 million litres of the total amount sprayed. Herbicides were used to strip the thick jungle canopy that helped conceal opposition forces, to destroy crops that enemy forces might depend upon, and to clear tall grass and bushes from around the perimeters of U.S. base camps and outlying fire support bases. Most large-scale spraying operations were conducted using airplanes and helicopters, but considerable quantities of herbicide were sprayed from boats and ground vehicles, as well as by soldiers wearing back-mounted equipment. Spraying began in 1962 and increased greatly in 1967. After a scientific report in 1969 concluded that one of the primary chemicals used in Agent Orange, namely 2,4,5-trichlorophenoxyacetic acid (2,4,5-T), could cause birth defects in laboratory animals, U.S. forces suspended use of this herbicide in 1970 and halted all herbicide spraying in Vietnam the next year.

Throughout the 1970s concern about possible long-term health consequences of Agent Orange and other herbicides heightened, fueled in particular by reports from growing numbers of Vietnam veterans that they had developed cancer or fathered handicapped children, which they attributed to wartime exposure to the herbicides. Along with the concerns of Vietnam veterans, public awareness increased because of reports of health concerns surrounding occupational and environmental exposure to dioxin—more specifically, 2,3,7,8-tetrachlorodibenzo-p-dioxin (2,3,7,8-TCDD), informally known as TCDD—a contaminant of 2,4,5-T. Thousands of scientific studies have since been conducted, numerous government hearings have been held, and veterans' organizations have pressed for conclusive answers, but the question of the health effects of herbicide exposure in Vietnam remained shrouded in controversy and mistrust. Indeed some veterans organizations, researchers, and public-interest organizations remained skeptical about whether the issue has received full and impartial consideration by the Department of Veterans Affairs (DVA; formerly the Veterans Administration) and other federal agencies.

Faced with this lingering uncertainty and demands that the concerns of veterans be adequately addressed, the Congress passed Public

Law (P.L.) 102-4, the "Agent Orange Act of 1991." This legislation codified DVA compensation policy for Vietnam veterans; decisions were to be made on a disease-by-disease basis, and any Vietnam veteran diagnosed with one of the "service-connected" diseases or conditions would be entitled to compensation without regard to exposure to herbicide. Compensation was to be based on preponderance of the evidence regarding a statistical association between exposure to the herbicides and disease. In particular, a condition would be determined to be "service connected" if the evidence for an association was stronger than the evidence against an association.

P.L. 102-4 also directed the Secretary of Veterans Affairs to request that the NAS conduct a comprehensive review and evaluation of available scientific and medical information regarding the health effects of exposure to Agent Orange, other herbicides used in Vietnam, and their components, including dioxin. In February 1992, the IOM agreed to review and summarize the strength of the scientific evidence concerning the association between herbicide exposure during Vietnam service and each disease or condition suspected to be associated with such exposure.

The IOM committee's review of the substantial data from basic research and animal studies established the biologic plausibility of health effects that may occur in humans after accidental or occupational exposure to herbicide and TCDD components. TCDD administered to laboratory animals interacts with an intracellular protein called the Ah receptor. This interaction appears to play a role in a number of health effects observed in animals. Because humans also have intracellular proteins that have been identified as Ah receptors, it is plausible that interactions between TCDD and these receptors could play a role in human health effects. In contrast to TCDD, the effects of the herbicides do not appear to be mediated through interactions with intracellular receptors. TCDD has also been shown to have a wide range of effects in laboratory animals on growth regulation, hormone systems, and other factors associated with the regulation of activities in normal cells. In addition, TCDD has been shown to cause cancer in laboratory animals at a variety of sites. If TCDD has similar effects on cell regulation in humans, it is plausible that it could have an effect on human cancer incidence. In contrast to TCDD, there is no convincing evidence in animals of, or mechanistic basis for, carcinogenicity or other health effects of any of the herbicides, although they have not been studied as extensively as TCDD.

In fulfilling its charge of judging whether each of a set of human health effects is associated with exposure to herbicide or dioxin, most of the committee's efforts concentrated on reviewing and interpreting epidemiologic studies. The committee staff carried out a systematic literature review, including both electronic searches and inspection of the bibliographies in scientific papers and government documents. In addition, the committee also convened three public hearings and other informal meetings at which veterans, veterans organizations, and others could bring material, including scientific studies, to the committee's attention.

In reviewing the literature, the committee discerned that the existing epidemiologic database is severely lacking in quantitative measures of individual exposure to herbicides and dioxin. Assessment of the intensity and duration of individual exposures is a key component in determining whether specific health outcomes are associated with exposure to dioxin or other chemicals found in the herbicides used in Vietnam. Although different approaches have been used to estimate exposure in Vietnam veterans and in others exposed occupationally or environmentally, each approach is limited in its ability to determine precisely the degree and level of individual exposure.

Although definitive data are lacking, the available quantitative and qualitative evidence about herbicide exposure suggests that Vietnam veterans as a group had substantially lower exposure to herbicide and dioxin than the subjects in many occupational studies. Due, in part, to the uncertain validity of exposure measurements in many of the studies of veterans, the committee decided to review studies of other groups potentially exposed to the herbicides used in Vietnam and TCDD—especially phenoxy herbicides, including 2,4-dichlorophenoxyacetic acid (2,4-D) and 2,4,5-T, chlorophenols, and other compounds. These groups include chemical production and agricultural workers, residents of Vietnam, and people exposed heavily to herbicide or dioxins as a result of residing near the site of an accident, such as occurred in Seveso, Italy in 1976. The committee felt that considering studies of other groups could help address the issue of whether exposure to these compounds might be associated with particular health outcomes, even though these results would have only an indirect bearing on the increased risk of disease in Vietnam veterans. Some of these studies, especially those of workers in chemical production plants, provide stronger evidence about health effects than studies of veterans, because exposure was more easily quantified and measured. Furthermore, the general level and duration of exposure to the chemicals were greater, and the studies were of sufficient size to examine the health risks among those with varying levels of exposure.

B. Committee's Conclusions

The committee's primary mandate was to determine whether there is a statistical association between the suspect diseases and herbicide exposure, taking into account the strength of the scientific evidence and the appropriateness of the methods used to detect the association. The committee addressed this charge by assigning each of the health outcomes under study into one of the following four categories on the basis of the epidemiologic evidence that it reviewed.

Sufficient Evidence of an Association. Evidence is sufficient to conclude that there is a positive association between herbicides and the outcome in studies in which chance, bias, and confounding could be ruled out with reasonable confidence. For example several small studies that are free from bias and confounding and that show an association that is consistent in magnitude and direction may be sufficient evidence for an association. The committee found sufficient evidence of association between exposure to herbicide or dioxin and soft-tissue sarcoma, non-Hodgkin's lymphoma, Hodgkin's disease, and chloracne (an acne-like skin disorder.)

Limited/Suggestive Evidence of an Association. Evidence is suggestive of an association between herbicides and the outcome but is limited because chance, bias, and confounding could not be ruled out. For example, at least one high-quality study shows a positive association, but other studies are inconsistent. The committee found limited or suggestive evidence of association between exposure to herbicide and respiratory cancer, prostate cancer, multiple myeloma, and spina bifida.

Inadequate/Insufficient Evidence to Determine Whether an Association Exists. Available studies are of insufficient quality, consistency, or statistical power to infer the presence or absence of an association. For example, studies fail to control for confounding, have inadequate exposure assessment, or fail to address latency. Most of the health outcomes fell into this category.

Limited/Suggestive Evidence of *no* Association. Several adequate studies, covering the levels of exposure that humans encounter, all fail to show a positive association between exposure to herbicides and outcome. A conclusion of "no association" is inevitably limited to the conditions, level of exposure, and length of observation covered by available studies. The committee found that a sufficient number and variety of well-designed studies exist to conclude that there is limited or suggestive evidence of no association between the herbicides or dioxin and gastrointestinal tumors, bladder cancer, and brain tumors.

The committee's characterization of the evidence can be seen schematically in Fig. 1. The horizontal axis represents the strength of the association, with a relative risk (RR) of 1.0 corresponding to no association, and larger, positive RRs corresponding to larger risks. The vertical axis represents the strength of the evidence, taken as a whole. The committee began its evaluation presuming neither the existence nor the absence of association, and many of the outcomes examined remained in the large oval, suggesting "inadequate/insufficient" evidence, and thus nothing about the presence or absence of an association. In some cases, the evidence was stronger, leading the committee to find "limited/suggestive" evidence of an association or of *no* association, and for some outcomes "sufficient" evidence of an association. This classification scheme is similar to the one used by the International Agency for Research on Cancer in evaluating the evidence for carcinogenicity of various agents (5), but the IOM focus was on evidence for particular outcomes rather than for "cancer" more generally, and on statistical association rather than evidence of causation. The classification scheme is also similar to the U. S. Preventive Services Task Force's rating system of quality of scientific evidence (6), apart from its focus on evidence of clinical benefits (as opposed to risk) and on randomized controlled trials.

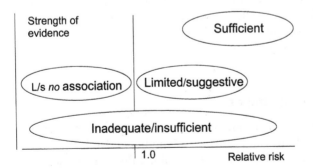

Figure 1 Schematic representation of the categories used in *Veterans and Agent Orange* (3). The horizontal axis represents the strength of the association, in terms of relative risk, and the vertical axis represents the strength of the evidence, taken as a whole.

The assignment of diseases to one of these categories required substantial judgment and group deliberation in which the committee compared diseases in the same group and considered whether the evidence was similar. These judgments have both quantitative and qualitative aspects. They reflect the nature of the exposure, health outcome, and population at issue; the characteristics of the evidence examined; and the approach taken to evaluate that evidence. The committee focused on the evidence as a whole rather than on each specific study, seeking to characterize and weigh the strength and limitation of the available evidence. In some instances, the committee felt that a formal meta-analysis could help summarize a subset of the available evidence. In most instances, the committee's judgment was based on a qualitative review of the existing studies.

In reviewing the committee's results, certain caveats must be borne in mind. First, as is consistent with the law governing DVA compensation decisions, the committee's conclusions are based on association rather than causality. The biological plausibility of the association was assessed, but this was kept separate from the committee's conclusions regarding association. The committee used the other components of the Bradford Hill criteria (7) to assess strength of the evidence regarding association, but used a lower threshold to determine its conclusions than if the question were causality. Second, because of poor exposure information, the committee determined that it was not possible to quantitatively estimate the excess risk of any health outcome for Vietnam veterans. This was not a major problem, since the law does not allow the DVA to consider attributable risks in making compensation decisions. Finally, the conclusions are drawn largely from occupational and other studies in which subjects were exposed to a variety of herbicides and herbicide components.

C. Detailed Examples

The following examples are presented to illustrate the committee's analysis. The first four—non-Hodgkin's lymphoma, Hodgkin's disease, multiple myeloma, and leukemia—are etiologically related and illustrate the differences between the first three evidence categories. The other three examples—respiratory cancer, prostate cancer, and spina bifida—illustrate the difficult judgments that must be made when the data are less than definitive.

The figures and some of the meta-analysis results in this section of the paper do not appear in the committee's reports. Although the committee used analyses of this sort in its review of the data, it was felt that presenting the data in this way could be misleading for three reasons. First, many epidemiological studies present multiple RR estimates, corresponding to different study groups, control groups, outcomes, exposure measures, and statistical models. These multiple estimates are presented in tabular form in the committee's report, but a judgment is required about which one to use in the type of summary figure used in this section. Second, graphical analyses of this sort do not convey important information about the nature of the exposure, within-study evidence of a dose–response relationship, or the quality of the individual studies, and are thus misleading. Third, some of the committee members felt that use of formal meta-analysis of the extremely heterogeneous study results available was unwarranted. The figures and formal meta-analyses presented in this section are the responsibility of the present author, and are included in this methodological paper only to explore the potential contribution of these techniques.

1. Non-Hodgkin's Lymphoma

Non-Hodgkin's lymphoma (NHL) includes a group of malignant lymphomas: that is, neoplasms derived from lymphoreticular cells in lymph nodes, bone marrow, spleen, liver, or other sites in the body. One large well-conducted case-control study in Sweden by Hardell et al. (8) examined NHL and Hodgkin's disease together and found an odds ratio of 6.0 (CI 3.7–9.7) based on 105 cases for exposure to phenoxy acids or chlorophenols, and these results held up under further investigation of the validity of exposure assessment and other potential biases (8). A more recent case-control study by Persson et al. (9) showed increased risk for NHL in those exposed to phenoxy acids (OR = 4.9, CI 1.0–27.0), based on a logistic regression analysis of 106 cases. Other studies of farmers and agricultural workers are generally positive for an association between NHL and herbicide/TCDD; however, only some are significant. All of the studies of U.S. agricultural workers reviewed showed elevated relative risks (although none were significant), and two NCI studies of farmers in Kansas and Nebraska (10, 11) show patterns of increased risk linked to use of 2,4-D. The CDC Selected Cancers Study found an increased risk of NHL in association with service in Vietnam; other studies of veterans, generally with small sample sizes, are consistent with an association.

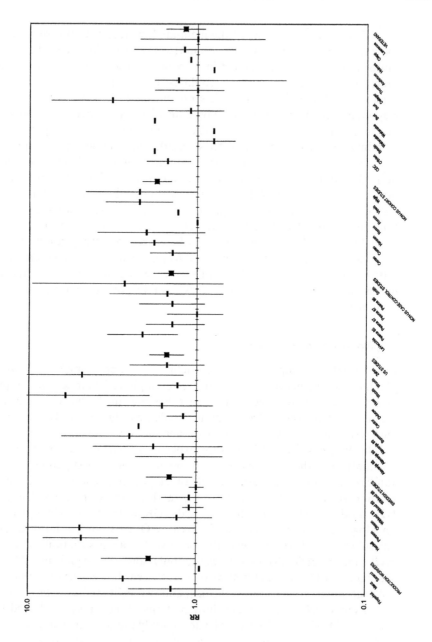

Based on this evidence, the committee concluded that there was sufficient evidence of an association between exposure to the herbicides used in Vietnam or their components and non-Hodgkin's lymphoma.

In contrast, studies of production workers, including the largest and most heavily exposed cohorts (12–15) indicate no increased risk. Thus, unlike most of the other cancers studied by the committee, for which the data do not distinguish between the effects of herbicides and TCDD, the available epidemiologic data suggest that the phenoxy herbicides, including 2,4-D, rather than TCDD may be associated with non-Hodgkin's lymphomas.

Figure 2, which indicates the primary RR estimate for each study with a horizontal line together with its 95-percent confidence interval where available, confirms the committee's finding that the estimated RRs are consistently greater than 1, and many significantly so. The studies are grouped in this figure by exposure category: production workers, Swedish case-control studies, U.S. cohort studies, non-U.S. case-control studies, non-U.S. cohort studies, and Vietnam veterans. For each group of studies, a formal meta-analysis was calculated using the DerSimonian–Laird (16) random-effect model, and the pooled RR and 95-percent confidence intervals shown on the figure with squares rather than bars. All of the pooled estimates, as well as the lower 95-percent confidence limits on the pooled estimates except for the Vietnam veterans, are greater than 1.0. The pooled estimate for production workers does not include the study by Saracci et al. (13), because no confidence intervals were published; if it had been included, then the pooled RR would be lower and not statistically significant.

2. Hodgkin's Disease

Hodgkin's disease (HD), also a malignant lymphoma, is a neoplastic disease characterized by progressive anemia and enlargement of lymph nodes, spleen, and liver. Fewer studies have been conducted of HD in

---

Figure 2 Summary of epidemiological studies relating Non-Hodgkin's lymphoma and exposure to herbicide and herbicide components. The primary RR estimate for each study is represented with a horizontal line, and the vertical line represents the 95-percent confidence interval where available. Meta-analysis results for each group of studies are represented by squares. For details on individual studies, see Ref. 3.

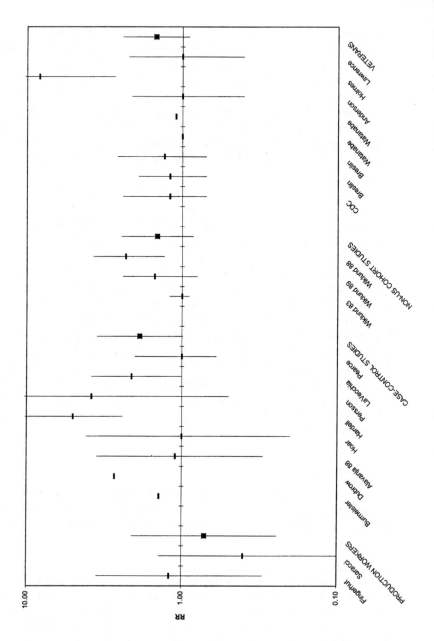

relation to exposure to herbicide or TCDD than have been conducted of NHL, but the pattern of results is strikingly consistent. The 60 HD cases in the study by Hardell et al. (8) were later examined by Hardell and Bengtsson (17), who found odds ratios of 2.4 (CI 0.9–6.5) for low-grade exposure to chlorophenols and 6.5 (CI 2.7–19.0) for high-grade exposures. Persson and colleagues' study (9) of 54 HD cases showed a large, but not statistically significant, OR of 3.8 (CI 0.5–35.2) for exposure to phenoxy acids. Furthermore, nearly all of the 13 case-control and agricultural-worker studies show increased risk for HD, although only a few of these results are statistically significant. As with NHL, even the largest studies of production workers exposed to TCDD do not indicate an increased risk. The few studies of HD in Vietnam veterans tend to show elevated risks; however, all but one are not statistically significant. Based on this evidence, the committee concluded that there was sufficient evidence of an association between exposure to the herbicide used in Vietnam or their components and Hodgkin's disease.

Graphically (Fig. 3), the Hodgkin's disease results look similar to those for non-Hodgkin's lymphoma, except that there are fewer published studies to include. The study by Saracci and colleagues is included in this analysis, and the pooled RR is less than 1.0.

3. Multiple Myeloma

Multiple myeloma (MM), like non-Hodgkin's lymphoma and Hodgkin's disease, is derived from lymphoreticular cells. Multiple myeloma has been less extensively studied than other lymphomas, but a consistent pattern of elevated risks appears in the studies that have been conducted, as can be seen in Fig. 4. Ten studies of agricultural and forestry workers provide information on MM risk in relation to herbicide or pesticide exposure. All demonstrated an odds ratio or SMR greater than 1.0; seven did so at a statistically significant level. This finding is made more specific for herbicide exposure by subanalyses in four of these studies (18–21) that

---

Figure 3 Summary of epidemiological studies relating Hodgkin's disease and exposure to herbicide and herbicide components. The primary RR estimate for each study is represented with a horizontal line, and the vertical line represents the 95-percent confidence interval where available. Meta-analysis results for each group of studies are represented by squares. For details on individual studies, see Ref. 3.

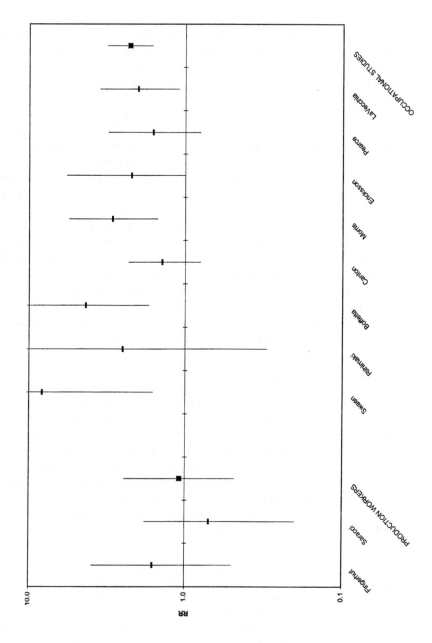

suggest higher risks for those exposed to herbicide, and higher risks for the studies of herbicide applicators (22, 23). The committee determined that the evidence for this association was limited/suggestive because the individuals in the existing studies—mostly farmers—have, by the nature of their occupation, probably been exposed to a range of potentially carcinogenic agents other than herbicide and TCDD.

4. Leukemia

The epidemiologic evidence for an association between exposure to herbicide and leukemia (all forms) comes primarily from studies of farmers and residents of Seveso, Italy. The observed overall relative risk for leukemia mortality and incidence in Seveso was elevated, but not significantly. As can be seen in Fig. 5, a number of studies of farmers that the committee found convincing for NHL, HD, or MM also show a consistently elevated risk of leukemia; but these results are not necessarily due to herbicide use, because (a) confounding exposures were not controlled for adequately in the analyses of these studies, and (b) when farmers are stratified by suspected use of herbicide, the incidence of leukemia is generally not elevated. Some studies of chemical workers found an increased risk of leukemia, but the number of cases was small in all of these studies. The available data on Vietnam veterans are generally not conclusive because the exposure data are inadequate for the cohort being studied. Small sample sizes weaken the studies of the Ranch Hands or Chemical Corps, where excesses are not likely to be detected. The committee thus determined that the evidence was inadequate/insufficient to determine whether an association exists between exposure to the herbicides used in Vietnam or their components and leukemia.

5. Respiratory Cancer

Among the many epidemiologic studies of respiratory cancers (specifically cancers of the lung, larynx, and trachea), positive associations were

---

Figure 4 Summary of epidemiological studies relating multiple myeloma and exposure to herbicide and herbicide components. The primary RR estimate for each study is represented with a horizontal line, and the vertical line represents the 95-percent confidence interval where available. Meta-analysis results for each group of studies are represented by squares. For details on individual studies, see Ref. 3.

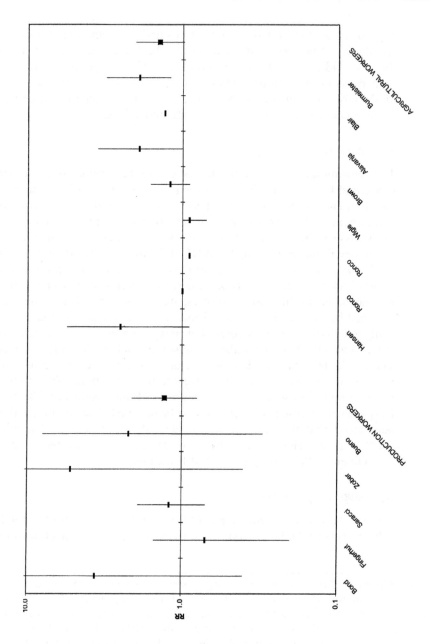

found consistently only in those studies in which TCDD or herbicide exposures were probably high and prolonged, especially the largest and most heavily exposed cohorts of chemical production workers exposed to TCDD (14, 12, 15, 13) and herbicide applicators (24–27). Studies of farmers tended to show a decreased risk of respiratory cancers (perhaps due to lower smoking rates), and studies of Vietnam veterans are inconclusive. The committee felt that the evidence for this association was limited/suggestive, rather than sufficient, because of the inconsistent pattern of positive findings across populations with various degrees of exposure and because the most important risk factor for respiratory cancers—cigarette smoking—was not fully controlled for or evaluated in all studies.

Figure 6 shows the results that contributed to this conclusion, including formal meta-analysis results which support the committee's conclusions. The pooled estimate for the production workers was included in the committee's report because the component studies were judged to be sufficiently similar to warrant a pooled analysis.

6. Prostate Cancer

Several studies have shown elevated risk for prostate cancer in agricultural or forestry workers. In a large cohort study of Canadian farmers (28), an increased risk of prostate cancer was associated with herbicide spraying, and increasing risk was shown with increasing number of acres sprayed. For the entire cohort (those spraying at least 250 acres), the relative risk for prostate cancer was 1.2 (CI 1.0–1.5). When the analysis was restricted to the farmers most likely to be exposed to phenoxy herbicides or other herbicides, and those with no employees or contractors to do the spraying for them, and aged between 45 and 69 years, the test for trend over increasing number of acres sprayed was significant. The risk was also elevated in a study of USDA forest conservationists (OR = 1.6, CI 0.9–3.0) (20), and a case-control study of white male Iowans who died of prostate cancer (18) found a significant association (OR = 1.2) that

---

Figure 5 Summary of epidemiological studies relating leukemia and exposure to herbicide and herbicide components. The primary RR estimate for each study is represented with a horizontal line, and the vertical line represents the 95-percent confidence interval where available. Meta-analysis results for each group of studies are represented by squares. For details on individual studies, see Ref. 3.

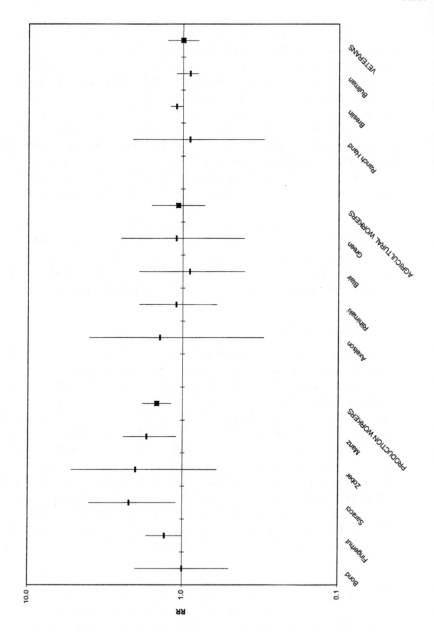

was not related to any particular agricultural practice. These results are strengthened by a consistent pattern of nonsignificant elevated risks in studies of chemical production workers in the United States and other countries, agricultural workers, pesticide applicators, paper and pulp workers, and the Seveso population. The largest recent study demonstrated a significantly increased risk of death from prostate cancer (RR 1.2, CI 1.1–1.2) in both white and nonwhite farmers in 22 of the 23 states that were studied (29). Studies of prostate cancer among Vietnam veterans or following environmental exposures have not consistently shown an association. However, prostate cancer is generally a disease of older men, and an increased risk among Vietnam veterans would be unlikely to show up in current epidemiologic studies. Because there was a strong indication of a dose-response relationship in one study and a consistent positive association in a number of others, as can be seen in Fig. 7, the committee felt that the evidence for association with herbicide exposure was limited/ suggestive for prostate cancer.

7. Spina Bifida

There have been three epidemiological studies that suggest an association between paternal herbicide exposure and an increased risk of spina bifida. In the Ranch Hand study (30), neural tube defects (spina bifida, anencephaly) were increased among offspring of ranch hands with four total (rate of 5 per 1,000), in contrast to none among the comparison infants (exact $p = 0.04$). The Centers for Disease Control and Prevention (CDC) VES cohort study (31) found that more Vietnam veterans reported that their children had a central nervous system anomaly (OR = 2.3; 95% CI 1.2–4.5) than did non-Vietnam veterans. The odds ratio for spina bifida was 1.7 (CI 0.6–5.0). In a substudy, hospital records were examined in an attempt to validate the reported cerebrospinal defects (spina bifida, anencephaly, hydrocephalus). While a difference was detected, its interpretation is limited by differential participation between the veteran groups

Figure 6 Summary of epidemiological studies relating respiratory cancer and exposure to herbicide and herbicide components. The primary RR estimate for each study is represented with a horizontal line, and the vertical line represents the 95-percent confidence interval where available. Meta-analysis results for each group of studies are represented by squares. For details on individual studies, see Ref. 3.

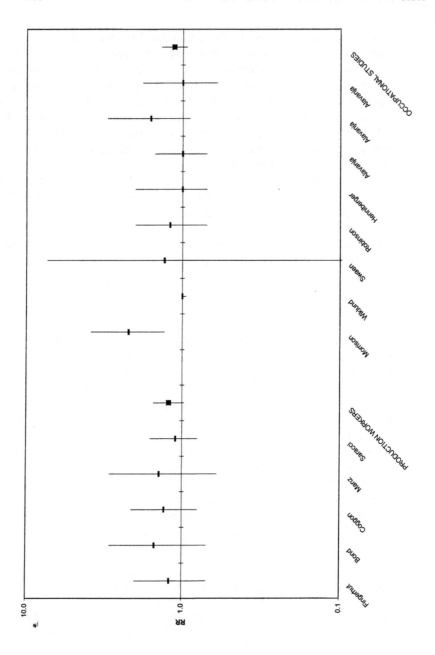

and failure to validate negatives reported; that is, the veterans not reporting their children having a birth defect. Thus, the issue of a recall bias is of major concern with this study. In the CDC Birth Defects Study, which utilized the population-based birth defects registry system in the metropolitan Atlanta area (32), there was no association between anencephaly (OR = 0.9, CI 0.5–1.7). However, the exposure opportunity index (EOI) based upon interview data was associated with an increased risk of spina bifida; for the highest estimated level of exposure (EOI-5), the OR was 2.7 (CI 1.2–6.2). There was no similar pattern of association for anencephaly. Thus, all three epidemiological studies (Ranch Hand, VES, CDC Birth Defects Study) suggest an association between herbicide exposure and an increased risk of spina bifida in offspring.

In contrast to most other diseases, for which the strongest data have been from occupationally exposed workers, these studies focused on Vietnam veterans. Although the studies were judged to be of relatively high quality, they suffer from methodological limitations, including possible recall bias, nonresponse bias, small sample size, and misclassification of exposure. For these reasons, the committee concluded that there is limited/suggestive evidence for an association between exposure to herbicide used in Vietnam and spina bifida in offspring.

D. Policy Impact

In a series of decisions, DVA decided to "service-connect" for all outcomes in the IOM "sufficient" and "limited/suggestive" categories. Prostate cancer was included in 1996, after the publication of second IOM report and the publication of the results by Blair et al. (29). Compensation for respiratory cancer is limited to cases occurring within 35 years of Vietnam service. In 1997, Congress passed legislation allowing the DVA, for the first time, to compensate veterans for a condition in their offspring: spina bifida.

Figure 7 Summary of epidemiological studies relating prostate cancer and exposure to herbicide and herbicide components. The primary RR estimate for each study is represented with a horizontal line, and the vertical line represents the 95-percent confidence interval where available. Meta-analysis results for each group of studies are represented by squares. For details on individual studies, see Ref. 3.

III. VACCINE RISKS

Childhood immunization has been one of the most effective public-health measures of the twentieth century. It prevents many diseases that in the past accounted for enormous morbidity and mortality. Serious neurological side effects and even death in some children, however, led some parents in the early 1980s to refuse vaccinations for their children. Some parents sued vaccine manufacturers, which led some companies to curtail vaccine research, development, and production. In response, Congress passed the National Childhood Vaccine Injury Act (NCVIA) in 1986 to encourage vaccine production by creating a no-fault compensation program for children who are injured by vaccines. To inform decisions about which injuries should be compensated, the NCVIA called upon the Institute of Medicine to study the evidence bearing on the adverse effects of childhood vaccines.

The charge to the vaccine safety committee concerned causality more than association. The committee judged whether each of a set of adverse events was caused by exposure to a particular vaccine. These judgments have quantitative and qualitative aspects. They reflect the nature of the exposures, events, and populations at issue; the characteristics of the evidence examined; and the approach taken to evaluate that evidence. The exposures—vaccinations—are widespread in the population, so absence of exposure may itself require explanation in interpreting studies. The adverse events under consideration by the committee are mostly rare in the exposed population, and can occur in the absence of vaccination. Some are clinically ill defined, or without known causes. Such features raise difficulties for the investigation and evaluation of the evidence.

The committee considered three kinds of causal questions about adverse events that may be caused by a vaccine:

1. *Can the vaccine cause the event, at least in certain people and circumstances?* This question was the main focus of the committee's work. The committee examined controlled epidemiological studies, using standard criteria for assessing causality: strength of association, analytic bias, biological gradient (dose response), statistical significance, consistency, and biological plausibility and coherence (7).
2. *In a vaccine recipient who developed the event, did the vaccine cause it?* This question came into plan rarely in the few

instances in which carefully prepared individual case reports provided convincing evidence, as illustrated below.

3. *Will the next vaccine recipient develop the event? Or, how frequently will vaccine recipients develop the event?* The committee addressed this question as the data permitted.

The committee assigned each vaccine-adverse event pair to one of the following five categories, according to the strength of the evidence that this relationship is causal:

- There is no evidence (other than biological plausibility).
- The evidence is inadequate.
- The evidence favors rejection of causal relation.
- The evidence favors acceptance of causal relation.
- The evidence establishes a causal relation.

The first vaccine safety report (1) used different wording for the categories, but the wording was changed in the second report (2) to clarify the meaning of the distinctions being made. The definitions of the categories remain the same in the two reports.

The committee found that available evidence established a causal relation between anaphylaxis and vaccines for pertussis, measles and mumps, hepatitis B, and tetanus; between protracted inconsolable crying and pertussis vaccine; between arthritis and rubella vaccine; between thrombocytopenia and the vaccine for measles, mumps, and rubella; and between live polio vaccines and polio (in those vaccinated or in the recipients' contacts). The committee also found evidence that favors the acceptance of a causal relation between diphtheria, pertussis, and tetanus vaccine (DPT) and acute encephalopathy, and Guillain–Barre syndrome (GBS) and tetanus toxoid. These adverse events were rare, however. The committee also found that data on DPT and sudden infant death syndrome (SIDS) favored rejection of a causal relation.

A. Detailed Examples

Despite the differences in the nature of the issues being addressed, the epidemiological and other evidence that was available, and the focus on causality rather than association, the methods used by the vaccine safety committee were similar in many respects to those used by the Agent Orange committee. This section illustrates some of the key differences, using two examples: the use of formal meta-analysis to confirm that there is no causal relationship between SIDS and DPT vaccine, and the use of case-study information suggesting a relationship between tetanus toxoid and GBS.

1. SIDS and DPT

Studies of DPT immunization and SIDS take one or both of the following approaches. Some investigators look for an association between DPT immunization status and SIDS in children. This can be done either through cohort studies of children vaccinated and not vaccinated, or through case-control studies comparing children who died of SIDS with other children, to see whether the SIDS cases were more likely to have received DPT in an interval before the death. Studies of this sort tend to find a decreased risk of SIDS in children who have received DPT immunization, but on closer examination it appears that this association is due to socioeconomic status, which is positively correlated with immunization and negatively associated with SIDS.

The second approach, which the committee favored, involves comparison of the timing of SIDS deaths relative to DPT vaccination, to see whether SIDS deaths are clustered in the few days following vaccination. Because this approach is limited to exposed cases, that is, those children who receive DPT vaccination and die of SIDS, the power to detect an elevated risk is lower than in the first approach. On the other hand, potential biases arising from inclusion of unvaccinated children are avoided.

The committee found nine studies that offer some information on the timing of SIDS relative to vaccination. Because only four of these studies correct for the age pattern of SIDS (which would lead one to expect more SIDS cases in the first few days after vaccination than a uniform distribution would predict), an additional step was needed to adjust data from the poorly controlled studies; see p 340 of Ref. 1.

To summarize these data in a meta-analysis, the committee chose to use the DerSimonian–Laird (16) approach to analyzing log odds ratios with appropriate sensitivity analyses for the major assumptions. The meta-analysis was based on the odds ratios that compared the number of deaths in the early period (0 to 3 days post vaccination) to the number in the early and late (8 to 28 days) periods combined. Deaths in the mid period (4 to 7 days) were excluded from the analysis because (a) one study (33) had no mid period, and (b) it was not clear whether mid-period deaths in the other studies should be aggregated into the early or late periods.

Two studies had sample sizes too small to include in the analysis (1 and 0 observed cases in the early period). Three alternative assumptions were made about the study of Hoffman et al. (33), which has two different control groups as well as a very different set of time breakpoints: (1) the results, with both control groups, are included as two separate studies; (2) only the results from the more highly matched control group B are included in the meta-analysis; and (3) the study is entirely excluded.

In four of the studies, the expected numbers of cases are based on calculated age distributions rather than on a sample of controls. In these cases, the committee assumed that only the observed proportion in the early period contributed to the standard error. This means that the confidence intervals from these studies understate the true uncertainty by an unknown amount. For the other three studies, the committee calculated standard errors without taking into account the matching and other variance-reduction techniques that were actually used in the study. This implies that the resulting confidence intervals overstate the true uncertainty, again by an unknown amount.

The odds ratios for the individual studies, as shown in Fig. 8, range from 0.2 to 7.3. As Fig. 8 shows, the 95-percent confidence intervals for these odds ratios differ markedly from study to study, and some of the confidence intervals do not overlap, suggesting that a random-effects model is appropriate. The results of the meta-analysis, shown in Fig. 9, reflect the three assumptions about the study of Hoffman and colleagues laid out above and show the impact of choosing a fixed-effects or random-effects approach. Both assumptions have an impact on the calculations, but not enough to change the qualitative results. The decision to include or exclude the three less well-controlled studies has somewhat more of an impact; if only the well-controlled studies are included in the meta-analysis, there is an almost significant inverse association between the vaccine and SIDS in the early period.

In summary, although the results depend somewhat on the statistical assumptions, in no case is there a significantly elevated average odds ratio for the early period. Including the less-well-controlled studies, the average odds ratio is slightly above 1.0. Inclusion of only the well-controlled studies leads to a lower average odds ratio, slightly below 1.0. Thus, the overall impression is that there is strong evidence that the odds ratio is not different than 1.0. These results contributed to the

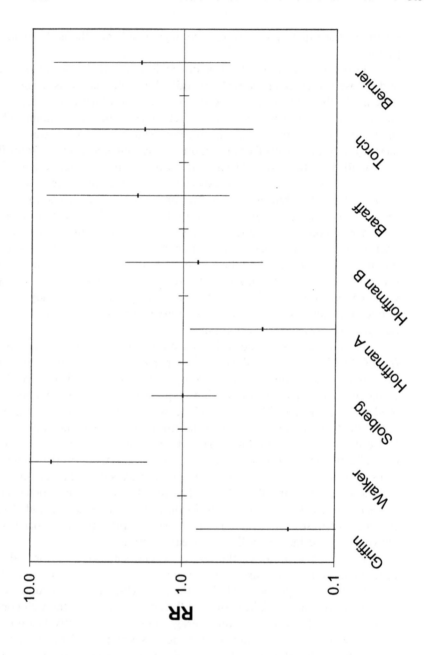

committee's conclusion that the evidence favors rejection of a causal relation.

2. Guillain–Barre Syndrome and Tetanus Toxoid

Guillain–Barre syndrome (GBS) is a neurological condition characterized by the rapid onset of flaccid motor weakness. As with many of the conditions that the vaccine safety committee analyzed, there are case studies, but no epidemiological studies, that suggest a relation between GBS and tetanus toxoid.

One particular case reported by Pollard and Selby (34) is particularly relevant. A 42-year-old male laborer received tetanus toxoid on three separate occasions over a period of 13 years, and following each vaccination a self-limited episode of clear-cut, well-documented GBS ensued, with latencies for each episode of 21, 14, and 10 days respectively. Since 1981, this man has experienced multiple recurrences of GBS, most following acute viral illnesses, suggesting a special sensitivity in this individual.

To consider systematically the implications for a causal relation of individual cases of this sort, the "did it?" question outline above, the committee addressed the following seven issues (35):

1. *Previous general experience with the vaccine.* How long has it been on the market? How many individuals have received it? How often have vaccine recipients experience similar events? How often does the event occur in the absence of vaccine exposure? Does a similar event occur more frequently in animals exposed to the vaccine than in appropriate controls?

2. *Alternative etiologic candidates.* Can a preexisting or new illness explain the sudden appearance of the adverse event? Does the adverse event tend to occur spontaneously (i.e., in the absence of known cause)? Were drugs, other therapies, or diagnostic tests and procedures that can cause the adverse event co-administered?

Figure 8 Epidemiological studies relating SIDS in the early post-vaccination period to receipt of pertussis vaccine. A horizontal line represents the primary RR estimate for each study, and the vertical lines represent the 95-percent confidence interval where available. For details on individual studies, see Ref. 1.

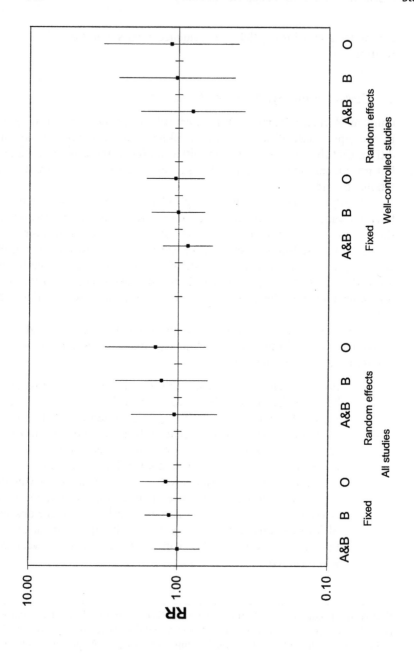

3. *Susceptibility of the vaccine recipient.* Has he or she received the vaccine in the past? If so, how has he or she reacted? Does his or her genetic background or previous medical history affect the risk of developing the adverse event as a consequence of vaccination?

4. *Timing of events.* Is the timing of onset of the adverse event as expected if the vaccine is the cause? How does that timing differ from the timing that would occur given the alternative etiologic candidate(s)? How does the timing, given vaccine causation, depend on the suspected mechanism?

5. *Characteristics of the adverse event.* Are laboratory tests that either support or undermine the hypothesis of vaccine causation available? For live attenuated virus vaccines, has the vaccine virus been isolated from the target organ(s) or otherwise identified? Was there a local reaction at the site at which the vaccine was administered? How long did the adverse event last?

6. *Dechallange.* Did the adverse event diminish as would be expected if the vaccine caused the event? Is the adverse event of a type that tends to resolve rapidly regardless of cause? Is it irreversible? Did specific treatment of the adverse event cloud interpretation of the observed evolution of the adverse event?

7. *Rechallenge.* Was the vaccine re-administered? If so, did the adverse event recur?

Considerations of this sort can be used in three ways: (a) global introspection (36), (b) algorithms or flowcharts (37), or (c) Bayesian analysis (38). The Bayesian approach explicitly calculates the probability that the event was caused by the vaccine from estimates of the prior probability

---

Figure 9   Meta-analyses of epidemiological studies relating SIDS in the early post-vaccination period to receipt of pertussis vaccine. The pooled RR estimate for each study is represented by a square, and the vertical lines represent the 95-percent confidence intervals. As a sensitivity analysis, the meta-analysis was carried out under various assumptions: (a) whether all studies or only well-controlled studies are included in the meta-analysis, (b) whether a fixed- or random-effect model is assumed, and (c) whether the meta-analysis includes results from the Hoffman study (33) based on age-matched controls and on controls matched for age, race, and birth weight (A and B); on controls matched for birth weight alone (B); or the study is entirely excluded (C).

(not including the particular facts of the current case) and a series of likelihood ratios for each pertinent element of the observed case. Each likelihood ratio is calculated by dividing the probability of observing what actually happened, under the hypothesis that the vaccine was the cause, by the probability of observing the same occurrence given non-vaccine causation.

The committee chose to adopt an informal Bayesian approach, focusing on the seven issues discussed above. In the Pollard and Selby case, for instance, the occurrence and subsequent resolution of GBS in three instances following tetanus toxoid (rechallenge/dechallenge), as well as the timing of the events, contributed to the committee's conclusion that the evidence favors a causal relation between tetanus toxoid and GBS. The apparent sensitivity of the individual in question to GBS, however, kept the conclusion from being stronger; the committee did *not* find that the evidence established a causal relation.

B. Policy Impact

In response to the two IOM reports and other clinical and scientific information, a number of diverse changes have been made to the "vaccine compensation table," the list of conditions covered by the Vaccine Injury Compensation Program. Many conditions studied in the reports were already on the compensation table, and did not change, although in some case the "qualifications and aids to interpretation" for these conditions did change to be more consistent with the way the conditions were characterized in the IOM reports. Chronic arthritis following rubella vaccine, thrombocytopenia (a blood disorder characterized by decreases platelet count) following vaccines for measles, brachial neuritis (a peripheral nerve disorder) following tetanus toxoid, vaccine-strain polio infection following live polio vaccine, and anaphylaxis following hepatitis B vaccine were all added to the vaccine compensation table. Shock-collapse or hypotonic-hyporesponsive collapse and residual seizure disorder following diphtheria, tetanus, and pertussis vaccines (in various combinations) were removed. Encephalopathy following vaccines containing tetanus toxoid (other than DPT) was also removed. Beyond the specific changes to the Vaccine Injury Compensation Program, the IOM reports removed a shadow of uncertainty about vaccine-related injuries that dated to the 1980s, and the evidence-based approach was important reassurance for the medical community and general public (G. Evans, personal communication, 1998).

IV. CONCLUSIONS

The two examples discussed in this paper illustrate the importance of systematic group judgment. The committees' process was characterized by a focus on the evidence as a whole rather than on individual epidemiological studies. Although the quality of individual studies were reviewed and assessed, and better studies were informally given more "weight" in the committees' synthesis, the committees did not attempt to reach definitive conclusions about each study (a task that had proven illusive in other groups' attempts to review these same studies). More important than the quality of individual studies was the careful review of differences among them—heterogeneity or systematic variation (39). The Agent Orange committee, in particular, looked for evidence of dose–response effects within and across studies. Both committees analyzed differences among studies and results in the kinds of exposure—TCDD vs herbicide, vaccine type—as clues to a more complete understanding of the evidence. In both cases, the systematic review of all of the data, with a focus on heterogeneity as well as quality, led to new insights and more informative conclusions.

The committees' review was facilitated by the decision to report the results in four or five standard categories (see Fig. 1). These judgments have both quantitative and qualitative aspects, and formal meta-analyses were helpful where they seemed to be appropriate. The judgments that had to be made with this approach had to do with whether the evidence for outcome X was more similar to the evidence for outcome Y (in the same category) or to the evidence for outcome Z (in another category). Although difficult, the committee members felt that categorical decisions of this sort were easier to make and more reliable than individualized conclusions for each outcome. Another factor in favor of this approach is that it does not make finer distinctions about the quality of the evidence than the data support; subtle differences in the wording between conclusions for different outcomes could be over-interpreted by readers. This approach was possible because the responsible government agencies had to make yes-or-no compensation decisions, and because the agencies were required to review all of the relevant information (including, but not limited to the IOM report) and make their own policy decisions.

The committees' approach is unique in that they adopted a neutral starting point in evaluating the evidence. There are two common approaches to risk assessments to guide public health policy. Scientists typically start from the assumption that there is no risk (the null hypothesis), and only reach a conclusion that there is a risk if the data "prove" it (the alternative hypothesis). Advocates, on the other hand, begin with the assumption that there is a risk, and ask if the data disproves it. The IOM committees took a different, more balanced approach. They viewed their task as evaluating the evidence, and thus started from a point between the two extremes. The committees then reviewed the data to see if the available studies either established or suggested a risk, or suggested that there was no risk. In both cases, many of the health outcomes remained in the "no evidence" category because the data simply was insufficient to make a case for either risk or no risk.

When presenting the results from their reviews, both committees found it important to carefully distinguish between (a) "no evidence" of an association/causal relation and (b) "evidence of no" association/causal relation. This is an important distinction for policy purposes, but not one with which many policy makers and non-scientists are not familiar. When the vaccine safety results were published, for instance, some of the adverse effects such as residual seizure disorder (epilepsy) that ended up in the "evidence is inadequate" category were already on the vaccine compensation table (reflecting decisions made by Congress when the compensation was established). Some argued that the IOM results meant these adverse effects should never have been put on the table. Others argued that the IOM results were not sufficient to remove the adverse effects. Both arguments are correct, and in this sort of example where the science is not clear, it *is* appropriate to decide on policy grounds.

Another important characteristic of the IOM approach was the use of "non-biased experts" on the committees. The IOM's institutional policies require all committees to be balanced in terms of the biases or perspectives of the committee members, and no committee members to have conflicts of interest. Because the issues were so controversial, the IOM decided to appoint committees for both studies in which no member had done research on the topic in question. The benefit was that the committee members came to the process in a sense naive, and everyone had the same scientific data on which to base their judgments. One result was that consensus was more easily achieved than in other bodies that had considered the same issues. This benefit came at a substantial cost, however, in terms of the amount of work needed by the committee members and staff to assess the literature.

A. Use of Formal Meta-analysis

Formal meta-analyses were used in both the Agent Orange and vaccine safety analyses, but only in some aspects of these questions. Statisticians might ask whether a more formal approach would have been feasible and more appropriate. For instance, could one formal meta-analysis have been carried out for each health outcome or adverse effect, incorporating all of the available information? One approach might be to formally model the sources of heterogeneity in the available studies (40). "Confidence profile" (41) or other Bayesian approaches (Stangl & Berry, Chap. 1) might also be applicable.

The committees chose not to adopt more formal approaches in part because of practical concerns. First, in both Agent Orange and vaccine safety, there was a striking amount of heterogeneity in the existing epidemiologic studies. These studies varied with respect to nature, level, and measurement of exposure; the definition and measurement of outcomes; and study design and quality. The committees attempted to learn from this heterogeneity, especially differences among the studies in the nature and level of exposure. Second, it was difficult to consistently extract data for a meta-analysis from published literature, as the SIDS example above illustrates. As is typically the case in epidemiology, many studies include multiple odds ratios or relative risks, corresponding to different study groups, control groups, outcomes, exposure measures, statistical models, and so on. It was simply not possible to choose a single measure from each study in a consistent way, and thus meta-analysis results would be subject to arbitrary decisions about data extraction. These practical difficulties made the committee members uncomfortable with using quantitative meta-analysis, except in the most limited circumstances.

Beyond these practical reasons, policy, political, and professional concerns also contributed to the committees' decisions not to use formal approaches. In both examples considered here, for instance, the policy question had to do with the existence of a statistical association or causal relation, not the relative risk, attributable risk, or other similar quantity. It is not clear how a quantitative estimate of a relative risk would or should be interpreted in these terms. Furthermore, in most cases, the risk depends on the nature and level of exposure, so a quantitative risk estimate requires an exposure specification. This is a major concern in the case of Agent Orange, since exposure levels were poorly characterized in Vietnam veterans, but they were surely lower, on average, than in many of the occupational studies on which the conclusions about statistical associations are based.

It is clear that scientific judgment is essential in any attempt to synthesize the information on the adverse health effects of herbicides and vaccines. The practical difficulties discussed above mean that any formal meta-analysis would require a substantial amount of scientific judgment about statistical models to use, which data to include and how, and so on. Given the contentious nature of both issues, there was substantial concern about personal judgments appearing to influence the reports' conclusions. The IOM addressed these concerns by adopting a less-formal group-process approach to synthesizing the data on these issues. Adopting a neutral starting point, giving fair consideration to all relevant data, giving interest groups a chance to bring data to the committee for consideration, and avoiding bias and conflict of interest in the choice of committee members. Although the final decisions were made by an informal committee process, all of the information on which the conclusions were based is laid out in a balanced way in the committee reports, together with a clear statement of the considerations that the committees used. This approach, the committee members believe, led to a better acceptance of the committee's conclusions than would have been possible, for the issues under study, than a formal quantitative meta-analysis.

References

1. Institute of Medicine (IOM). Adverse Effects of Pertussis and Rubella Vaccines. Washington: National Academy Press, 1991.
2. IOM. Adverse Events Associated with Childhood Vaccines: Evidence Bearing on Causality. Washington: National Academy Press, 1993.
3. IOM. Veterans and Agent Orange: Health Effects of Herbicides Used in Vietnam. Washington: National Academy Press, 1994.
4. IOM. Veterans and Agent Orange: Update 1996. Washington: National Academy Press, 1996.

5. International Agency for Research on Cancer (IARC). Some Fumigants, the Herbicides 2,4-D and 2,4,5-T, Chlorinated Dibenzodioxins and Miscellaneous Industrial Chemicals. IARC Monographs on the Evaluation of the Carcinogenic Risk of Chemicals to Man, Vol. 15. Lyon: IARC, 1977.

6. U.S. Preventive Services Task Force. Guide to Clinical Preventive Services. Baltimore, MD: Williams & Wilkins, 1989.

7. AB Hill. Principles of Medical Statistics, 9th ed. New York: Oxford University Press, 1971.

8. L Hardell. Relation of soft-tissue sarcoma, malignant lymphoma and colon cancer to phenoxy acids, chlorophenols and other agents. Scandinavian Journal of Work, Environment, and Health 7:119–130, 1981.

9. B Persson, A-M Dahlander, M Fredriksson, HN Brage, C-G Ohlson, O Axelson. Malignant lymphomas and occupational exposures. British Journal of Industrial Medicine 46:516–520, 1989.

10. SK Hoar, A Blair, FF Holmes, CD Boysen, RJ Robel, R Hoover, JF Fraumeni. Agricultural herbicide use and risk of lymphoma and soft-tissue sarcoma. Journal of the American Medical Association 256:1141–1147, 1986.

11. SH Zahm, DD Weisenburger, PA Babbitt, RC Saal, JB Vaught, KP Cantor, A Blair. A case-control study of non-Hodgkin's lymphoma and the herbicide 2,4-dichlorophenoxyacetic acid (2,4D) in eastern Nebraska. Epidemiology 1:349–356, 1990.

12. MA Fingerhut, WE Halperin, DA Marlow, LA Piacitelli, PA Honchar, MH Sweeney, AL Greife, PA Dill, K Steenland, AJ Suruda. Cancer mortality in workers exposed to 2,3,7,8-tetrachlorodibenzo-p-dioxin. New England Journal of Medicine 324:212–218, 1991.

13. R Saracci, M Kogevinas, PA Bertazzi, BH Bueno De Mesquita, D Coggon, LM Green, T Kauppinen, KA L'Abbe, M Littorin, E Lynge, JD Mathews, M Neuberger, J Osman, N Pearce, R Winkelmann. Cancer mortality in workers exposed to chlorophenoxy herbicide and chlorophenols. Lancet 338:1027–1032, 1991.

14. A Zober, P Messerer, P Huber. Thirty-four-year mortality follow-up of BASF employees exposed to 2,3,7,8-TCDD after the 1953 accident. International Archives of Occupational and Environmental Health 62:139–157, 1990.

15. A Manz, J Berger, JH Dwyer, D Flesch-Janys, S Nagel, H Waltsgott. Cancer mortality among workers in chemical plant contaminated with dioxin. Lancet 338:959–964, 1991.

16. R DerSimonian, N Laird. Meta-analysis in clinical trials. Controlled Clinical Trials 7:177–188, 1986.

17. L Hardell, NO Bengtsson. Epidemiological study of socioeconomic factors and clinical findings in Hodgkin's disease, and reanalysis of previous data regarding chemical exposure. British Journal of Cancer 48:217–225, 1983.

18. LF Burmeister, GD Everett, SF Van Lier, P Isacson. Selected cancer mortality and farm practices in Iowa. American Journal of Epidemiology 118:72–77, 1983.

19. KP Cantor, A Blair. Farming and mortality from multiple myeloma: a case-control study with the use of death certificates. Journal of the National Cancer Institute 72:251–255, 1984.

20. MC Alavanja, S Merkle, J Teske, B Eaton, B Reed. Mortality among forest and soil conservationists. Archives of Environmental Health 44:94–101, 1989.

21. P Boffetta, SD Stellman, L Garfinkel. A case-control study of multiple myeloma nested in the American Cancer Society prospective study. International Journal of Cancer 43:554–559, 1989.

22. V Riihimaki, S Asp, E Pukkala, S Hernberg. Mortality and cancer morbidity among chlorinated phenoxyacid applicators in Finland. Chemosphere 12:779–784, 1983.

23. GMH Swaen, C van Vliet, JJM Slangen, F Sturmans. Cancer mortality among licensed herbicide applicators. Scandinavian Journal of Work, Environment, and Health 18:201–204, 1992.

24. O Axelson, L Sundell. Herbicide exposure, mortality, and tumor incidence. An epidemiological investigation on Swedish railroad workers. Scandinavian Journal of Work, Environment, and Health 11:21–28, 1974.

25. V Riihimaki, S Asp, S Hernberg. Mortality of 2,4-dichlorophenoxyacetic acid and 2,4,5-trichlorophenoxyacetic acid herbicide applicators in Finland: first report of an ongoing prospective cohort study. Scandinavian Journal of Work, Environment, and Health 8:37–42, 1982.

26. LM Green. A cohort mortality study of forestry workers exposed to phenoxy acid herbicides. British Journal of Industrial Medicine 48:234–238, 1991.

27. A Blair, DJ Grauman, JH Lubin, JF Fraumeni. Lung cancer and other causes of death among licensed pesticide applicators. Journal of the National Cancer Institute 71:31–37, 1983.

28. H Morrison, D Savitz, R Semenciw, B Hulka, Y Mao, D Morison, D Wigle. Farming and prostate cancer mortality. American Journal of Epidemiology 137:270–280, 1993.

29. A Blair, D Mustafa, EF Heineman. Cancer and other causes of death among male and female farmers from twenty-three states. American Journal of Industrial Medicine 23: 729–742, 1993.

30. WH Wolfe, JE Michalek, JC Miner, AJ Rahe, CA Moore, LL Needham, DG Patterson. Paternal serum dioxin and reproductive outcomes among veterans of Operation Ranch Hand. Epidemiology 6:17–22, 1995.

31. CDC. Health status of Vietnam veterans. Vietnam Experience Study. Atlanta: U.S. Department of Health and Human Services, 1989, Vols. I–V, Supplements A–C.

32. Centers for Disease Control (CDC). Health status of Vietnam veterans. III. Reproductive outcomes and child health. Journal of the American Medical Association 259: 2715–2717, 1988.

33. HJ Hoffman, JC Hunter, K Damus, J Pakter, DR Peterson, G van Belle, EG Hasselmeyer. Diptheria–tetanus–pertussis immunization and sudden infant death: results of the National Institute of Child Health and Human Development Cooperative Epidemiological Study of Sudden Infant Death Syndrome Risk Factors. Pediatrics 79:598–611, 1987.

34. JD Pollard, G Selby. Relapsing neuropathy due to tetanus toxoid: report of a case. Journal of Neurological Science 37:113-125, 1978.

35. MS Kramer. Difficulties in assessing the adverse effects of drugs. British Journal of Clinical Pharmacology 11:105S–110S, 1981.

36. DA Lane. A probabilist's view of causality assessment. Drug Information Journal 18:323–330, 1984.

37. J Venulet. Assessing Causes of Adverse Drug Reactions with Special Reference to Standardized Methods. London: Academic Press, 1976.

38. DA Lane, MS Kramer, TA Hutchinson, JK Jones, C Naranjo. The causality assessment of adverse drug reactions using a Bayesian approach. Pharmaceutical Medicine 2:265–283, 1987.

39. JA Berlin. The benefits of heterogeneity in meta-analysis of data from epidemiological studies. Invited commentary. American Journal of Epidemiology 142:383–387.

40. JA Berlin, MP Longnecker, S Greenland. Meta-analysis of epidemiological dose-response data. Epidemiology 4:218–228, 1993.

41. DM Eddy, V Hasselblad, R Schacter. Meta-analysis by the Confidence Profile Method: The Statistical Synthesis of Evidence. San Diego: Academic Press, 1992.

15
Meta-analysis in Practice: A Critical Review of Available Software

Alexander J. Sutton, Paul C. Lambert, Keith R. Abrams, and David R. Jones
University of Leicester, Leicester, England

Martin Hellmich
University of Cologne, Cologne, Germany

I. INTRODUCTION

Meta-analysis, the science of combining quantitative estimates from a number of different studies, was named such by Glass in 1976 (1). In the twenty years since its inception, an array of different methods, often of increasing sophistication, have been developed for the pooling of estimates from primary studies. While a brief overview of the most common methods is given below, the reader who is unfamiliar with any of the methodology is strongly advised to seek other sources (such as Refs 2 and 3) for clarification.

The simplest of the methods for combining the results of studies include vote counting, where the number of studies showing beneficial and harmful effects are tallied. Alternatively the combination of p-values of the treatment effect for each study has also been used. Neither method yields a pooled estimate of effect size, and both are recommended as a last resort when, due to limitations in information available from study reports, other methods are not possible (4).

A method known as the fixed-effect model produces a pooled treatment estimate by taking an average of the primary study estimates, weighting each study estimate by the inverse of its variance (5). This method assumes that all the studies estimate the same underlying treatment effect, and that no between-study heterogeneity is present. The assumption of no heterogeneity may not be realistic; formal tests for its presence exist (4) but they have low power.

Random-effect models can incorporate some aspects of heterogeneity (6), and are popular alternatives to the fixed-effect method. Though the two approaches may give similar results, the confidence interval obtained by the random-effect method will generally be wider than that for the fixed-effect model. To explain between-study heterogeneity rather than simply incorporate it into the analysis (as random-effect models do), study-level covariates can be included in the model. If they are included in a fixed-effect analysis, the resulting analysis has been termed meta-regression (7), whereas if a random-effect term is included to account for residual heterogeneity not explained by covariates, this analysis is referred to as a mixed model (8).

In addition to the classically derived methods mentioned above, Bayesian approaches to meta-analysis exist (9). The Bayesian approach considers a priori beliefs regarding the overall intervention or exposure effect and the level of between-study heterogeneity, and allows pertinent background information to be included in the analysis.

All the methods previously described can be applied to binary outcome scales (including odds ratios, relative risks), and continuous outcome scales (directly or standardized). In addition to these general methods, specialist methods specific to certain study designs and data types exist, some of which will be discussed further in a later section of this chapter. Also, several graphical techniques have been used to communicate the findings of a meta-analysis. These include Forrest plots used to display the study and pooled estimate (see Fig. 1); and the radial plot (10), an alternative display of the treatment estimates. For more information on this topic, see DuMouchel and Normand (Chap. 6). A final important issue touched upon later is publication bias; see also Smith et al. (chap. 13). This bias occurs when significant results are more likely to be submitted and published (or published quickly) than work with null or non-significant results (11). Among the methods for dealing with this problem are the funnel plot and Rosenthal's "file drawer" method (12). The former helps identify publication bias, while the latter helps assess its impact.

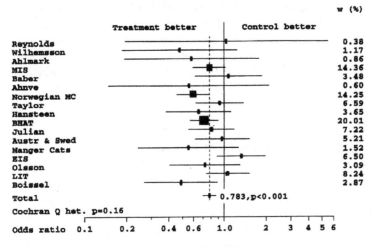

Figure 1   Forrest plot produced by EasyMA.

Many meta-analysis techniques, including many described and used in the other chapters of this volume, are not specifically supported by leading statistical software packages. This can make their implementation problematic for researchers who are not statistical experts. Fortunately, specialist software is becoming increasingly available to the meta-analyst, and code routines for several standard statistical packages are being made available through the published literature. The latter often implement new techniques which the specialist packages cannot. The aim of this paper is to inform the researcher of the software packages and routines currently available, and describe the capabilities of each. This chapter is not intended as a tutorial on how to carry out a meta-analysis, nor does it suggest appropriate analyses for a given situation. Rather, its aim is to assist the researcher who has decided on the form of the analysis required, but wishes to know how to implement it most conveniently. It will also help researchers who want to carry out a meta-analysis find packages that will give them some guidance in choosing analysis type.

The remainder of this chapter has four sections. Section II reviews existing software packages and outlines the capabilities of each. Section III discusses code routines published, or referenced in the literature, many of which implement specialist or recently developed applications. Section IV discusses the use of standard software functions to carry out meta-analysis, and presents concise code for SAS to implement regression models. We suspect that these functions are little used—primarily because researchers do not know how to implement them in a meta-analytic setting. Finally, *Sec. V* is a critical overview of the previous sections, and highlights some meta-analysis methods for which no software is presently known to be available.

II. REVIEW OF EXISTING PACKAGES

Six pieces of software are reviewed below; three of them are stand-alone programs written exclusively to carry out meta-analyses (Review Manager, EasyMA (13), MetaGraphs (14, see Chap. 5)); two are collections of macros to run on the statistical packages SAS—a registered trademark of the SAS Institute Inc., Cary, N.C. (15)—and Stata (16–22) respectively; and one (ARCUS) is a general statistical analysis package with meta-analysis options. Three other commercial packages are known to exist; namely DSTAT, TRUE EPISTAT, and FAST*PRO (23), as well as one which is available for no charge (Meta-Analyst), and one currently in development (Descartes).

All three of the commercial packages are somewhat older than the packages reviewed here, and all have been reviewed elsewhere (24–26). DSTAT is somewhat limited, performing only fixed-effect analyses. However, it does include options for combining p-values, z statistics, t statistics, F statistics, χ^2 statistics, and correlation coefficients, which none of the five packages under review here provide. As previously mentioned, however, combining studies using test statistics is not recommended. It is a last resort and might be used when it is impossible to combine effect estimates (27). TRUE EPISTAT was improved after the criticisms in a previous review (24), with an updated version being released, which performs fixed- and random-effect analyses for a wide variety of outcome measures. FAST*PRO implements a conceptually different approach to meta-analysis, called the Confidence Profile Method (28). It has been noted previously that the current version (1.0) produces erroneously wide confidence intervals (26). Meta-Analyst is limited to binary outcomes, but carries out both fixed-effect and random-effect analysis.

Of the software under review here, MetaGraphs (29) and ARCUS (30) are commercially available, while Review Manager (31), EasyMA, and the SAS and Stata macros are all freely distributed. However, support for Review Manager is only available to members of the Cochrane Collaboration. Appendix A details where each package can be obtained, and a website link from which free software/code can be downloaded. To simplify the description of the capabilities of the various packages two summary tables have been produced: Table 1 highlights the main statistical features of each, while Table 2 describes the graphical plots available.

Arcus. This Windows-based statistical software package is not specific to meta-analysis; however, it does include several menu-driven options for carrying out fixed- and random-effect analyses on both continuous and binary outcomes. The program is very easy to use, with data entered interactively, or in spreadsheet form. The graphical output is limited with only Cochrane plots supported (similar to that produced by EasyMA in Fig. 1). A feature unique to this package is that it produces exact confidence intervals for each of the individual studies effect-size estimates by Gart's method (32).

Review Manager. This suite of software was developed by the Cochrane Collaboration to assist their reviewers in carrying out systematic reviews. The software is unique in that it manages the whole systematic review process, as well as carrying out simple statistical analyses (implemented in a unit called MetaView). Once all the study data are entered, the supported fixed- and random-effect analyses, for binary and continuous outcomes, can all be carried out very easily using the user-friendly menu driven interface. Restrictions for the reviewer wanting to do more than produce a Cochrane study report are that the software and databases created within it are designed to stand alone, so no facilities are available for printing/exporting results/graphs. Additionally, it is not possible just to enter raw data; several forms have to be completed within the package before one can proceed with the analysis.

SAS Macro Suite. This is not a program but a collection of macros for the SAS system, written by Kuss and Koch (15). The macros perform various simple analyses and produce a variety of graphical figures. Versions of the macros and output annotated in either English or German are available. It should be stressed that they were written for use with binary outcome data only (such as odds ratios, relative risks etc.),

Table 1 Summary of Analyses Covered by Each Piece of Software

Software Package	Arcus	Review Manager (Meta View)	SAS Macro Suite	Stata Macros	Easy MA	MetaGraphs
Version reviewed	1.0	3.0	Slightly updated since Ref. 15	As described in Refs 17–22	97b	1.1
Fixed-effect						
Continuous outcomes						
Mean differences	✓	✓		✓		✓
Standardised mean differences	✓	✓		✓		✓
Binary outcomes						
Odds ratios	✓	✓	✓	✓	✓	✓
Inverse variance weighted		✓	✓	✓	✓	✓
Peto method		✓	✓		✓	
Mantel Haenszel method	✓		✓		✓	
Maximum likelihood			✓		✓	
Relative risk	✓	✓	✓	✓	✓	✓
Risk difference	✓	✓	✓	✓	✓	✓
Adjustment for zeros in cells	✓		✓		✓	
Random-effect						
Continuous outcomes						
Mean differences	✓	✓		✓		✓
Standardised mean differences	✓	✓		✓		✓

Binary outcomes						
Odds ratios	✓	✓	✓		✓	✓
Inverse variance weighted	✓	✓	✓		✓	✓
Peto method						
Relative risk	✓	✓	✓	✓	✓	✓
Risk difference	✓	✓	✓	✓	✓	✓
Test(s) for heterogeneity	✓	✓	✓	✓	✓	✓
Cumulative analysis		✓	✓	✓		
Publication bias assessment		✓		✓		
Subgroup analyes				✓		
Study level covariates (regression models)	✓		✓			
Hierarchical Bayesian modeling	✓			✓		

Table 2 Summary of Graphical Plots Produced by Each Piece of Software*

Software	Arcus	Review Manager	SAS Macros	Stata Macros	EasyMA	MetaGraphs
Cochrane/Forrest plot	✓	✓				
Galbraith/Radial plot (1)			✓	✓	✓	✓
Funnel plot (41)			✓	✓	✓	†
Thompson sensitivity plot (34)			✓			
Cumulative plot (42)				✓	✓	
Covariate plot (43)						✓
L'Abbé's plot (44)					✓	✓
Shrinkage plots (43)						✓
Residual plots (43)						✓

*This figure is not exhaustive, certain packages do other plots, these are mentioned in the text. †A similar plot, described as a trace plot, is produced (43).

but several of the plots can also be produced for continuous outcomes. Figure 2 shows a radial plot produced using these macros. Users familiar with the SAS language will be able to extend and customize these macros for specific needs. They are useful for processing raw data as the first stage of more sophisticated analyses such as the method of Biggerstaff and Tweedie (33), discussed in Sec. III, and the implementation of mixed models in SAS covered in Sec. IV. All standard fixed- and random-effect models are covered. Additionally the sensitivity plot advocated by Thompson (34) is also implemented. To our knowledge, this is not possible in any other software reviewed here. Two minor criticisms are that, unlike the sensitivity plot in the paper (15), no confidence intervals are plotted by the distributed code, and there is a lack of subgroup representation on the radial plot. However, improvements to other plots have been made in the most recent version of the macros. As basic knowledge of the SAS system is required for their use, and the output is in a very concise but minimally annotated form, this option is less user-friendly than the others reviewed here.

Stata Macros.   Several macros have been written for Stata which implement different aspects of meta-analyses. Sharp and Sterne (17, 18) have written a standard meta-analysis procedure for combining study results implementing fixed- and random-effect and empirical Bayes estimates; it produces numerical and graphical output in the form of a

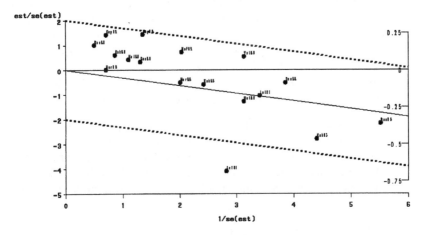

Figure 2   Radial plot produced by the SAS® macro suite.

Forrest plot (similar to, but of better quality, than Fig. 1). A recent modification to this macro allows one to enter confidence intervals, as an alternative to standard errors for each primary study being combined. A macro to implement mixed models has also been written (19), allowing extensions to the standard random-effect analysis. Several iterative algorithms for calculating parameter estimates, as well as moment-based methods, are available and clearly documented. Cumulative meta-analysis (35) is possible via the macro *metacum*, which allows both fixed- and random-effect analyses. A macro which implements tests for publication bias is also available (21). It performs an adjusted rank correlation test (36), and a regression asymmetry (37) test. This appears to be the first software implementation of the latter technique. In addition, it allows the user to produce funnel (Fig. 3) and regression asymmetry plots. Finally, a macro to produce a Galbraith plot to assist the assessment of heterogeneity has been written (22). In combination, these macros provide a powerful and easy-to-use analysis system.

EasyMA. EasyMA is a stand-alone program which runs under MS-DOS® 5.0 or later. It is restricted to meta-analysis of studies with binary outcomes. As Tables 1 and 2 show, it possesses a wide range of options, many of which can be customized, and all of which can be produced by the intuitive menu-driven interface. An example Forrest plot produced by this program is provided in Fig. 1. Sensitivity analyses should be carried out routinely in a meta-analysis (2); EasyMA facilitates this by allowing subgroup and cumulative analyses to be performed, in addition to assessments (12, 36, 38) of publication bias. Several endpoints from the same studies may be conveniently entered for analysis. It would seem that for simple (i.e. non-regressive or Bayesian) analyses, this package incorporates the widest range of features of those reviewed. A possible limitation for some users is that the printing of plots is restricted to PostScript format. For more details on this software, see Ref. 13.

MetaGraphs*. This recently developed package deals with both continuous and binary outcomes. It supports standard fixed-effect, random-effect, and mixed models, in addition to Bayesian hierarchical models (39). This is the only package in which a fully Bayesian analysis is possible, allowing both informative and non-informative specification of

---

*At the time of writing, this software was only available in beta-test form, and its range of functions may change on commercial release.

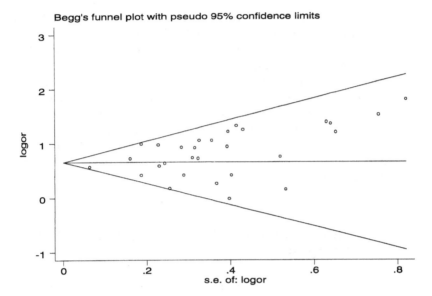

Figure 3 Funnel plot produced by Stata macro

prior distributions for model parameters. A plot showing the shrinkage of individual estimates produced by the Bayesian analysis is available. An innovative aspect of this package is that it recommends the most appropriate model for a given analysis (using Bayes factors to select between models (40)). The range of features of MetaGraphs is greater than any of the others reviewed here—though facilities for cumulative meta-analysis, or the updating of a Bayesian analysis, and tests for publication bias are notable omissions—and many of them are customizable by the user. Emphasis is placed on exploratory data analysis as well as the synthesis, with many types of exploratory plot available for different types of data (note that these are not included in Table 2). Other useful features are that it is possible to "drill down" by clicking on points in any of the plots to view the underlying data; and assessment of model fit is addressed through two sorts of residual plot. MetaGraphs runs under all Windows platforms and requires a minimum of 16Mb of memory. This package was used to produce the graphs and the computations presented by DuMouchel and Normand (Chap. 6).

Summary. It is now possible to carry out simple fixed- and random-effect analyses, for binary and continuous outcomes, using several reasonably user-friendly programs. Four of the programs (SAS macros, Stata macros, EasyMA, and MetaGraphs) produce several types of plot, of moderate to good quality; however, the SAS macros and EasyMA deal only with binary outcomes. EasyMA and the Stata macros are the only packages that make an assessment of publication bias, perform cumulative meta-analysis, and produce subgroup analyses. In addition, both these programs are freely available over the internet, which makes them an attractive choice for basic analyses. MetaGraphs supports both binary and continuous outcomes, is capable of more complex analyses, supports many exploratory plots, and has the most powerful and flexible interface.

III. CODE AVAILABLE IN PAPERS, AND SPECIALIST METHODS

This section reviews code described (and in many cases reproduced) in the meta-analysis literature, but excludes code reviewed in most other chapters of this volume. The code was identified through an extensive review of the meta-analysis methodology literature generally, which is described elsewhere (2, 45). Where the code was not fully reproduced in print, it was obtained by contacting the authors. As well as referencing the paper where the software code is reproduced or described, we have also listed, in Appendix A, websites at which the code is available, where authors' permission has been granted. In many instances the code implements new non-standard techniques which have not been included in the software packages reviewed in Sec. II. They should be seen as add-ons rather than alternatives to the packages reviewed above. The different code routines have been collated under several broad technique groups below.

A. Incorporating More Parameter Uncertainty into a Random-effect Analysis

Biggerstaff and Tweedie (33) present a method of incorporating into the standard random-effect model the uncertainty introduced by estimating the between-study variation; previous random-effect methods assumed this variance term to be known. SAS code is provided in their paper to implement the method. An estimate for the between-study variance, with confidence interval, and a revised pooled estimate and confidence interval for the overall pooled effect estimate are provided. Code is given for the new part of the analysis only. However, the SAS macros in Section A can be modified to do the rest. The paper also notes that similar routines for carrying out this analysis have been coded in C and are available from the first author.

Hardy and Thompson (46) present a likelihood approach for incorporating the same uncertainty into the analysis as the method of Biggerstaff and Tweedie. An S-Plus (47) routine is available to carry out this analysis, which includes three-dimensional displays of the likelihood functions being evaluated; sample output is presented in Fig. 4. Senn has also demonstrated how this approach can be implemented very easily in the package Mathcad, and gives full details of the code required (48).

B. Regression Models

Few software packages described in Sec. II support regression models for meta-analysis. Such models include covariates to explain systematic variation between studies being combined. Random-effect terms can be included in regression models to account for variation not explained by covariate terms; models of this type are commonly known as mixed models. Code written to fit classically based regression/mixed models for meta-analysis is summarized below.

Raudenbush (8) presents straightforward code for the package Minitab to fit mixed-effect models with one covariate (though the author points out that this could be extended to multiple covariates by the user familiar with matrix notation). Three different estimation procedures are included, namely approximate and exact method of moments, and a maximum-likelihood-based approach.

Berkey et al (49) present a SAS routine for implementing regression models, with and without a random-effect term, for combining studies with either continuous or binary outcomes. The code presented incorporates one covariate, but the authors explain how this can be easily extended to include several. An empirical Bayes approach is presented. An iterative weighted-least-squares approach is used which iterates between estimating the regression coefficients via weighted least squares and estimating the between-study variance. It should be noted that, with improvements to the built-in procedure PROC MIXED within SAS (implemented since this code was written), mixed models for meta-ana-

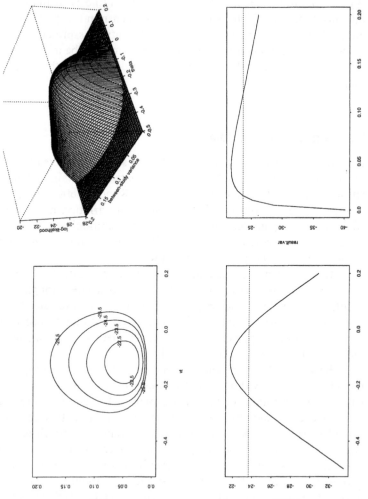

Figure 4 Likelihood surfaces for the profile likelihood technique of Hardy and Thompson (46) produced by their S + macro.

lysis can now be calculated without the need for macros (see Sec. 4 for details of regression and mixed models using SAS).

Code for S/S-Plus, similar to that given above for SAS is presented by Hellmich et al (50). This code is based on the empirical Bayes methodology developed by Carlin (51) and Morris (52), and implements a random-effect regression model via maximum likelihood and restricted-maximum-likelihood methods.

Lambert and Abrams (53) discuss and demonstrate how meta-analyses can be carried out using the software package ML3/MLn/MLwiN. This approach uses the same algorithms as the S-Plus code of Hellmich et al (50) discussed above. Starting with a standard random-effect approach with no covariates, they build the model up to a full mixed-effect model capable of including multiple covariates. Although the explicit code used is not included in the paper, it is available from our website (see Appendix B).

Stram (54) discusses a very general form of mixed models for meta-analysis. The author comments that programs have been written in GAUSS to implement this method; additionally these procedures have also recently been written into Epilog Plus (55).

DuMouchel (56) also presents a very general mixed model, which is capable of incorporating repeated measure outcomes, multiple intervention groups, and multiple time periods or crossover states. This flexibility allows the inclusion of information which would normally be omitted in a meta-analysis. The model is fitted, allowing estimates and their respective confidence intervals to be obtained, using the SAS PROC MIXED statement. The necessary code and a worked example are provided in the paper.

The above code should make the implementation of a regression/mixed model possible for researchers with a basic understanding of standard least-squares regression. Further details on code for including covariates into a meta-analysis are also given in Sec. IV.

C. Code for Bayesian Analyses

With the exception of MetaGraphs, reviewed in Sec. II, all the software discussed implements classical or empirical Bayes statistical methods. The alternative (fully) Bayesian formulation of the meta-analysis model has recently aroused much interest. For a review of Bayesian methods in meta-analysis, see Ref. 2 and Stangl and Berry (Chap. 1). These methods are often more difficult to implement and compute than their classical counterparts (though they do have advantages); fortunately software for their implementation is becoming available.

DuMouchel has written a series of programs in S-Plus for implementing hierarchical Bayes linear models (57), which are available via the internet. These functions fit and graph the results from a mixed-effect three-level hierarchical Bayes linear model, which allows the inclusion of covariates. The routines also support predictive crossvalidation of a meta-analysis, and several complementary plotting functions are provided. These functions form the basis of part of the commercial package MetaGraphs, reviewed in Sec. II;* see Chap. 6.

An effective way of applying a Bayesian approach to modelling generally is through the use of Markov chain Monte Carlo simulation techniques (see examples in Chaps 1, 4, 8, 11, and 12). A program which implements a specific form of this approach, known as Gibbs sampling, has been developed by Spiegelhalter et al. (58). The availability details of this program, BUGS (Bayesian inference Using Gibbs Sampling), are given in Appendix B. By using this program, it is possible to perform a fully Bayesian meta-analysis. Several clear accounts of its implementation within BUGS are available (9, 58, 59). Care should be exercised in the use of this software package because failure to establish convergence of methods may lead to wrong answers.

D. Assessing Underlying Risk as a Source of Heterogeneity in Meta-analysis

Thompson et al. (60) present a method for assessing underlying risk as a source of heterogeneity in meta-analysis. The meta-regression models outlined previously should not be used for this since, attenuation of the relationship between baseline risk and outcome may result due to regression to the mean (60,61). A Bayesian approach, implemented using the package BUGS, is presented (see subsection C above).

---

*Personal communication.

E. Likelihood Techniques

Van Houwlingen et al. (62) present a likelihood-based method for the estimation of the parameters in the random-effect model, which avoids the use of approximating normal distributions. They have written a program for the package GAUSS to implement this method for binary outcomes. The program produces several illustrative plots in addition to carrying out the analysis.[†]

Aitkin and Francis (63) present macros for the package GLIM which fits generalized linear variance-component models by non-parametric maximum likelihood. In their paper, they apply this technique to a typical meta-analysis dataset.

F. Cross-design Synthesis

The concept behind cross-design synthesis is the combination of evidence from studies of differing designs; capturing the strengths of the studies and reflecting the full range of available data. The idea can be traced back to the confidence-profile method described by Eddy (28) and implemented via the software program FAST*PRO (23). More recently, Smith et al (64) have described a fully Bayesian hierarchical modelling approach to combine studies of different types. Code for the software package BUGS, described under the Bayesian analysis section, is available for this type of analysis.

G. Plots

Table 2 outlines the capabilities of the software packages for creating graphical plots. The plots produced are often of a rigid format, not allowing much modification by the user. In our experience, if high-quality fully-customizable plots are desired, then the S-Plus language is very powerful, though considerable time is required to learn the language

---

[†] It is important to note that we could not get this program to run initially on a new version of GAUSS (3.2.15), due to a modification of the language; a small amount of modification to the code is needed. The code runs unmodified on older versions of Gauss; we know that it works on 3.1.4, for instance.

and produce the plots. However, a flexible S-Plus function is available[‡] which emulates the graphical output of MetaView. Being based on estimates and standard errors, it is applicable to any type of data. A welcome feature is that it allows the results of Bayesian and classical analyses to be displayed on the same plot for comparison.

H. Miscellaneous Methods

For completeness, other references in the literature concerning software used for meta-analysis are given below. These are for analysis of very specific situations, and are not reviewed in full.

The first two of these discuss the practical application of meta-analysis to survival data.

> Dear (65) reports that he has written code in SAS/IML for the joint analysis of survival proportions reported at multiple times in published studies, to be combined in a meta-analysis. Additionally, this method can incorporate single-arm studies into the analysis.[§]
>
> Messori and Rampazzo (66) discuss a DbaseIII program for the meta-analysis of clinical trials based on censored endpoints.
>
> Moreno et al. (67) discuss the analysis of matched and unmatched case-control studies, giving GLIM 4 macros to carry out the analysis via composite-link models.
>
> Hellmich et al. (68) present BUGS and S-Plus code for a Bayesian meta-analysis of ROC curve data.
>
> Berkey et al (69) present a method of combining multiple outcomes in a meta-analysis of clinical trials, and comment that they have written code in SAS/IML to carry out the analysis.
>
> To the authors' knowledge no software is available for meta-analysis when outcomes are measured on ordered scales. However, Whitehead and Jones (70) mention the use of the SAS procedure PROC LOGISTIC for carrying out a meta-analysis of studies using ordinal scales.

[‡]Written by Julian Higgins, Systematic Reviews Training Unit, Department of Epidemiology and Public Health, University College London Medical School. It is available from our website.
[§]FAST*PRO is also capable of doing this.

Finally, this chapter is concerned with methods for pooling results aggregated at the study level. An alternative, which is becoming increasingly common, is to perform meta-analyses combining individual-patient data. Many of the statistical techniques used in this situation are similar to those used in multi-centre trials. However, Higgins (71,72) has developed methodology for including both patient-level and study-level covariates, which he implements using BUGS.

IV. USING STANDARD SOFTWARE FUNCTIONS TO CARRY OUT META-ANALYSES

It is sometimes possible to use standard functions present in common statistical packages to carry out certain meta-analyses. However, the appropriateness of the functions and/or the model specification may not be obvious; two such examples are given below.

A.   Meta-regression (Fixed-effect)

Logistic regression is available in a variety of statistical packages (including Minitab, SPSS, SAS, S-Plus, Arcus, Glim, EGRET, and Stata). This can be used to fit fixed-effect regression models when the (log) odds ratio is used as the outcome (which is often the case with clinical or epidemiological data).

To perform a fixed-effect meta-regression with a continuous outcome variable is just as easy; any package that can perform weighted regression will suffice. If one is combining binary outcomes, such as relative risks or odds ratios, these can be modelled using the same procedure, if one is willing to make the assumption of normality. A log transform of the data is usually carried out prior to synthesis to improve the approximate normality of the data. In these situations, the model is specified in the same way as for normal ordinary least-squares regression—only an extra line specifying the weights given to each observation is needed. Figure 5 gives the routine necessary for implementation in SAS. The data column supplying the outcome of interest is included in the ⟨outcome⟩ field. The covariate(s) under investigation are inserted in the ⟨x1⟩ field. If a single continuous covariate is being included, simply specify the name of the variable here. Categorical covariates can also be modelled: it is possible (as with other types of regression) to include them

```
proc glm;
model <outcome> = <x1>;
weight <wt>;
run;
```

Figure 5   SAS code to fit a meta-regression (fixed-effect model).

using dummy variables to ease interpretation. Finally, a variable giving the weights associated with each trial are included in the ⟨wt⟩ field (often taken to be the inverse of the variance of the study estimate).

B.   Mixed Models (Random-effect Regression)

The second example illustrates how the SAS procedure PROC MIXED can be used to fit mixed-effect regression models. Mixed models have been mentioned in the previous sections. With the exception of Bayesian modelling, they are the most sophisticated type of model regularly used for meta-analysis. Several routines were given in Sec. III to fit them, the most flexible method being the implementation in MLn. However, if a specialist multilevel package is not available however, PROC MIXED in SAS can be used for modelling continuous or binary outcomes instead.¶ Recent refinement of PROC MIXED allows the necessary model specification (version 6.12 works). This possibility is little publicized. By using this standard procedure, it is now possible to fit the model of Berkey et al. (49), eliminating the need for the PROC IML routines discussed in the previous section.

    An outline of the necessary code is given in Fig. 6. The computation method (either maximum likelihood or restricted maximum likelihood) is stated in the ⟨ML/REML⟩ field. As above, a variable numbering each of the studies being combined is included in the ⟨idnumber⟩ field. The model is specified in the same way as the fixed-effect regression model given in Fig. 5. The outcome can be either continuous or a log-transformed binary outcome such as the odds ratio. The same variables are included in the

---

¶Note however, a mixed logistic model for directly modelling binary data on the odds ratio scale is not possible in SAS (Smith comments that this is, however, possible in EGRET (9)). Recently written macros should also allow random-effect logistic regression in MLn. However, we had difficulty getting these to work.

```
proc mixed  method= <ML/REML>;
class <idnumber>;
parms (0.5) (1)/eqcons=2;
model <outcome> = <x1> /solution;
random intercept/subject=<idnumber> solution;
weight <wt>;
run;
```

Figure 6 SAS PROC MIXED routine for carrying out meta-analysis using mixed models.

⟨wt⟩ field as described above for Fig. 4. The *parms* subcommand is required to restrict variance terms appropriately.

The PROC MIXED procedure now supports limited Bayesian analysis. Although this does not have the same flexibility as using BUGS, the Bayesian option within the function allows specification of flat and Jefferys (the default) priors for the model parameters. The code described here can be expanded upon to produce more complex models. The paper by DuMouchel (56) provides details of how this can be achieved.

V. SUMMARY AND BRIEF RECOMMENDATIONS FOR USERS AND FURTHER SOFTWARE DEVELOPMENT

This paper has reviewed and brought together the disparate literature concerning software routines for carrying out meta-analysis. Many of the developments discussed are recent, and clearly the options available to the researcher analysing a dataset using pre-written software are growing daily—cf. an earlier review (24). Now both fixed- and random-effect analyses can be carried out on virtually any dataset with binary or continuous outcomes using very simple menu-driven software. Implementation of more sophisticated techniques, such as regression and mixed modelling, is now also possible. These techniques have been advocated because it is advantageous to explain systematic variation between studies, if it exists, rather than simply accounting for it, as the random-effect models do (49). One of the recommendations reported in the review by Sutton et al. (2) on meta-analysis methodology was the development of experience with practical applications of mixed models.

This current review should bring the practical implementation of such models within reach of all statistically competent researchers.

Normand, in her previous review of meta-analysis software (24), commented that a feature lacking from all packages was model-checking facilities; it would appear that, with the exception of MetaGraphs and the DuMouchel S-Plus routines, which enable one to calculate normalized predictive residuals, this is still a neglected area. Other notable omissions from the existing software include integrated routines for assessing publication bias. Although several pieces of software will draw funnel plots, only EasyMA and the Stata macros will carry out the associated formal tests recently devised (36,37). Though still at the experimental stage, several authors have discussed the role of selection models using weight functions to adjust an analysis suspected to be affected by publication bias (73, 74). We are not aware of any software that is available to do this yet.

Key assumptions made in any analysis should be assessed through the use of sensitivity analysis (2). The lack of a formal framework for doing this in the literature is reflected by a lack of features in the software. It would be desirable (a) to be able to omit certain studies from an analysis, to assess their impact, or the impact of varying inclusion criteria; (b) to vary the relative weights given to the individual studies, possibly in light of quality assessments; and (c) to perform cumulative meta-analyses, not only chronologically but also ordering using covariates of interest (as advocated by Lau et al. (42)). While certain factors such as the omission of particular studies can be assessed by changing the data set, others are problematic and time-consuming. No package facilitates all these forms of sensitivity analysis; however, EasyMA does offer a certain amount of flexibility in this respect.

Companion Website

All relevant links, plus further information/code are to be found at the following companion resource site:

http://www.prw.le.ac.uk/epidemio/personal/ajs22/meta/

If further websites concerning software for meta-analysis are created, we will be happy to provide links from our site (please contact the first author of this chapter regarding this).

Acknowledgments

Part of the research on which this paper is based was funded by the NHS Research and Development Health Technology Assessment Programme methodology project 93/52/3.

The authors are indebted to Calvinda Shields for assistance in the use of GAUSS, and Stephen Sharp for supplying detailed information on the Stata macros reviewed. Warm appreciation goes to the authors mentioned herein who took time to supply and discuss their code with us and allowed us to place it on our website.

Disclaimer

The authors have made every effort to make this review as complete and accurate as possible as at February 1998. Inevitably revisions and developments of software will have taken place since then.

While the results obtained from each piece of software were checked, the authors of this chapter accept no responsibility for mistakes in the original code. The software packages may use slightly different formulae for carrying out certain methods. It is beyond the scope of this article to investigate and describe full details of every method in every package.

The authors have tried all the software/code routines described in this paper with the exception of "miscellaneous methods" (subsection III.H) and those needing Epilogue to run them. All software/code was run on a PC Pentium100Mhz with 8Mb of RAM running Windows 3.1/DOS 6—with the exception of that requiring S-Plus, which was run using a Silicon Graphics UNIX platform, and MetaGraphs which was run on a more powerful PC.

Appendix A: Contact Addresses for Software Packages Reviewed in Part A

Packages Reviewed in Part A

ARCUS.

http://www.camcode.com

At the time of writing, a fully working demoonstration version was available to download for a one-month evaluation period

Review Manager. This software is only available from:

http://hiru.mcmaster.ca/cochrane/

SAS macros. These can be obtained from the authors, by email from:

xarmin@imbi.uni-heidelberg.de

or by post from Institut für Medizinische Biometrie und Informatik, Abteilung Medizinische Biometrie, Universität Heidelberg, Im Neuenheimer Feld 305, D-69120 Heidelberg, Germany. They are also available at our website.

Stata & macros. Information on obtaining Stata can be obtained from Stata Corporation, 702 University Drive East, College Station, TX 77840, USA, or

e-mail stata@stata.com

The macros can be obtained from the Stata website:

www.stata.com

See issues 38, 41, and 42 of the Stata newsletter downloads for the various macros described.

EasyMA. The program and users guide are available from two internet sites:

www.spc.univ-lyon1.fr/citccf/easyma/
www.spc.univ-lyon1.fr/~mcu/easyma/

MetaGraphs. The version reviewed was a beta-test version. A commercial version should be out by the time this chapter appears. Information can be obtained from:

Belmont Research, Inc., 84 Sherman Street, Cambridge, MA 02140, USA.
e-mail: metagraphs@belmont.com

Other Packages Reviewed Elsewhere (24, 26)

DSTAT. Available from Lawrence Erlbaum Associates, Inc., 365 Broadway Hillside New Jersey, 07642, USA

TRUE EPISTAT. Available from Epistat Services, 2011 Cap Rock Circle, Richardson, TX 75080, USA; telephone: 214-680-1376; fax: 214-680-1303.

FAST*PRO. Available from Academic Press, Inc., 955 Massachusetts Avenue, Cambridge, MA 02139, USA; telephone: 800-321-5068; fax: 800-336-7377. A limited demonstration version is given away with each new copy of the book by Eddy et al. (28).

Meta-analyst. Available from Dr J Lau, New England Medical Center, Box 63, 750 Washington St, Boston, MA 02111, USA.

e-mail: joseph.lau@es.nemc.org

Software Under Development

Descartes. This program is produced by Update Software Ltd, whose web address is

http://www.update-software.com/

SCHARP. SAS software for meta-analysis of individual-patient survival data, which plots survival curves and hazard ratios. Developed by the MRC Clinical Trials Unit, Cambridge, UK and the Istituto "Mario Negri," Milan, Italy. More information at

http://www.mrc-cto.cam.ac.uk/scharp.htm

Appendix B: Further Information on the Code Routines Reviewed

Table 3 shows where the different pieces of code are available. Contact addresses of authors can be obtained from the papers. By the time this chapter appears, it is hoped that permission will have been granted from more authors allowing us to put the code on our website.

Table 3 Contact Details for Meta-analysis Code

Author(s) of code and source of description	Brief description	Package code written for	Reproduced in paper	Available directly from author	Available/link from our website
Biggerstaff & Tweedie (33)	Incorporating more parameter uncertainty	SAS & C	√ (SAS only)	√ (code in (C)	
Hardy & Thompson (46)	Incorporating more parameter uncertainty	S-Plus		√	
Senn (48)	Incorporating more parameter uncertainty	Mathcad	√		
Raudenbush (8)	Mixed regression models	Minitab	√		
Berkey et al. (49)	Mixed regression models	SAS	√		
Hellmich, Abrams & Sutton (50)	Mixed regression models	S-Plus			√
Lambert & Abrams (53)	Mixed regression models	ML3/MLn/ MLwiN			√
DuMouchel* (57)	Bayesian analysis	S-Plus		√	√
Smith et al. (9, 59)	Bayesian analysis	BUGS	√		
Thompson et al. (60)	Assessing underlying risk	BUGS	√		
Van Houwligen et al. (62)	Likelihood techniques	GAUSS		√	
Aitkin and Francis* (18)	Likelihood techniques	GLIM	√		√
Smith et al. (64)	Cross-design synthesis	BUGS	√		
Hellmich et al. (68)	ROC curves	BUGS			√

*See below for further details.

Further Information on Code. GLIM macros by Aitken and Francis (63) for fitting generalized linear variance-component models by non-parametric maximum likelihood are available from

http://www.mas.ncl.ac.uk/~ nma9/allvc.txt/

DuMouchel's S-Plus macros are available at two sites:

http://research.att.com/~dumouchel/
ftp://ftp.research.att.com/dist/bayes-meta/

Information on BUGS. The software packages BUGS (Bayesian inference Using Gibbs Sampling) (75), used for several of the Bayesian methods covered here, together with CODA (Convergence Diagnosis and output Analyses software for Gibbs sampling output) (76) are freely available at

http://www.mrc-bsu.cam.ac.uk/bugs/

A Windows version—WinBUGS—is also freely available.

Addendum

Since this review was written, several important developments have occurred in this fast moving field. These are summarized below. The package *Descartes*, which was under development when the review was completed, has been released under the title *metaxis* (see *http://www.up-date-software.com/metaxis.htm*). (Note: *Meta-graphs* is still at the beta test stage). Another commercial software package, *Comprehensive Meta-analysis*, has been developed (*http://www.Meta-Analysis.com/*); as has a free software package called *Epi Meta*, which can be downloaded at: *http://www.cdc.gov/epo/dpram/epimeta/epimeta.htm*. A specialized program to carry out meta-analysis of diagnostic test studies has been written. Several new macros for STATA have been written (77) increasing the range of analyses possible; these include: i) *Metan* (78) which provides an alternative command for basic analyses, ii) *Metap* (79) which combines *p*-values, iii) *Metainf* (80) which allows assessment of the influence of each individual study in the analysis, and iv) *Metabias* (81) has been modified. Finally, a book describing how to carry out meta-analysis using SAS(R) software has been written (82).

References

1. GV Glass. Primary, secondary and meta-analysis of research. Educ Res 5:3–8, 1976.

2. AJ Sutton, KR Abrams, DR Jones, TA Sheldon, F Song. Systematic reviews of trials and other studies. Health Technology Assessment 2 (19), 1998.

3. H Cooper, LV Hedges, eds. The Handbook of Research Synthesis. New York: Russel Sage Foundation, 1994.

4. LV Hedges, I Olkin. Statistical Methods for Meta-analysis. London: Academic Press, 1985.

5. V Hasselblad, DC McCrory. Meta-analytic tools for medical decision making: A practical guide. Med Decis Making 15:81–96, 1995.

6. R DerSimonian, N Laird. Meta-analysis in clinical trials. Controlled Clin Trials 7:177–188, 1986.

7. LV Hedges. Fixed effects models. In: H Cooper, LV Hedges, eds. The Handbook of Research Synthesis. New York: Russell Sage Foundation, 1994, pp 285–300.

8. SW Raudenbush. Random effects models. In: H Cooper, LV Hedges, eds. The Handbook of Research Synthesis. New York: Russell Sage Foundation, 1994, pp 301–322.

9. TC Smith, DJ Spiegelhalter, A Thomas. Bayesian approaches to random-effects meta-analysis: a comparative study. Stat Med 14:2685–2699, 1995.

10. RF Galbraith. A note on graphical presentation of estimated odds ratios from several clinical trials. Stat Med 7:889–894, 1988.

11. PJ Easterbrook, JA Berlin, R Gopalan, DR Matthews. Publication bias in clinical research. Lancet 337:867–872, 1991.

12. R Rosenthal. The file drawer problem and tolerance for null results. Psychol Bull 86:638–641, 1979.

13. M Cucherat, JP Boissel, A Leizorovicz, MC Haugh. EasyMA: a program for the meta-analysis of clinical trials. Comput Methods Programs Biomed 53:187–190, 1997.

14. W DuMouchel, D Fram, Z Jin, SL Normand, B Snow, S Taylor, R Tweedie. Software for exploration and modeling of meta-analysis (abstract). Controlled Clin Trials 18:181S, 1997.

15. O Kuss, A Koch. Metaanalysis macros for SAS. Computational Statistics & Data Analysis 22:325–333, 1996.

16. STATA [computer program]. 702 University Drive East, Texas: Stata Press, 1985.

17. S Sharp, J Sterne. Meta-analysis. Stata Technical Bulletin 38(sbe16):9–14, 1997.

18. S Sharp, J Sterne. New syntax and output for meta-analysis command. Stata Technical Bulletin 42(sbe16):6, 1998.

19. S Sharp. Meta-analysis regression. Stata Technical Bulletin 42(23):16–22, 1998.

20. J Sterne. Cumulative meta-analysis. Stata Technical Bulletin 42(22):13–16, 1998.

21. TJ Steichen. Tests for publication bias in meta-analysis. Stata Technical Bulletin 41(sbe20):9–15, 1998.

22. A Tobias. Assessing heterogeneity in meta-analysis: the Galbraith plot. Stata Technical Bulletin 41(sbe20):15–17, 1998.

23. FastPro: Software for MetaAnalysis by the Confidence Profile Method [computer program]. Eddy DM &Hasselblad V. San Diego, California: Academic Press Inc, 1992; including 3.5-inch disk for PC.

24. SLT Normand. Metaanalysis software—a comparative review. DSTAT, version 1.10. American Statistician 49:298–309, 1995.

25. D Behar. FastPro—software for metaanalysis by the confidence profile method. JAMA 268:2109, 1992.

26. M Egger, JAC Sterne, G Davey Smith. Meta-analysis software. Br Med J 316 (1998). Published on website only: http://www.bmj.com/archive/7126/7126ed9.htm

27. BJ Becker. Combining significance levels. In: H Cooper, W Hedges, eds. The Handbook of Research Synthesis. New York: Russell Sage Foundation; 1994; 15, Combining significance levels. p. 215–30.

28. DM Eddy, V Hasselblad, R Shachter. Meta-analysis by the Confidence Profile Method. San Diego: Academic Press, 1992.

29. MetaGraphs [computer program]. 1.0. Cambridge, MA: Belmont Research Inc, 1998.

30. Arcus QuickStat [computer program]. Buchan IE. 1.0. Cambridge: Research Solutions, 1997.

31. Reference Manager [computer program]. 3.0. Oxford: Update Software Ltd, 1996.

32. H Sahai, A Kurshid. Statistics in Epidemiology: Methods, Techniques, and Approaches. CRC Press, 1996.

33. BJ Biggerstaff, RL Tweedie. Incorporating variability in estimates of heterogeneity in the random effects model in meta-analysis. Stat Med 16:753–768, 1997.

34. SG Thompson. Controversies in meta-analysis: the case of the trials of serum cholesterol reduction. Stat Methods Med Res 2:173–192, 1993.

35. J Lau, EM Antman, J Jimenez-Silva, B Kupelink, SF Mosteller, TC Chalmers, J JimenezSilva, B Kupelnick, F Mosteller. Cumulative meta-analysis of therapeutic trials for myocardial infarction. New Engl J Med 327:248–254, 1992.

36. CB Begg, M Mazumdar. Operating characteristics of a rank correlation test for publication bias. Biometrics 50:1088–1101, 1994.

37. M Egger, GD Smith, M Schneider, C Minder. Bias in meta-analysis detected by a simple, graphical test. Br Med J 315:629–634, 1997.
38. TR Einarson, JS Leeder, G Koren. A method for meta-analysis of epidemiological studies. Drug Intelligence Clin Pharm 22:813–824, 1988.
39. W DuMouchel. Hierarchical Bayes linear models for meta-analysis. Technical Report #27, National Institute of Statistical Sciences, PO Box 14162, Research Triangle Park, NC 27709; 1994.
40. KR Abrams, B Sanso. Model discrimination in meta-analysis—a Bayesian perspective. Technical Report 95–03, Department of Epidemiology and Public Health, University of Leicester, 1995.
41. CB Begg. Publication bias. In: H Cooper, W Hedges, eds. The Handbook of Research Synthesis. New York: Russell Sage Foundation, 1994, pp 399–410.
42. J Lau, CH Schmid, TC Chalmers. Cumulative meta-analysis of clinical trials: Builds evidence for exemplary medical care. J Clin Epidemiol 48:45–57, 1995.
43. Anonymous. MetaGraphs Manual. 1.1 ed. Cambridge, MA: Belmont Research Inc 1998.
44. KA L'Abbe, AS Detsky, K O'Rourke. Meta-analysis in clinical research. Annals of Internal Medicine 107:224–233, 1987.
45. AJ Sutton, DR Jones, KR Abrams, TA Sheldon, F Song. Systematic reviews of randomized trials. In: N Black, J Brazier, R Fitzpatrick, B Reeves, eds. Health Services Research Methods: A Guide to Best Practice. London: BMJ Books, 1998, pp. 175–186.
46. RJ Hardy, SG Thompson. A likelihood approach to meta-analysis with random effects. Stat Med 15:619–629, 1996.
47. S-Plus [computer program]. 4.0. Statistical Sciences Inc, 1700 Westlake Ave. North, Suite 500, Seattle, WA 98109, 1997.
48. S Senn. Meta-analysis with Mathcad. ISCB News 20:4–5, 1996.
49. CS Berkey, DC Hoaglin, F Mosteller, GA Colditz. A random-effects regression model for meta-analysis. Stat Med 14:395–411, 1995.
50. M Hellmich, KR Abrams, AJ Sutton. Bayesian approaches to meta-analysis of ROC curves: A comparative review. Medical Decision Making; 19:252–264, 1999.
51. JB Carlin. Meta-analysis for 2 × 2 tables: a Bayesian approach. Stat Med 11:141–158, 1992.
52. CN Morris. Parametric empirical Bayes inference: theory and applications. J Am Statist Assoc 78:47–65, 1983.
53. PC Lambert, KR Abrams. Meta-analysis using multilevel models. Multilevel Modelling Newsletter 7(2):17–19, 1996.
54. DO Stram. Meta-analysis of published data using a linear mixed-effects model. Biometrics 52:536–544, 1996.
55. Epilog Plus, Statistics Package for Epidemiology and Clinical Trials [computer program]. Epicenter. Pasadena, California: Epicenter Software, 1994.

56. W DuMouchel. Repeated measures meta-analysis. Bulletin of the International Statistical Institute, Session 51,Tome LVII, Book 1:285–288, 1998.

57. W DuMouchel. Predictive cross-validation in Bayesian meta-analysis. In: Bernardo et al., eds. Proceedings of Fifth Valencia International Meeting. Valencia, Spain, 1994

58. D Spiegelhalter, A Thomas, W Gilks. BUGS Examples 0.30.1. Cambridge: MRC Biostatistics Unit, 1994.

59. TC Smith, D Spiegelhalter, MKB Parmar. Bayesian meta-analysis of randomized triles using graphical models and BUGS. Bayesian Biostatistics 1995, pp 411–427.

60. SG Thompson, TC Smith, SJ Sharp. Investigation underlying risk as a source of heterogeneity in meta-analysis. Stat Med 16:2741–2758, 1997.

61. S Senn. Importance of trends in the interpretation of an overall odds ratio in the meta-analysis of clinical trials. Stat Med 13:293–296, 1994.

62. HC Van Houwelingen, KH Zwinderman, T Stijnen. A bivariate approach to meta-analysis. Stat Med 12:2273–2284, 1993.

63. M Aitken, B Francis. Fitting generalized linear variance component models by nonparametric maximum likelihood. GLIM Newsletter 1999. Issue 28.

64. TC Smith, KR Abrams, DR Jones. Using hierarchical models in generalised synthesis of evidence: an example based on studies of breast cancer screening. Technical Report 95-02, Department of Epidemiology and Public Health, University of Leicester, 1995

65. KBG Dear. Iterative generalized least squares for meta-analysis of survival data at multiple times. Biometrics 50:989–1002, 1994.

66. A Messori, R Rampazzo. Metaanalysis of clinical-trials based on censored end-points—simplified theory and implementation of the statistical algorithms on a microcomputer. Computer Methods And Programs In Biomedicine 40:261–267, 1993.

67. V Moreno, ML Martin, FX Bosch, S De Sanjose, F Torres, N Munoz, S Desanjose. Combined analysis of matched and unmatched case-control studies: Comparison of risk estimates from different studies. Am J Epidemiol 143:293–300, 1996.

68. M Hellmich, KR Abrams, DR Jones, PC Lambert. A Bayesian approach to a general regression model for ROC curves. Medical Decision Making 18:436–443, 1998.

69. CS Berkey, JJ Anderson, DC Hoaglin. Multiple-outcome meta-analysis of clinical trials. Stat Med 15:537–557, 1996.

70. A Whitehead, NMB Jones. A meta-analysis of clinical trials involving different classifications of response into ordered categories. Stat Med 13:2503–2515, 1994.

71. JPT Higgins, A Whitehead. Inclusion of both patient-level and study-level covariates in a meta-analysis (abstract). Controlled Clin Trials 18:84S, 1997.

72. JPT Higgins. Exploiting information in random effects meta-analysis. PhD Thesis, Department of Applied Statistics, The University of Reading, 1997
73. LV Hedges. Modeling publication selection effects in meta-analysis. Statistical Science 7:246–255, 1992.
74. GH Givens, DD Smith, RL Tweedie. Publication bias in meta-analysis: a Bayesian data-augmentation approach to account for issues exemplified in the passive smoking debate. Statistical Science 12:221–250, 1997.
75. DJ Spiegelhalter, A Thomas, NG Best, et al. BUGS: Bayesian inference Using Gibbs Sampling, Version 0.50 (version ii). Cambridge: MRC Biostatistics Unit, 1996.
76. MK Cowles, NG Best, K Vines. CODA—Convergence diagnostics and output analysis software for Gibbs samples produced by the BUGS Language. Version 0.30. Technical Report, MRC Biostatistics Unit, University of Cambridge, England, 1994.
77. Meta-Test [computer program]. J. Lau version 0.6. New England Medical Center, Boston: 1997. (download from: http://som.flinders.edu.au/FUSA/COCHRANE/cochrane/sadt.htm.)
78. MJ Bradburn, JJ Deeks, DG Altman. Metan – an alternative meta-anlaysis command. Stata Technical Bulletin, STB 44(24):4–15, 1998.
79. A Tobias. Meta-analysis of p-values. Stata Technical Bulletin 49(28):15–7, 1999.
80. A Tobias. Assessing the influence of a single study in the meta-analysis estimate. Stata Technical Bulletin, 47(26):15–27, 1999.
81. TJ Steichen, M Egger, J Sterne. Modification of the metabias program. Stata Technical Bulletin 44:(19.1):3–4, 1998.
82. MC Wang, BJ Bushman. Integrating results through meta-analysic review using SAS(R) software. Cary, North Carolina: SAS Institute Inc., 1999.

Index